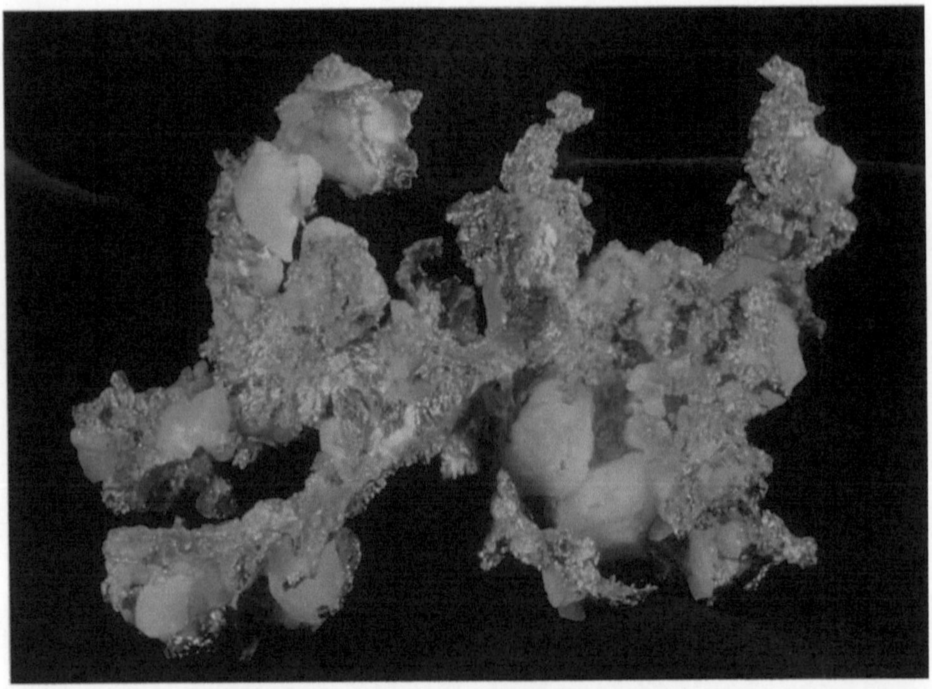

Photograph courtesy of the Department of Mineralogy
Royal Ontario Museum, Toronto, Canada

Claudia
Gasparrini

GOLD and Other Precious Metals

From Ore to Market

With Forewords by
G. C. Amstutz and P. M. J. Gray

With 66 Figures
in 121 Separate Illustrations
Mostly in Color

Springer-Verlag
Berlin Heidelberg New York
London Paris Tokyo
Hong Kong Barcelona
Budapest

Dr. Claudia Gasparrini
Minmet Scientific Inc.
6651 N. Campbell Ave. # 102
Tucson, AZ 85718, USA

ISBN-13:978-3-642-77186-6 e-ISBN-13:978-3-642-77184-2
DOI: 10.1007/978-3-642-77184-2

Softcover reprint of the hardcover 1st edition 1993

Typesetting: Camera ready by author

32/3145-5 4 3 2 1 0 – Printed on acid-free paper

FOREWORD I

The literature on gold deposits has increased *exponentially* during the past eight years. The books on gold geology and exploration cover in most cases certain geographical areas (for example, **Gold in Canada**, etc.), or they contain papers presented at symposia and congresses.

The approach often missing in other texts is the one which combines geology with mineralogy and the recovery problems and which addresses the earth scientist **as well as** the extractive metallurgist. Dr. Claudia Gasparrini has managed to fill this gap very well. She builds a bridge from the ore geology aspect to the mineralogical parameters and properties, specifically for the ore dressing engineer and, although to a lesser extent, for the exploration geologist. With her book, she helps to stop blind, one-sided recovery attempts, and points aptly to the importance of mineralogical and largely physical properties, such as the mineral intergrowth and the role of associated minerals. The intergrowth determines important steps of the metallurgical processes, and some of the associated minerals may mess-up the recovery procedure.

I am convinced that engineers and scientists will recognize and acknowledge that the approach presented in the present book needs to be introduced everywhere and not only in very few places. As a matter of fact, the small company needs it just as badly, if not more than the large one.

I wholeheartedly support the author's intention *to prove, by use of the appropriate reasoning supported by practical examples of actual case studies, that recoveries from most precious metal deposits can be planned in a way to yield appreciably higher values, if a clear representation of the distribution of the metal of interest is available. Logic suggests that, if blind attempts extract a certain amount of the metal, carefully designed projects based on the knowledge of the distribution of that same metal within the ore, must extract larger amounts. And in most situations, such amounts are sufficiently large to extensively justify the revised approach.* (page 2 in this book).

I am also convinced that her *approach will prove that, being the highest valued metals and therefore the most important source of wealth for the country that produces them, the precious metals need more attention, at least as much as is presently given to the base metals. After all, if gold is the most significant means of international payment and if it is often used as a backing for paper money, more systematic research for a better understanding of its occurrence and production is a very logical expectation.* (page 3 in this book).

Every exploration geologist and manager knows that the success of an exploration and mining venture depends on a number of factors, **imponderable** and **ponderable**. In order to move **imponderable** factors into the classification of **ponderable** ones, specialist knowledge is needed. The knowledge of a specialist has a similar function as additional drill holes, to move the ore body from the class of **possible** to the **probable**, or to the **calculated** or **proven** class of ore.

Dr. Gasparrini is one such rare specialist. She has put an enormous amount of knowledge and experience into her book and that will be of benefit to the whole spectrum of professionals active in gold exploration, the prospector, the geologist, the mineralogist and the metallurgist. Of course, it is also a good compendium for any manager who has to make decisions on factors involved in gold exploration or the search and recovery of any other mineral.

Dipl.Ing.Geol.ETH
Dr.Dr.h.c.mult.**G.C.Amstutz,**
Professor and **Director,**
Institute of Mineralogy and Petrology
University of Heidelberg

May 10 1990

FOREWORD II

MINERAL PROCESSING AND MICROANALYSIS. Mineral processing technology exists to prepare mined ore for subsequent processes which ultimately yield marketable products as metals or oxides.

In economic terms therefore, the mineral processing steps are required to fulfill two purposes:

- To insure that as high a proportion of the values in the ore as is economically possible report to the marketable products.
- To limit as far as possible the adverse effects on processing costs of ore components which wastefully consume reagents or energy, or contaminate the end products.

In terms of tonnage of ore processed, the mineral processing steps most widely used are based on achieving a physical separation of value from non-value minerals as they occur.

Very few mineral deposits (mineral sands are one obvious example of an exception) have been formed by purely physical processes of agglomeration of distinct mineral grains. In the majority of deposits, even if they were originally sedimentary, successive phases of alteration by chemical, thermal or mechanical (e.g. brecciation) forces have resulted in a texture in which most of the mineral grains are highly interconnected, chemically and physically, with their neighbors.

It appears therefore somewhat illogical to try to isolate by purely physical methods, mineral grains in a texture that has been created by chemical and thermal forces. From a scientific point of view, it would make more sense to use chemical or thermal methods to unravel what Nature has put together - i.e. to, so to speak, put the processes of ore genesis in reverse.

Nevertheless, mineral separation is widely used and is successful economically. Its limitations are in the complexity of the texture and in the fine size of some individual grains that may be generated by comminution. Historically, the number of economic ore deposits which are amenable to really efficient separation of valuable minerals (say a recovery of 90 per cent or more of the values into a concentrate comprising 90 per cent of that mineral) has never been great, has declined throughout the past century and is still declining. The boundary between the field of upgrading by physical methods and of chemical processing of minerals has of necessity moved nearer to the ore composition. Today smelters have to treat lower grade concentrates than they would like and concentrators have to sacrifice too much of their values into tailings in order to meet even the smelter's minimum specifications.

One may note a third consequence of the increasing textural complexity of mined ores which is of considerable economic importance. To find ore deposits amenable to processing by well established methods to mineral concentrates, even to minimum smelter specifications at recoveries which repay the mine owner, deposits are being exploited in more remote and undeveloped parts of the world and underground mining has gone to greater and greater depths. This has markedly put up the cost of ore exploration and exploitation through transport, infrastructure and site costs.

It is broadly true to say that scientific and engineering developments over the past 100 years in mineral liberation and separation have not kept pace with the world's need to treat deposits of increasing textural complexity, although many of the advances are in themselves impressive.

Because the processes of unravelling a *chemical* ore texture by physical means are inevitably imperfect, the application of any higher technology has now to be directed to gaining the extra percents of recovery or costs that are needed. One may instance as an example in a recent introduction of Hi-Tech, the application of methods of continuous analysis and computerized control systems. These have paid for themselves, even if the gains have measured in a few per cent.

The equipment and methods available for microanalysis of ore have advanced greatly in the last few years. We are now able to observe and describe with precision the texture and constitution of an ore. The challenge is both to get mineral engineers to take note of these descriptions, and to develop a new science of translating those descriptions into process practices which will eventually be considerably more efficient than they would be without access to the microanalytical observations.

How does one translate, for example, the visual or x-ray picture of an ore into quantitative data that will enable the engineer to foresee or explain the outcome of the application of the mechanical forces applied in comminution or the differential forces used in mineral separation? It is easy to see in the microcomposition, and that is independently of whether or not physical separation can be achieved by the *commonly used* mineral separation technology. This should save the engineer needless time and money in the test laboratory, and it may represent the first step towards the adoption of a new technology developed as a result of the knowledge of the ore microcomposition.

What the mineral engineer most seeks from the analyst of ore microcompositions is guidance on how the known physical properties of minerals (grain size, density, hardness, magnetic susceptibility, etc.) will apply in a particular texture. At the end of the day, processing ores by physical methods is always going to mean an application of mechanical forces for size reduction followed by exploitation of differential physical properties for separation, but there is a wide variety of methods and equipment to do this. To be systematic in selecting equipment in the first instance and optimizing the use of that equipment involves a great amount of trial and error test work and operation which is both time consuming and costly. The science of interpretation of ore texture to permit this systematic progression to production to be carried out through more rapidly and with less risk of being wrong can now be built on a solid basis of measurement of naturally occurring mineral textures.

If chemical or thermal ore processing is, scientifically and geologically, a more logical method of unravelling a natural association of minerals, it is also a less developed science than mineral separation. Perhaps the correct knowledge of the ore microcomposition will promote and help its use by future generations.

Hydrometallurgy is used only where mineral separation is impracticable. In terms of ore tonnage, the two biggest uses are for gold (because most gold occurs in grains too fine to be collected by gravity separation) and aluminum (because the aluminous and waste components are chemically bound).

Iron ore provides the only significant example of ore treated by thermal fusion processes, albeit with some physical separation. Ore processing by fusion, as pointed out in chapter 6 (page 63), makes microcomposition almost irrelevant since the structure is totally destroyed and the end products are largely predictable from thermodynamic and kinetic data. For more complex base metal ores it may become more widely employed since there is significant development of real promise going on at the present.

A scientific prediction of hydrometallurgical ore processing is more difficult even than it is for mineral separation for two main reasons:

- If a physical mineral separation process achieves only, say 70 per cent recovery of a value, it is generally possible to explain with some precision why the 30 per cent have been lost in tailings. The 30 per cent may be identified in particles too fine to be collected, or in mineral composites. For leaching processes, it is **seldom** possible to find the comparable answer, **unless** one is prepared to develop an entirely new approach, based on the knowledge of the ore microcomposition (see comments on pages 276-278). Theoretically, a chemical reaction between a liquid and a solid which can attain 70 per cent completion, should continue in time, to 100 per cent completion. Some mineral may be shrouded from contact with solution by insufficient liberation or formation of impervious reaction products, but this physical explanation is usually only partial for the *lost* recovery.

- Mineral chemistry is a much less well known science than is the measurement of physical properties of minerals. Even where measurements of reaction chemistry and kinetics have been made on samples of a pure mineral, their applicability to a mineral grain in its own ore texture is not always precise.

The data in the book indicate that the studies by Dr. Claudia Gasparrini of the relationship of gold leaching to ore microcompositions, have probably reached further than for other ores, and they may be taken as a model for the interpretation of hydrometallurgical processes: Gold however, being nearly always present as the same and chemically unbound mineral, is a relatively simple metal to follow in leaching. More difficult are the associated cyanicide minerals. The mineral chemistry of silver, uranium, copper, nickel and other metals which can be present in a multiplicity of mineral forms and environments are as yet more difficult to describe in quantitative and scientific terms, but the knowledge of gold leaching is providing a platform for advances. Perhaps, by showing the importance of the correct knowledge of ore microcompositions, this book may represent the first step towards the systematic characterization of any product to be treated for metal extraction, before and during the metallurgical test work.

The science of hydrometallurgy is different from that of mineral separation, but the challenge to microanalysis is similar: to assist process testing by providing quantitative data on the behavior of minerals in their ore texture. Also, as with physical separation, the aim will be to provide the metallurgist with the means of getting better extraction and saving in reagent and energy costs.

Philip M.J. Gray F. Eng.

London, December 21 1988

PREFACE

Gold and Other Precious Metals - From Ore to Market has the unique distinction of being the only scientific book in recent years to be conceived, researched, written, illustrated, typeset, brought to a complete professional looking form and financed by one person alone. Although many people provided help and advice during the final stages of the preparation of the hard bound version, the original paperback version, circulated during 1990, was produced entirely by myself in my United States office.

Such an effort proved well invested, because the response that I received, after I made available the paperback version, was so positive and so sincere, that it more than compensated for the limited support during its preparation stages. Orders came from countries such as Australia, Austria, China, Chile, Denmark, Finland, France, Germany, Great Britain, Greece, Haiti, India, Ireland, Italy, Namibia, Saudi Arabia, Spain, Sweden, South Africa, Switzerland and Uganda, in addition of course to the United States and Canada. The orders were followed by letters and telephone calls expressing approval of what they considered as a very ambitious enterprise to be initiated by one person alone. The encouraging words that they conveyed may very well be responsible for the finished hard bound copy that has now finally become available: I would have probably given up the project, had it not been for those letters and telephone calls. The advice and suggestions that came with them are responsible (at least partly) for the many text changes that can be recognized in this new edition.

The most provocative and significant quality of the book is probably in its format designed to include subjects that are usually published separately. There is presently no other book **written by one author only** that addresses topics belonging to both the earth and the metallurgical sciences. I do not know how successful my effort to bridge the gap between the two fields was, but I am fully aware that, given its pioneering quality, the book is far from perfect. First attempts often are. For such reason, the book's main quality should be recognized in the emphasis that it places on the approach to the study and the actual planning of the precious metal extractions. Having shown the potential, through the appropriate reasoning, as well as through practical demonstrations, it is now up to the newer generations to accept the importance of such an approach, then to develop and perfect it. I will consider my effort successful, if those new generations do recognize this quality and use the book accordingly, for the understanding of their precious metal occurrences.

Having discussed the manuscript extensively, I am confident that many errors have been recognized and eliminated. And, being all the people with whom I have spoken eminent representatives within their fields, I know that the book itself now meets with the general approval of the mining community. Still, should the readers discover inaccuracies that were overlooked, I would be more than grateful if they were to point them out to me. All constructive criticism will be taken into consideration and acknowledged in future editions. After all, **constructive** criticism is what makes progress possible.

Some colleagues raised objections regarding the base metals and their definition. They suggested that the base metals should be taken out, because they do not belong in a book on precious metals. They pointed out that I have included within the category elements that are not always classified as base metals. I did not consent to making any major changes regarding the inclusion and handling of the base metals and have given reasons why in several occasions (for example, on pages 3, 7, 46, 227). I have discussed their complete definition on pages 4 and 226 (footnote). The general term of *base metals* is is adequate for the purposes of the book, whenever reference is made to elements other than precious; and, having the readers been made aware of what and why I refer to under the general heading of *base metals*, the scientific accuracy has been maintained.

Among the colleagues who provided the most substantial help and advice, my very special and sincerest thanks go first of all to Prof. Dr. G.C. Amstutz, Director of the Institute of Mineralogy and Petrography of the University of Heidelberg, Germany, and one of the foremost representatives within the field of Applied Mineralogy. Prof. Amstutz technically reviewed several versions of the manuscript, wrote the flattering foreword and provided much support in the actual publication by Springer-Verlag.

Other individuals who extensively discussed portions of various drafts or helped in other ways are, in alphabetical order: Bart Cannon, President of Cannon Microprobe in Seattle, Washington; Greg Davison, Mineralogist with Lakefield Research of Canada Limited, in Lakefield, Ontario; Philip M.J. Gray, Consulting Metallurgist in London, England; Arnold F. Griffin, Director of Micron Research (W.A.) in Osborne Park, Australia; David A. Lowrie, President of Argyle Ventures Inc., in Toronto, Ontario; David E. Lundy, Executive Director, and the staff of the International Precious Metals Institute, in Allentown, Pennsylvania; Maurice Magee, Geology Consultant in Tucson, Arizona; Thomas P. Mohide, Consultant and Retired Director of Mineral Resources, Ontario Government, in Toronto, Ontario; Roger Runck, Editor of **Precious Metals News and Review** in Columbus, Ohio; John C. Taylor, President of Ian Reimers and Associates in Oakville, Ontario; Edmund von

Pechmann, Mineralogist with Uranerz in Bonn, Germany. Additional assistance was provided in a less formal manner by numerous other colleagues who phoned or wrote with various suggestions. Although these are too many to mention individually, I do feel deeply grateful to each one of them.

Barbara A. Chappell, Chief Librarian with the US Geological Survey in Reston, Virginia; Rosemary Swan, Head of Reference and Circulation with the Geological Survey of Canada Library in Ottawa, Ontario; and the Mintek Library in Johannesburg, South Africa, all provided literature and other information. Ms. Chappell was patient enough to discuss some technical topics on the telephone.

The photograph on the front cover, representing magnified crystals of gold, was used with the kind permission of Frank L. Fisher, Director of External Affairs, BP Minerals America, formerly Kennecott Minerals Company in Salt Lake City, Utah. The photograph on the dedication page, representing incrustations of native gold on quartz, was provided by J. A. Mandarino, Curator of Mineralogy with the Royal Ontario Museum in Toronto.

Permission to publish excerpts of consistent length from various books was requested and granted by the Society of Mining Engineers of AIME (Marianne Snedeker), the Metallurgical Society of AIME (Judy Parker), MET-CHEM Research Inc. (David C. Schneller), the Colorado School of Mines Press (Marty Jensen), the Ministry of Energy, Mines and Resources - Canada (R.G. Blackadar), the Encyclopaedia Britannica (Helen L. Carlock). As for the data taken from scientific reports prepared by myself for the mining industry, much of such data was obtained over a period of approximately 12 years, starting from 1976. Permission to publish was not obtained, due to the often lack of availability of the persons and even of the companies that submitted the projects. In view of such difficulty, only data were used from reports where no written or verbal or in other way understood agreement as to confidentiality existed, or was stipulated prior to the author's acceptance of the project. Use of the data is within the laws of the United States and of Canada.

Finally, last but certainly not least, I would like to thank Dr. Wolfgang Engel and Ms. Monika Huch, both with the Geoscience Editorial Section of Springer-Verlag in Heidelberg, Germany, for their assistance during the preparation of the final hard-bound version and for the suggestion and use of color photography. Perhaps the difference between the original and the present versions of **Gold and Other Precious Metals - From Ore to Market** gives a good indication of the importance of an efficient team work. Their kind hospitality while I visited their offices, and their patience in answering all my questions during the several telephone calls that preceded the actual submission of the finished manuscript will also never be forgotten.

Dr. Claudia Gasparrini

September 19 1991

CONTENTS

INTRODUCTION

This is the age of computers and of facsimile machines, the age of super-fast airplanes and of space exploration. An age when in many fields of technology, nothing seems impossible. An age when progress keeps on happening at a faster pace than it ever has in recorded history. Against such a background of major scientific advances, it becomes difficult to explain why recoveries of precious metals are today, for many deposits, not significantly better than they were 50 years ago. Despite the incredible progress of modern technology within the theoretical fields of the earth sciences and metallurgy, many properties which were considered uneconomical before the two major increases in the price of gold (during the seventies, and then again during the early eighties), still remain just as uneconomical today. In fact when trying to asses the progress within the field of precious metal mining\extractive metallurgy against most other scientific fields, one cannot help observing that:

- Techniques for extraction have remained essentially unchanged for the past 100 .years, at least based on the open admission by many metallurgists, as well as on much of the available literature.
- As a consequence, the approach to the base and precious metal extractions, during the planning stages of recovery, is still very similar, despite the many differences, most important of which being the market value and the abundance in the ore.

The low quantities of the precious metals within their ores require different more specialized methods for their observation and study, than those used for the base metals. Instruments and techniques for the characterization of fine particles occurring in trace amounts have been available for at least three decades. It is against all principles of progress and economic growth to ignore such techniques, while not exploiting a deposit to its full potential.

The more conservative among us insist that progress **has** actually taken place within the applied fields of the earth and mining/metallurgical sciences. Perhaps it has. And perhaps, when planning precious metal extractions, well equipped laboratories **are** used whose mineralogical/chemical data are produced in a manner which interacts closely with the metallurgical test work. But as long as recoveries from many properties remain as unsatisfactory as, or little more satisfactory than they were before the recent technical advances, there is in fact no evidence proving the existence of such progress. Even if studies **are** conducted, it is difficult to come to any conclusion other than that they are either poorly planned or not interpreted correctly. Despite the many arguments to the contrary, it may be difficult to deny that presently the common North American approach to recovery still often involves a few attempts at extraction with little understanding of the actual processes taking place. If the amount extracted is sufficiently high to justify the investment, the property is exploited. If not, the property is abandoned, while exploration continues for more suitable areas of development.

Exploration and, more so, discovery of new deposits, will not go on forever, and the need for a more efficient exploitation of the deposits that have already been discovered may soon become apparent.

Reasons for the limited *visible* progress within the precious metal industry are obviously complex. One however cannot help pondering that efficiency and quality of recovery would be considerably higher, if the above mentioned studies, were conducted **systematically** by use of the newer techniques for the characterization of ores and in a manner that gives complete and clear representations of the precious metal distributions; and if such studies were then **interpreted correctly.**

The purpose of this book is to prove the full extent to which representations of the precious metal occurrences in their ores are possible and the manner in which these help improve their recoveries by the already existing extraction methods. It is the author's intention to demonstrate, by use of the appropriate reasoning supported by practical examples of actual case studies, that recoveries from most precious metal deposits **can** be planned in a way to yield appreciable higher values, if a clear representation of their distribution is available. Logic suggests that, if blind attempts extract a certain amount of the metal, carefully designed projects based on the knowledge of the distribution of that same metal within the ore, **must** extract larger amounts. And in most situations such amounts are sufficiently large to extensively justify the revised approach.

* * * * * *

The subjects described, such as mineralogy, mineral deposits, mineral chemistry, extraction processes and techniques, are usually found in books or articles which are either related to the earth sciences or to the mineral processing/metal extractions, two commonly separate fields of University training. These subjects were included here together because, for the purpose of the recovery process, they belong together and should be taught and studied together. A metallurgist will not be in a position to design the most suited and most efficient extraction plan for a certain metal from a certain ore, unless he is familiar with the occurrence of that metal within that ore. An exploration geologist will be unable to evaluate a newly discovered deposit, unless he understands the processes involved in the successful extraction of all of its metals. Hopefully the author's approach will be an incentive for future scientists and mining engineers to include both the earth sciences and the extractive metallurgy in their research projects and study plans.

The other area where the author did not conform to the more traditional formats, is in the amount of space given to the precious and base metals. Many more pages are dedicated to gold, silver and platinum group metals, than they are to most of the base metals. Readers may feel that the treatment of the base metals is superficial and far from complete when compared to other sources in the available literature. There was a reason why the base metals were included in a book whose main subject is precious metals, and only treated superficially. Hopefully those same readers will realize that such is the manner in which gold and other precious metals are usually treated in many metallurgical texts: in a superficial and incomplete manner. For example, the **Encyclopaedia Britannica - Macropaedia** in an article entitled **Industries, Extraction and Processing** (1988) dedicates approximately 28 pages to iron, 7 to aluminum, 4 to copper 4 to zinc and 3 to lead, all of which are treated in separate chapters. The same reference dedicates 2 and 1/2 pages to gold, 1 and 1/2 to silver and 1 and 1/2 to all of the platinum group metals in subsections of a chapter entitled **Production of other industrial minerals**. The **SME Mineral Processing Handbook** (1985) dedicates approximately 34 pages to iron, 20 to aluminum, 147 to copper, and 53 to lead and zinc together, but only about 35 to all six of the precious metals (29 to gold, 3 to silver and 2 and 1/2 to all of the platinum group metals).

Perhaps the author's approach will prove that, being the highest valued metals and therefore the most important source of wealth for the country which produces them, the precious metals need more attention, at least as much as is presently given to many of the base metals. After all, if gold is the most significant means of international payment and if it is often used as a backing for paper money, more systematic research for a better understanding of its occurrence and production, is a very logical expectation.

* * * * * *

Precious (or noble) metals are gold and the platinum group elements, namely ruthenium, rhodium, palladium, osmium, iridium and platinum. They are so called because they do not frequently combine with other elements, but rather occur in their ores in the metallic state or in the form of compounds with a limited number of other elements, which seldom succumb to secondary alteration and weathering. Therefore their occurrence in alluvial or placer deposits. In general terms, their definition is based on their ability to remain unaltered under conditions which would affect all other metals. Precious metals trade at the highest market values. Their abundance in the Earth's crust and in the economic deposits is low (in the ppm range, for gold and platinum). Silver, although unstable under certain conditions (it tarnishes to the sulfide phase in the presence of some of the sulfur compounds), is usually regarded as a precious metal along with gold and the platinum group elements, because of its scarcity and high price, when compared to many of the base metals. Silver is considered a precious metal in this book, because the microanalytical studies and extraction techniques require an approach and an amount of accuracy which are similar to those used for the precious metals. Although modern metallurgical classifications subdivide the non-precious metals/elements/compounds into *base metals* (high tonnage heavy metals), *light metals* (aluminum, beryllium, magnesium and titanium), *rare elements* (zirconium, yttrium, rare earths, gallium, germanium) (P.M.J. Gray, personal communication), and *industrial minerals* (barite, boron minerals, clay, feldspar, etc.), in this book the general term of **base metals** is used when referring to all the economic elements characterized by lesser stability and resistance to alteration. The market value is generally lower than that of the precious metals, often considerably lower. For many of them, the abundance in the Earth's crust and in the economic deposits is higher than that of the precious metals by factors of up to more than 1 million (representative of 1 per cent and more).

* * * * * *

Before the actual description of the technical topics, a discussion of the incorrect practices, misconceptions and improper or incomplete teaching and training was considered necessary. This was by no means intended as a destructive criticism of the mining industry or of the universities as they operate today. Any criticism, if taken as such, has a constructive purpose: no system will improve unless the faults that keep it from running at its full potential are pointed out and corrected. In this context perhaps it should be made clear that any reference to misconceptions or incorrect practices refers to situations in which the author was directly involved in Canada and the United States only. Therefore, although prevalent, such misconceptions and incorrect practices are not representative of the views, opinions or way of operating of all the earth scientists, metallurgists and mining engineers in the two countries, or in the rest of the world.

Development of the author's ideas and techniques for the study of precious metals was affected by her two year experience in South Africa where studies of precious metal ores appear to be extensive and systematic.

The information presented is the result of the author's experimental studies of ores and metallurgical samples, of discussions with colleagues and of literature search. Some material is derived, in an appropriately revised manner, from a number of her earlier articles. In those instances where her knowledge on the subject was limited, or where she felt that the information already available needed no interpretation, excerpts were included *as published elsewhere* without changes or comments. The purpose of such an approach, that is of republishing selected data already in print, is to provide the reader with information which all belongs in one book, thus avoiding tedious referrals to books and other publications which may not always be readily available. A large amount of this information has been taken from the **SME Mineral Processing Handbook** (1985), probably the most comprehensive volume on the subject. Because of the author's interest in conveying clarity, rather than information which may be too technical to maintain the interest of several categories of scientists, some data were taken and used, in a revised format, from the latest edition of the **Encyclopaedia Britannica** (1988), one of the most accurate references addressing the widest variety of readers.

The data, including the case studies, photomicrographs and figures were selected or prepared in a manner which, if appropriately interpreted, makes them of practical value. They are representative of situations which are general enough to apply to the individual readers' extraction problems. The case studies were all taken from actual projects conducted on samples originated from precious metal and other deposits, although the specific location of the deposit is not usually mentioned. Such policy was adopted because the book is not about deposits as much as it is about the techniques used for the study and understanding of deposits. Names and localities therefore do not generally matter. And perhaps more important than that, because many of the studies were conducted on limited numbers of samples, not necessarily representative of the bulk of the ore. Had the name of the locality been mentioned, conclusions drawn about the deposit itself and its composition could have been erroneous and/or misleading.

A strong effort was made to present the information in as clear a form as possible for at least two reasons. One is the obvious one, and that is to maintain the reader's interest and attention. The other reason results from the author's concern to reach a wider audience than the one usually addressed by this type of scientific treatise and which includes individuals with a training in both the earth sciences and metallurgy. Because of such an effort, many concepts were described in a very comprehensible manner and definitions were included which might have been unnecessary, if writing for a more restricted number of readers.

* * * * * *

The book is divided into three parts which together include twelve chapters and the conclusions. Part I addresses theoretical topics which serve as an introduction to the following section. Part II describes gold, silver, the platinum group elements and uranium in detail, and the base metals in very general terms. Part III includes a few considerations on the importance of the knowledge of the microcomposition of the ore in mineral exploration and the conclusions.

- *Chapters 1 and 2.* These are the chapters which discuss the limitations in the present approach to precious metal recovery planning. They provide a general view of the state of the art of the mineral industry, as interpreted by the author (perhaps not necessarily shared by all of the readers), and include an attempt at elucidating the reasons why precious metal recoveries are not always planned in a manner which would produce optimum recoveries. The chapters elaborate on the two concepts mentioned above, that is the slow progress of the precious metal industry during the past 100 years, when compared to other fields of technology, and the superficial approach to precious metal ore characterization. A description follows of the mineralogical studies of the precious metals, as conducted today, with an analysis of the most obvious reasons for their limited contribution to the understanding of the recovery problems. That is the lack of communication between metallurgists and earth scientists. It is assumed that such lack of communication is largely due to insufficient or improper university training. The two chapters are not of direct scientific or technical nature. Their importance is in the emphasis which they place on the flaws of the present approach to the study and characterization of the precious metal ores, **before** any attempt at recovery is made, and on some suitable revisions.

- *Physical and chemical character of ores* lists the different variables which affect recoveries and which should all be accurately determined or kept in mind when extractions are planned. Although essential for the treatment of any metal, these variables are more complex and their accurate determination more critical, if successful recoveries of the precious metals are expected.

- *Methods and techniques for the study of ores* describes how to obtain or determine the above variables, and covers the advances within the field of microanalysis up until about the mid 80's. Progress has taken place since that time, but is not discussed because the book was conceived and written in the mid 80's. And because it is assumed that the average laboratory (in North America or in other parts of the world) does not have the most recent equipment at its disposal. The basic design of microanalytical instruments such as the electron microprobe and the scanning electron microscope (SEM) is outlined. Emphasis is placed on the importance of the microanalytical techniques for the understanding of the precious metal occurrences, more so than for the understanding of the base metal occurrences, prior to any recovery attempts.

- *General principles of mineral processing* describes the basic processes used for metal extractions. The chapter provides a very limited amount of new information, and has been included, as an alternative to a glossary, relegated to the end of the book, and essential to the earth scientist for the comprehension of the more technical metallurgical terms and practices. Should any of the readers with metallurgical training or background find this chapter too concise, or perhaps even incorrect, they may want to keep in mind the just as concise and often incorrect short chapter on ores and on minerals that is always included in the metallurgical texts.

- The chapters addressing the *gold, silver and platinum group elements* include material previously published by the author (Gasparrini 1983-A, Gasparrini 1984-A, Gasparrini 1984-B), presented in an **extensively** revised and updated form. The reader will find that many more pages are dedicated to gold than to the other precious metals. That is largely because there was in the 80's considerably more interest in gold than there was in any of the other precious metals, therefore more information has become available.

- *Differences between gold and silver.* The two metals are often treated by using similar techniques despite their numerous differences. A detailed separate list of such differences may help explain the differential gold and silver recoveries sometimes experienced from ores where gold and silver occur together.

- *Uranium.* Although not classified as a precious metal, uranium is described in a separate chapter and in more detail than the other base metals, because it is a complex element the accurate study of which is often necessary for proper extractions. An extensive description was also given because, due to its radioactive nature, the techniques for the study of the uranium ores are in some ways different from those used for the ores of other metals. The chapter includes revised material from a previously published article (Gasparrini and Williamson 1981).

- The chapter *on base metals* describes the mineralogy and metallurgy of many of the *non precious* economic metals/compunds (as defined on page 4), as well as their more common applications, year of discovery and most important countries producing them. The chapter provides concise, but adequate information, not usually available within one book. One of the reasons why the chapter was considered necessary (to emphasize the importance which should be given to the precious metals), has already been mentioned (page 3). Descriptions of the base metals and of the techniques used for their study were also included to demonstrate the differences between the two groups of metals and to point out the necessity for differential approaches to their study, when planning the recovery process.

- The chapter on *mineral exploration* gives examples where the knowledge of the microcomposition of the ore may help in the discovery and evaluation of newly found deposits.

- Finally, in the *conclusions*, a short discussion is presented of how and to what extent the microanalytical techniques can actually assist in the planning and understanding of the recovery process. Topics of research are suggested, to provide some much needed data that will contribute to the optimization of the precious metal extractions.

* * * * * *

The book introduces the word *microcomposition*, not commonly in use among earth scientists or metallurgists. The word refers to the mineralogical and chemical components of an ore, their distribution and variation, all on a **microscopic** level. A complete knowledge of the composition of ores on a microscopic scale has only recently become possible through the availability of very advanced microanalytical instruments and techniques. A new terminology, related to these new techniques seems like a logical expectation. Because related to the field of microanalysis, microcomposition appears to be an appropriate word within such a context.

Mineralogy, or *applied mineralogy*, the closest available terms for the representation of the same concept, were not considered sufficient for a complete description of an ore on a microscopic scale and down to the limits made possible by the microanalytical techniques. These terms were in use long before the introduction of the microanalytical instruments, therefore, they do not describe the ore to the same extent. Mineralogy, traditionally defined as the *science and study of minerals*, does not cover all the variables which need to be determined for the planning of an efficient process of extraction, particularly if dealing with gold and other precious metals. It is rather intended for minerals of coarser, often macroscopic sizes or sizes which can be resolved by microscopes with lower power magnifications. The word mineralogy does not extend to the chemistry of the ore or of the minerals within the ore on a microscopic scale and down to the micron size. Because gold and other precious metal ores require more detailed examinations involving the extensive use of the microanalytical techniques, when compared to the base metals, we can probably say that the knowledge of the microcomposition of the ore is necessary where planning recoveries of the precious metals, while the determination of the mineralogy is sufficient where studying most of the base metals. After all, the more common of the base metal minerals were discovered, observed and defined long before the microanalytical techniques became available. But many precious metal minerals were only recognized and classified as such, following the advent of the electron microprobe and SEM. For all of these reasons, the word microcomposition is used throughout the book when describing the precious metals and uranium, the word mineralogy is used when describing the base metals. And by the same token, the word *microanalyst* is used when referring to a scientist who is competent and involved in the study of ore microcompositions, while the word *mineralogist* is used to indicate an expert in mineralogy, as intended in its more traditional sense.

Terms such as *free gold, locked, encapsulated gold particles*, and *refractory/refractoriness*, commonly used within the gold mining industry, were not used, other than for their definition (on pages 66, 104 and 105). These terms are applied without the direct observation of the actual ore or gold on a microscopic level and therefore are not accurate. **Ideally**, if the ore microcomposition is known and the correct extraction technique for that microcomposition is employed, no ore is refractory.

The more theoretical among the mineralogists may find that some of the mineral names are not identical to those favored by the International Mineralogical Association, for example **niccolite** instead of **nickeline.** The names chosen for this book are those most commonly used by the mining industry in North America. Given the prevalently practical purposes of the book, the author assumed that actual names make no difference, as long as the nature of the compound is made clear.

Finally, an *overbearing consultant in gold* is mentioned in some of the case studies in connection with the improper handling of metallurgical problems. Such is a fictitious character and the result of the author's view of the attitude of a small number of metallurgical consultants and other scientists whose advice is sought and valued by the average mining company. The character **does not exist,** neither is it representative of the way of operating of **all** of the metallurgical people. The case studies however are real. The character is used for an indication of the type of hurried reasoning and approach, sometimes employed while planning a recovery, and of how misleading such an approach can be in the light of the findings of accurate microanalytical studies. The character is also used to prove that incorrect interpretations of the microanalytical data may be just as damaging as their lack of use. The representation of an attitude and interpretation of data by an actual person, even if fictitious, can be more convincing and have a stronger impact than just the use of abstract descriptions of incorrect practices.

PART I

CHAPTER 1

THE MINERAL INDUSTRY - PRESENT STATE OF THE ART

During the past 30 years and largely following the advent and development of computers, an incredible advance has taken place in many sectors of modern technology, for example medicine, dentistry, the automobile industry, the TV and stereo, the appliances industries. This advance has happened so quickly that a wide intellectual gap has developed between the older and younger generations. Many people over 40 cannot believe that a child under 10 understands and operates computers better than they can, if they can at all.

In the theoretical fields of the earth sciences and metallurgy, the same rapid development has taken place through the availability of high magnification/high resolution microscopes, image analyzers, electron microprobes, ion probes and SEM's, all capable of giving detailed descriptions of mineral occurrences and particles down to 1 or 2 microns (1 millionth of a meter) in size. Computers attached to such instruments provide for the data to be processed within relatively short times, thus resulting in the production of large quantities of accurate information which only a few decades ago would have been considered impossible. The speed at which everything has been changing has been remarkable. For example, the information on the microanalytical instruments in chapter 4 of this book, up-to-date during the 1980's, became partly outdated by the time the manuscript went to the presses.

In the midst of such rapid scientific progress, the mineral industry, particularly in the field of precious metal extractions, has generally developed while maintaining a slower pace. Changes in metal recovery techniques during the past 100 years do not appear to be as dramatic as those registered within other fields of technology. And most important improvements, such as those related to the development and use of flotation for mineral beneficiation, to an increase in size and capacity of the process equipment and to the recognition of several new hydro and pyrometallurgical processes (Anthony 1988), apply largely to the extraction of some of the base metals. The precious metals seem to have been neglected. For example, the cyanide process, the most common technique for gold extraction, has remained basically unchanged for more than 100 years, with the exception of the introduction of a few variations, such as heap leaching, pressure leaching, etc. (pages 120-121). Amalgamation is a much older technique, as it is reported to have been practiced as far back as 1000 BC (Anthony 1988). T.K. Rose's book **The Metallurgy of Gold,** whose first edition was published in 1894 and whose 7th edition was out of print in 1945, was recently reprinted by Met-Chem Research Inc. (1986) in Boulder, Colorado and is still regarded as one of the most comprehensive treatises on gold. Less information is available on the metallurgy of silver and the platinum group elements. Fire assay, used for the analysis of precious metals world-wide, has been practiced since Roman times (Rose, 1986). Textbooks on fire assay by Bugbee (1981) and Smith (1987) are recent poorly revised reprints of books the first editions of which were published in 1922 and 1913. Very little research has been and is being conducted on the subject, despite the awareness within the mining industry that there are situations where the technique may be faulty (pages 92-95).

Progress does not appear to have proceeded at a faster pace within the applied field of the earth sciences, at least not if we judge by the available literature. Boyle's book **GOLD - History and Genesis of Deposits** (1987) includes a collection of the most important papers on gold deposits, and more than two thirds of them were published before 1975. Many were published in the 1800's. Although probably the most recent and comprehensive reference on gold within its field, the book lacks significant information on deposits such as Hemlo in Ontario, Mesquite in California, and Pueblo Viejo in the Dominican Republic, all reported to be some of the potentially largest gold producers in the Americas.

The above observations are confirmed and extensively demonstrated by Anthony (1988) who expresses concern about the methods used within the mineral industry, hardly different from those current in 1900, while at the same time drawing attention to the astonishing rates of technical development in most other fields. Anthony's conclusions are supported by detailed comparisons of the present state of the mineral industry with that of many other technical fields. They are illustrated by practical examples of the pace at which various instruments and equipment belonging to different fields of technology have developed (information and paper provided by P.M.J. Gray).

Anthony also states that investment in research and development in mining is presently smaller than in any other sector of technology. It is three times as low as it is in the chemical industry which is one of the lower spending industries.

When observing such slow progress, one could actually speculate that a more conservative attitude within the mineral industry could be a consequence of the manner in which nature takes care of our needs on this planet. Unlike products from the animal and vegetable worlds, the reproduction of which can be controlled, minerals come in limited supply. Were we to extract our metals too fast and too soon, there would be little left for the future generations to exploit. Extent of need may also be an explanation: if we only extract the metals in amounts related to our needs at a particular time and at the proper pace, efficiency and modernization may not always be necessary. Perhaps it is more than a coincidence that many of the light metals were discovered and/or isolated shortly before the advent of the space age and space exploration, where they found extensive applications in the construction of spacecrafts, airplanes and missiles. These are however philosophical considerations which should not affect the purpose of our book: that is to point out the errors and to open the way to the modernization of an industry the present state of which may not be as advanced as potentially possible.

The other oddity, again possibly the consequence of the precious metal industry not keeping up with the latest developments, relates to the study of the precious and base metals: very much in the same manner, when characterizing the ore in preparation for the recovery process. Such an approach could be a consequence of the fact that many of the United States geologists, metallurgists and mining engineers (or at least those with whom the author has had significant exchanges), have limited knowledge or understanding of the gold and other precious metal occurrences in their ores on a microscopic level. Most of them appear to believe that gold occurring in amounts below 1 oz/ton, or 34.2 parts per million (ppm), cannot be seen or detected, therefore gold occurrences in economically feasible ores, cannot be studied in detail, and certainly not in any better detail than the base metals can. Such belief is not as strong and irritatingly persistent in Canada, yet literature from other countries, for example Russia (Smirnov 1951), suggests that a similar conviction may not be limited to some of the mining people in the United States.

Given the differences between precious and base metals (pages 3-4), the conclusion that a complete knowledge of the microcomposition of the ore is essential for the recovery of the precious metals, far more so than for the recovery of the base metals, is logical. And the need for comprehensive studies by techniques different than those used for the base metals and prior to any recovery attempts, becomes obvious: because the precious metals are so valuable, we want to be able to extract as much of them as possible; because they occur in low quantities and in a finely disseminated state, separation, particularly during the beneficiation and leaching stages, can be complex. Strangely enough however, much has been and is being published on the study of the applied mineralogy of the base metals (for example, Amstutz 1961, 1972, 1982, and 1983, Henley 1981, AIME publications of the **Process Mineralogy Committee**, including one symposium volume per year since 1981, Amstutz et al 1987). Yet, when it comes to the precious metals, *generally* very little emphasis is put on the need for differential techniques for the systematic characterization of the ores, before any recovery attempts are made. When information on precious metal deposits (gold deposits most commonly) and recovery techniques is published, actual data on the metal occurrence and distribution is limited and usually represented by a very small number of identified particles, not sufficient for the description of the entire ore. Many studies are conducted on the high grade portions or on concentrates derived from high grade portions, neither of which are representative of the occurrence throughout the majority of the deposit. Lately (mid 80's) **The Metallurgical Society (TMS) of AIME** suggested the formation of a new committee in which the precious metals were being dumped in the same group as nickel, cobalt and copper, possibly because the precious metals are sometimes recovered as byproducts of these metals. A **TMS** committee on copper, nickel, cobalt and precious metals was actually formed and lasted for at least one year, before a separate committee on precious metals finally came into existence. This trend is consistent with the observations made in the introduction, regarding the available literature on the metallurgy of the base and precious metals (page 3).

There are organizations, such as the **European Precious Metals Industry**, the **European Association of Metals (Euromeatux)** and the **International Precious Metals Institute (IPMI)** that encourage the exchange of technology and information on precious metals. For example, the **IPMI**, whose headquarters are in Allentown, Pennsylvania, does so through the distribution of a newsletter titled **Precious Metals News and Review** and through the patronage of one conference and several seminars every year. During such events, miners, refiners, producers, users, research scientists, mercantilists, government and other representatives of the precious metal community have the opportunity to present their new findings by delivering lectures, as well as by participating in informal gatherings and discussions. The **Institute** has been publishing the proceedings of these conferences and seminars in hardbound volumes since 1979 (details in the references). The contribution to such volumes on the part of representatives of the mining industry, and of the mineralogical sciences in particular, is however rather limited.

There is presently no accepted systematic technique for the study of the precious metals in their ores. The method used is essentially the same as the one applied to the study of the base metals. In many laboratories it is assumed that the availability and use of a microscope (it does not matter what make and with what magnifications) is sufficient to provide answers to most metallurgical problems, whether related to the precious or the base metals. In order to collect reliable data on precious metal ores, a microanalyst, alias a person with a special training in the use of the microanalytical instruments and techniques (page 8), is necessary. And yet many average and even some of the larger North American mining/exploration companies do not employ a microanalyst on a full time basis. In fact they do not even employ a mineralogist/petrographer. When characterizing an ore containing either precious or base metals, a selected number of samples are sent to a mineralogist or petrographer to provide thin section studies of the transparent minerals on a contract basis. Ore microscopy for the description of the metal occurrence and distribution, and therefore equally, if not more important than the thin section work, is requested on a smaller number of samples, if at all. These samples may or may not be the same as those of which thin section descriptions were obtained. The company expects to pay similar fees and to have the results within similar amounts of time, whether dealing with a base or a precious metal bearing ore. Many mining companies believe that accurate microanalytical work is too expensive and too slow and that they can obtain just as good results from faster more reasonably priced sources. When viewed against the high costs involved in the development of a mining property, such an approach is illogical.

Not all mineralogists today are trained and experienced in the use of the ore microscope in combination with the electron microprobe and SEM. Fewer are familiar with gold (and other precious metal) occurrences and distributions within their ores. Consequently, when contacted by the mining company, they use a very similar approach, whether dealing with copper, zinc, or lead, occurring in amounts in the range of 1 per cent and up to 20 or 30 per cent, or with gold occurring in amounts more than a million times smaller (below one ounce per ton or 34.2 ppm). They prepare and look at one polished mount (two at the most) per sample. The study does not take longer than half an hour, perhaps one hour, if conducted by *conscientious* investigators. Sometimes, hoping for positive results by different techniques, these mineralogists have someone else (often a technician with very little knowledge of the earth sciences, and none of the gold occurrences in ores) analyze the same polished mount in the SEM, in the electron microprobe or even in the ion probe. Naturally not being completely clear with what he needs to find, the technician spends even less **productive** time searching for gold and finds just as little as the mineralogists did. Finally they all conclude and report that the gold is in solid solution in one of the major minerals *in amounts that elude detection.* In pyrite, if pyrite is present, or in some other sulfide. Unfortunately they neglect to report or probably even to consider that perhaps there was no gold on the surface of that single polished mount examined. The chances of ppm amounts occurring within a two-dimensional space 1 inch/2.5 cm in diameter are extremely small. Naturally not every mineralogist functions in such a disorganized superficial manner, but the author has met a sufficiently large number of them in influential positions, to want to describe and comment on their way of operating. We will call such "experts" *unaware* mineralogists.

In addition to the expertise, an important requirement for a microanalyst (and mineralogist) is that he communicates effectively with the mine geologist, with the metallurgist and with the decision making staff of the company that has engaged his services. Such is the only way in which he can become familiar with the actual nature of the deposit, then convey the correct information back to all of those individuals. Being the expert, it is logical to expect that he is the one to decide on the choice of the type and number of samples that are representative enough for the proper description of the precious metal distributions. Or at least that he is the one to discuss such choice. He may need to visit the exploration or mine area, to observe the natural conditions and the location where the sampling took place. When solving problems of beneficiation in cooperation with the metallurgist, the microanalyst may need to examine the mine sections from which the unresponsive ores originate. Insurance of constant communication of the microanalyst with his colleagues in exploration, mining/beneficiation, and chemical separation, is a must for any mining company interested in obtaining the best recoveries from a property. But communicative skills are not very common, neither are they generally sought in a microanalyst (or mineralogist).

The situation is certainly not encouraging, and yet it is not hopeless either. Although very few, there are individuals capable of successful interaction. There are occasionally some very progressive mining companies (large and small alike) that are aware of the important function of an experienced microanalyst who is both knowledgeable and communicative. There are universities (within and outside North America) that recognize the importance of the training of *aware* microanalysts who are academically competent and at the same time useful to the mining industry. And papers (for example, Baum and Gilbert 1989) stressing the importance of systematic studies of gold and silver ores, are in fact published. All such evidence suggests that, despite the appearances, the need for modernization **is** felt within the precious metal industry. Hence, the author is confident that, being the awareness of the need for a new approach already existent; and being such approach occasionally applied for the teaching and understanding of the precious metal ores, it is only a question of time, before the use of systematic studies by the more modern techniques becomes accepted by the mining community at large.

* * * * * *

A book entitled **The Extractive Metallurgy of Gold** by J.C. Yannopoulos was published in 1991 by Chapman & Hall in London. According to the review in the August 1991 issue of **Precious Metals News and Review**, this book emphasizes the methods employed during the past 15 years, although some historical methods are also mentioned. The book also includes a very wide bibliography with 240 references. Another book, entitled **Mineria del Oro** by Hagel (1991), is a review in Spanish of the world gold mining, which extends from the classification and description of the major deposits to the actual extraction techniques. Unfortunately, because **Gold and Other Precious Metals** had already been delivered to the publisher, the author did not have an opportunity to examine such new references in any detail for inclusion and more extensive discussion at the beginning of the chapter.

CHAPTER 2

REASONS FOR POOR PLANNING OF PRECIOUS METAL RECOVERIES

Instruments and techniques for the study of gold and other precious metals in their ores are available. Nobody ever denies the importance of the knowledge of the microcomposition of the ore for proper recovery planning. Yet it seems as if *generally* very little is done towards the use of systematic studies before separation is attempted. Let us discuss in some more detail how some metallurgists and mineralogists function and try to understand why their way of operating is largely responsible for such an apparent contradiction.

Presently, a person in charge of the recovery process, who understands the intricacies of the ore microcompositions, as well as of the metallurgical processes is seldom available. There are earth scientists and there are mining engineers/metallurgists. And there is a nearly total lack of communication between the two groups: very few are familiar and comfortable with both fields. In other words, very few metallurgists appreciate the complexity of the earth sciences. Very few earth scientists are prepared to accept and satisfy the needs of the metallurgists. We have seen that communication is an essential requirement for the successful study of precious metal ores (page 15).

The average metallurgist/mining engineer is not aware of the chemical and mineralogical complexity of rock formations. The attitude towards the studies of precious metal ores has already been mentioned (page 14). Besides wanting answers from relatively quick, superficial examinations, many engineers expect the solution to metallurgical problems from the composition of a small number of samples only. The samples are sometimes chosen at random, in which case they are in no way representative even of portions of the ore studied. Sometimes they are chosen from the high grade sections of the ore, upon the assumption that the gold or other precious metals will be easier to see. These samples again are not representative of the ore under study. No matter how accurate the microanalytical data, they will provide little insight into the ore microcomposition.

Case study 1. Use of data collected from unrepresentative samples chosen from the high grade portions of the ore. This is a case where *the overbearing consultant in gold* was involved in the planning of a recovery project related to a gold deposit in the Western United States. A microanalytical report was available describing some coarse gold concentrated from a small number of samples derived from some very high grade portions of the ore, and containing 50 to 200 times as much gold as the rest. The very small number and their origin made the samples oddities rather than anything else and definitely unsuitable for use in the understanding of the general microcomposition of the ore. Yet the consultant provided very detailed descriptions of the gold extractions to be expected from such deposit, based on the data in the report. Needless to say, further systematic studies of the rest of the ore proved his comments and recommendations incorrect.

Studies of a few unrepresentative samples to understand and describe the composition of the entire deposit could be compared to the detailed observation of small portions of a human body, for example, an eye, a kidney and a toe (taken from the top, middle and bottom) and then to the assumption that the in-depth knowledge of these three parts is sufficient for the understanding of the functioning of the entire body.

There are situations where sufficient microanalytical data are available, yet their interpretation, often provided by consultants to the metallurgical laboratory conducting the test work, is incorrect. This results in additional lack of communication.

Case study 2. Incorrect interpretation of microanalytical data. *The overbearing consultant in gold* read a report describing silver losses (yes, he consults in silver as well) due to impermeable coatings which become more numerous with fine grinding, because of the larger amount of available particle surface (see page 171 for details). He praised the report extensively as one of the best he had ever read, then advised the metallurgical laboratory to use as fine a grinding as possible. Results of his recommendation, naturally, were decreased silver recoveries and loss of interest on the part of the company behind the project, in spite of a perfectly sound and accurate report.

There are other situations where the microanalytical studies are conducted after attempts at recovery have been made and the results of the test work are unsatisfactory. In many of these situations, the samples submitted often relate poorly to those tested, or the project is planned in a hurried way which seldom provides the necessary answers.

Case study 3. Use of samples which relate poorly to the samples tested. A practical example of how the microanalytical studies should have been conducted and how they were actually conducted during the course of a project to understand poor recoveries, is given in charts 1 and 2, at the end of the chapter. Purpose of the study was to determine the proportions of coarse native gold associated with the sulfide and of invisible (probably colloidal) gold associated with the clay. Such data would have provided leads as to the fraction of the ore and amount of gold which needed special attention, due to the clay interference.

ference. During the microanalytical study, the percentage of invisible gold could not, for obvious reasons, be determined from direct microscope determinations. However, its calculation would have been possible, based on the observation of the fire assay/cyanide soluble gold data and of the microanalytical results, had the study been conducted in an organized manner, and on the correct samples. Chart 2, showing the incorrect approach, indicates that, although the microanalytical work was actually conducted, this was conducted on the wrong samples, and not in a systematic way. Insufficient material was submitted for the project to be successfully completed. It produced little positive results. Charts 1 and 2 are discussed again in chapter 4, pages 57-58.

The input provided by what we decided to call the *unaware* mineralogist (page 15) contributes very little to improved communications. Perhaps it should be mentioned that answers based on assumptions, such as those regarding the gold solid solutions in pyrite, are less helpful than no answer at all. They are actually damaging, because they confuse the metallurgist by concentrating his attention on the wrong mineral and on the wrong manner of occurrence. The assumption that gold occurs in solid solution in the pyrite is probably one of the most widespread misconceptions regarding gold in North America. We shall see that solid solutions are very rare, if existent at all, and pyrite, although a common host to gold, is not always the host, when present in a gold-bearing ore.

In addition to data which are superficial and sometimes inaccurate, many such *unaware* mineralogists give little advice on the choice of the samples or on the necessity for the study of representative portions of the deposit. We have seen (page 15) that such advice is seldom expected from them. Consequently, they only limit their services to the examination of whatever samples are submitted. It is not very often that they produce mineralogical reports giving answers to metallurgical problems, as they tend to include an excessive amount of academic detail, often related to some of the major minerals, rather than to the precious metal occurrences.

Results of this lack of communication among scientists belonging to such closely related fields, are skepticism on the part of the metallurgist and mining engineer in the usefulness of studies of the ore on a microscopic level, and frustration on the part of the earth scientist who finds his work unappreciated.

* * * * * *

The above mentioned lack of communication could be interpreted as the result of the type of knowledge acquired at the university level by earth scientists and extractive metallurgists.

At the risk of being repetitive, this **is not** intended to be a criticism of the type of teaching offered by some modern universities, it is only an interpretation of the facts, as viewed by the author. Representatives of universities may disagree or have different opinions on the subject.

The correct or incorrect training is almost totally in the hands of the universities. It is easier to point a young person in the right direction before he/she has had any opportunity to develop false ideas or to make important decisions in the treatment of ores, than it is to convince someone like *the overbearing consultant in gold*, endowed with more than average self confidence, that it could be his techniques that cause the less than satisfactory recoveries.

There is a considerable amount of theoretical geology/mineralogy taught in geology departments and just as much theoretical metallurgy taught in metallurgy departments. All of these subjects go into so much detail that when the average earth scientist and extractive metallurgist complete their university training, they are each skilled in subjects which are seemingly unrelated to those of the other. The subjects are so specialized that sometimes neither of them understands the language of the other. The terminology may be different, even when referring to the same subject. Whatever each knows about the subjects within the other's field is incomplete and, from the author's experience derived from attending university courses in extractive metallurgy, often incorrect.

A professor of mineralogy describes the 32 classes of crystallographic symmetry, lists the several hundred minerals existing in nature and illustrates their physical and chemical properties. This is all very important information, yet one cannot help pondering that such importance would better manifest itself, if that same professor were to make his students aware of the significance of the function of some of the mineral properties, when related to the extraction processes. For example, he might mention the relation between anisotropism and physical/chemical behavior in processes such as crushing and grinding, leaching, etc. Mineralogy courses offered by the average instructor make little mention to the importance of mineral abundance and grain size during the process of extraction. They place considerably more emphasis on the identification of the silicate minerals in thin sections than on the identification of the sulfide/sulfosalt minerals in polished mounts. This is obviously a consequence of the larger amount of available data on the *rock forming minerals*, the most abundant minerals in the Earth's crust. And yet, if the sulfide/sulfosalts bear so many economically important elements, it seems logical to expect that the student should become familiar with these minerals, at least as much as he/she is with the silicate minerals.

A professor of economic geology, training students in the metal occurrences in their natural form, describes the textural and structural features of a mineral deposit, and lists all the hypotheses and possible explanations for the origin of such deposit. This information is essential for the understanding of the exploration techniques leading to the **discovery** of the metal deposits. Very seldom does he refer to the manner in which the occurrence and distribution of the metal of interest relate to its amenability to mining and metallurgical recovery. Yet such information is just as essential for the understanding of the **extraction processes** to be applied to the metal in question.

Many geology departments in the United States have no adequate sample preparation facilities. They rely on outside technicians hired on a contract basis for their thin sections and polished mounts. Some expect the students to prepare the samples. Observations of mineral particles in a mount the surface of which has been prepared by an inexperienced person, and usually observed under an average to subaverage microscope, can and often give an incorrect description of the mineral occurrences.

Techniques related to the use of the SEM or electron microprobe, essential for the study of fine particles, when taught, cover many theoretical topics such as the Montecarlo techniques or topics which are not related to the study of two-dimensional particles in polished mounts. For example, the stereo images. The instructor is often a physicist or other scientist not trained in the earth sciences or metallurgy and is generally unfamiliar with the study of fine particles occurring in trace amounts and with their significance in relation to the metallurgical processes.

A professor of extractive metallurgy proves obvious information such as the one related to the notion that a small particle dissolves faster in a fluid than a large one, or, as he expresses it, *the reaction rate is proportional to the inverse of the particle radius.* When teaching about the hydrometallurgy of metals, for example silver, he explains in detail the chemical processes which lead to the silver dissolution in a particular solvent (sometimes not the one commonly used during the actual extractions), but he mentions very little about the number of silver minerals existing in nature and their different solubilities. Or about the necessity to grind down the ore to a fine enough size for liberation or exposure of the silver particles, before the process of dissolution can take place. He mentions very little about the complexity of ores and the necessity for a good understanding of the metal occurrence, before any treatment is applied. In this light, the time spent showing the processes leading to the silver dissolution is almost totally wasted, in that it confuses the student by making him/her believe that recoveries by hydrometallurgical techniques are simply a question of using the correct chemicals, chemical strength, pH, and other variables which are only **partly** related to the actual metal separation. Unfortunately such is the common teaching system today, and perhaps more unfortunately, such is the approach to precious metal extractions in many metallurgical laboratories.

Some of the readers with metallurgical backgrounds may object that it is the present book that makes it appear as if extractions depend entirely on the microcomposition of the ore. In fact one reviewer of an earlier draft already made such an observation. Well don't they? If the gold is completely enclosed in an impermeable host mineral, thus **never** coming into contact with the dissolving agent, then what advantage would derive from the use of faster or more effective dissolving agents? And if the silver occurs in minerals which tend to tarnish, thus becoming unreactive, how does the understanding of the reaction processes during silver dissolutions help in the development of a successful recovery plan?

A few among the larger universities and mining schools are already aware of the necessity for a different type of training, such as the one suggested in this chapter. Hopefully a revised teaching system and reorganization on the part of the majority of the universities, will in time lead to improved communication among the different scientists responsible for the extractions of the precious metals, therefore to a better understanding and a more efficient handling of the recovery processes.

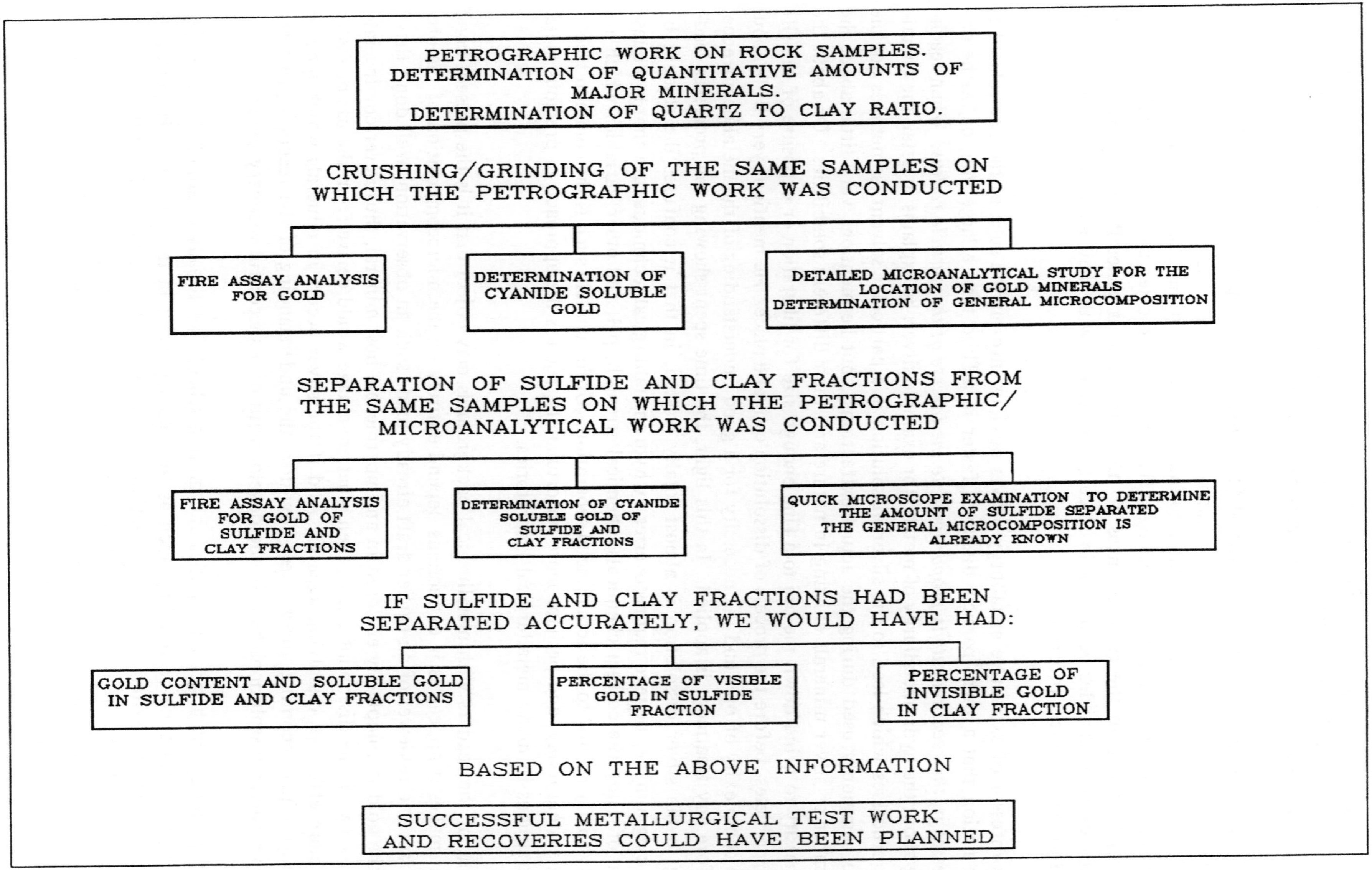

Chart 1. *Steps that should have been followed to determine the gold distribution in some clay rich samples.*

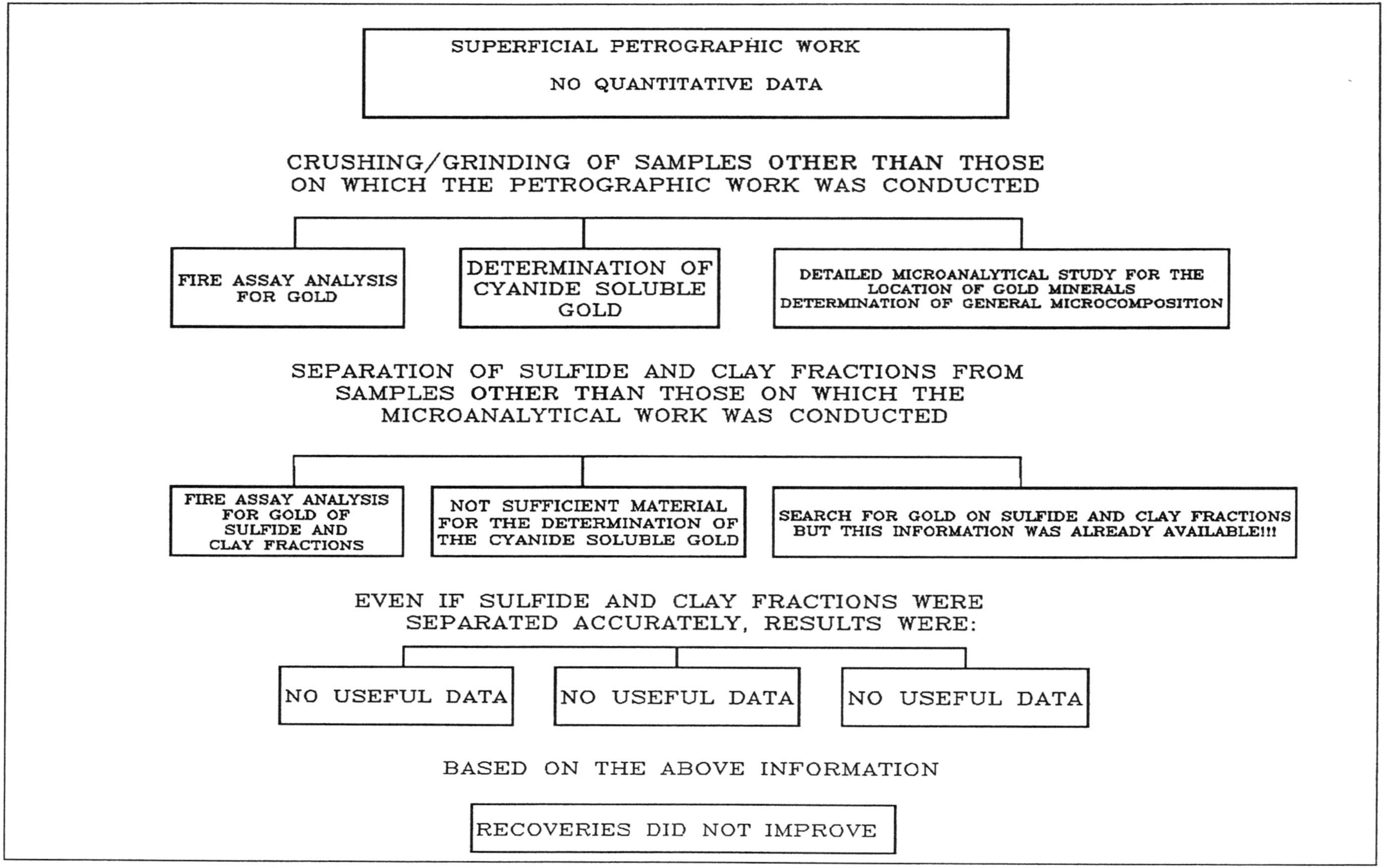

Chart 2. *Steps that were followed during the study of the gold distribution in some clay rich samples.*

READERS PLEASE NOTE. The photomicrographs shown with the text were all taken by using an **SM-LUX-POL Leitz** microscope and reflected light, except for those in figures 48, 49B, 51 and 53 (uranium chapter), 65 (page 266) and 66 (page 272), which were taken by using transmitted light.

All references to the "Soviet Union" or the "USSR" should be read as the "former Soviet Union" or the "former USSR".

CHAPTER 3

MARKET VALUE AND OTHER VARIABLES AFFECTING RECOVERY DECISIONS AND RECOVERY COSTS

INTRODUCTION

So precious metal deposits are more complex than metallurgists believe. Let us examine what makes the applied study and understanding of these deposits complex.

When an economic metal is discovered in concentrations above background, there are many variables that affect first the decision to go ahead with the extraction, then the actual recovery. In this chapter we will try to classify the different variables which affect the decision to exploit first, then the planning of an efficient recovery strategy.

Exploitation of an ore may be economically justifiable at one time, when the market value of the metal of interest is high, but not so when the market value falls below a certain level. Recovery costs will always need to be estimated to be lower than the total earnings derived from the sales of the final concentrate or purified metal. The first and most important variable therefore, affecting the decision as to exploitation of an ore, is the **market value** of the metal of interest.

Market value varies with the metal and with time, based on availability and demand. In general, "demand for precious metals is determined by people's want, demand for base metals by people's needs." (T.P. Mohide 1988, personal communication).

Other information necessary for estimates of recovery costs first, and then for the planning of efficient recovery strategies, extends to a variety of chemical and mineralogical data which could be summarized to include the **metal geochemistry**, **the general ore chemistry**, the **type and physical character of the host rock** and the **mineralogy/ore microcomposition**. Detail varies, when determining such information, depending on whether we are dealing with base or precious metals: the precious metals require higher accuracy which extends well into the microscopic level, and which is provided only by the more advanced microanalytical instruments. Such higher accuracy is needed during the planning stages as well as during the actual recovery.

It follows then that familiarity with the geochemistry of the metal, ore grades and a *reasonably accurate* mineralogical report conducted with an average microscope, are usually sufficient for a good indication of ore exploitability and costs, when dealing with most base metals. The same information provides limited insight, when dealing with precious metals. Let us find out why.

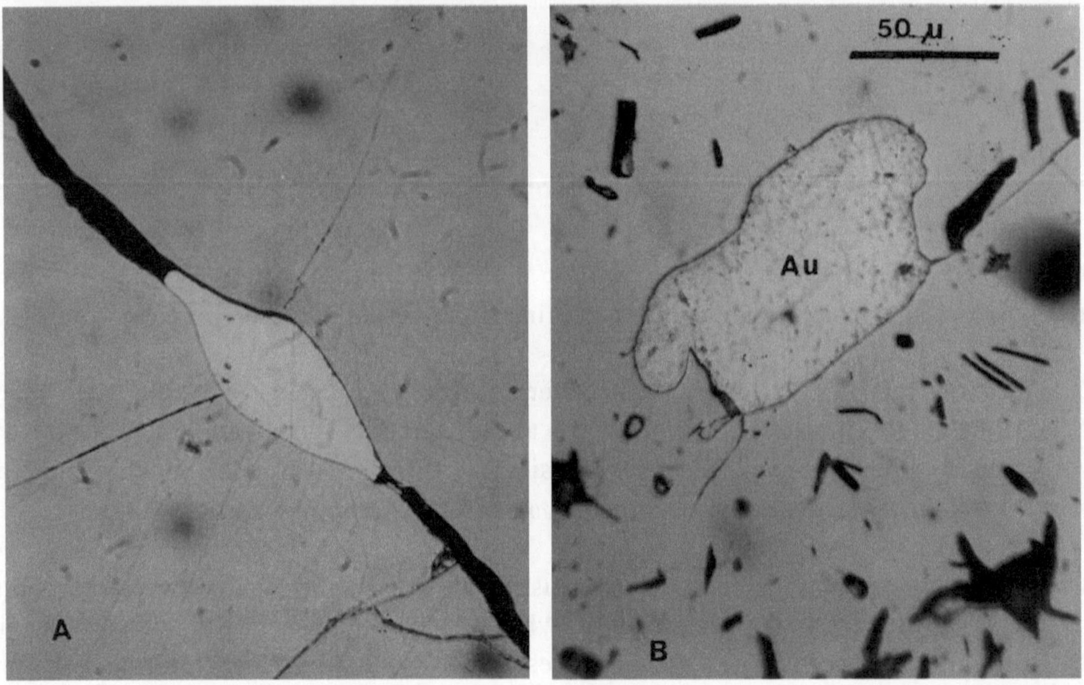

Fig. 1. *Native gold in pyrite requiring different accuracy of approach for satisfactory recoveries. The gold in A occurs in a fracture and a coarse grinding is sufficient to expose, or to make it in other ways available to the cyanide solution. The gold in B is totally enclosed and fine grinding sizes (similar to those of the native gold) will be necessary, if satisfactory recoveries are expected.*

It is generally known that, when extracting the same or similar amounts of gold from a number of different ores, recovery costs and techniques are not necessarily the same. Special processes involving higher expenditures such as very fine grinding, chemical techniques such as roasting before the actual separation step, may be necessary for the treatment of one ore, but not for the treatment of another. Figure 1, showing photomicrographs of occurrences representative of two ores in which the native gold is distributed predominantly in fractures (1A) and in a totally enclosed state in the pyrite (1B), is a good practical example. Although both ores contain similar amounts of gold, the one represented by figure 1A is likely to give the easier recoveries by the standard techniques. The abundant fractures will promote: (i) breakage and therefore particle exposure (in general fractures and gaps are directions of lesser resistance), and (ii) cyanide penetration. Successful gold recoveries therefore will only require relatively coarse grinding sizes and lower expenditures. The ore represented by figure 1B, where there are no fractures or gaps to ease breakage and particle exposure, or to allow cyanide penetration and contact with the gold, requires grinding sizes as fine as those of the gold, therefore higher expenditures. A particularly expensive extraction would be the one where the average size of the native gold is finer than the finest to which the ore can be ground (10 to 15 microns), because chemical methods will then be necessary to obtain particle exposure before any leaching is attempted.

Acquisition of the above type of information requires the observation of the ore on a microscopic level which often extends down to the micron size range: much of the economic gold today occurs in particles which are considerably finer than those shown in figure 1. Consequently determinations and knowledge of all the variables affecting recovery decisions and costs **on a microscopic level**, are essential when dealing with gold and other precious metals much more so than when dealing with base metals. The *average* mineralogical report **does not** provide the needed answers. In order to be of assistance, such answers **need to be** collected through detailed observations and chemical analyses of representative samples by the microanalytical techniques.

Awareness of the necessity for accurate determinations of the variables affecting recovery costs and techniques, leads to the planning and execution of appropriately detailed studies for the understanding of the ore. A thorough knowledge of the ore then allows for proper budgeting and for correct experimentation with techniques chosen to apply to that particular ore, or to various individual sections within, if the ore is complex.

METAL GEOCHEMISTRY

Geochemistry is the study of the occurrence (**abundance** and **distribution**) of the elements in the Earth's crust. The abundance and distribution of an element depends on its physical and chemical properties which in turn are related to the electronic configuration of the atom and vary with the composition of its nucleus. Recurring characteristic properties of elements with increasing atomic number, allow for the classification within the periodic table. If we regard the deposit as a small scale reflection of the Earth's crust (as suggested by the theories of fractal geometry, page 289 and Mandelbrot 1982), then knowledge of the element's geochemistry within the Earth's crust may give a good indication of the geochemistry of that same element within the deposit. Knowledge of the chemical behavior of elements in the natural environment, assists in the understanding of their behavior during the metallurgical processes.

Abundance. The correct abundance of the metal of interest in the ore is an important variable when making decisions regarding ore exploitability. Each metal is characterized by a minimum exploitable amount which is related to its market value, thus, the higher accuracy required for the extraction of the precious metals. The abundance in economic ores is for many metals directly related to the abundance in the Earth's crust. Low quantities, if economic, are often coupled with high market values.

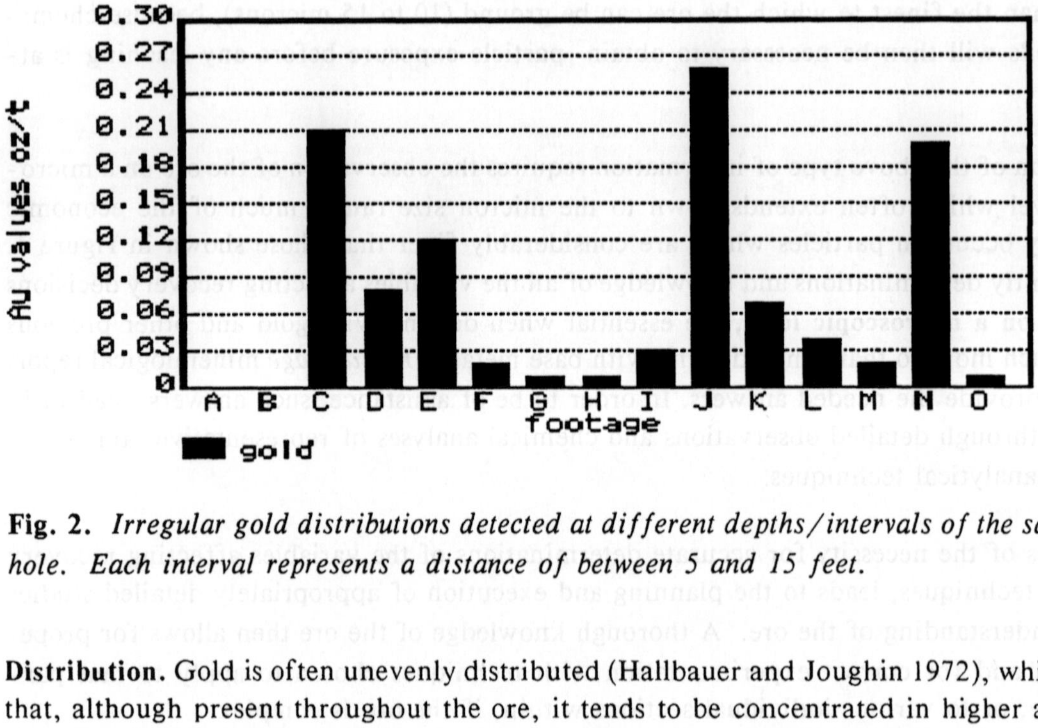

Fig. 2. *Irregular gold distributions detected at different depths/intervals of the same drill-hole. Each interval represents a distance of between 5 and 15 feet.*

Distribution. Gold is often unevenly distributed (Hallbauer and Joughin 1972), which means that, although present throughout the ore, it tends to be concentrated in higher and lower quantities within intervals that can be as small as a few feet apart (figure 2). High and low quantities may vary between the limits of detection of the analytical technique used, and more than 1 ounce per ton (34.2 ppm). Consequently, knowledge of the correct gold distribution becomes particularly important, as it may assist in decisions whether to exploit the bulk of the ore, or the high grade portions only. It may then lead to the choice of the portions of the same ore to be exploited. A similar reasoning applies to the study of the other precious metals, when unevenly distributed.

Exploitation of the high grade portions of the ore only may in fact give better results by consuming less of the recovery agent and by saving time. Sometimes perhaps even by producing higher gold extractions. This last alternative would be particularly suitable for ores where the higher concentration levels are coupled with higher solubility in cyanide, like in the following case study.

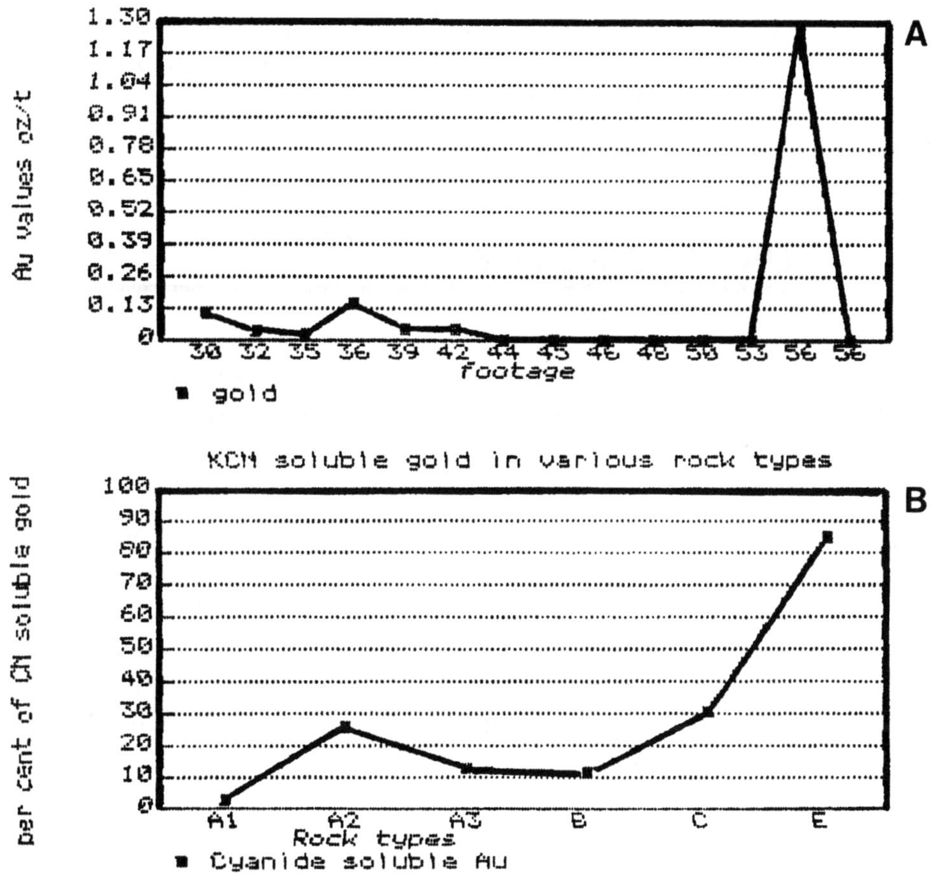

Fig. 3. *Diagrams showing a very high concentration of gold (top diagram, level 56) which is coupled with a much higher cyanide solubility (bottom diagram, rock type E).*

Case study 4. High gold concentrations/high cyanide solubility. During the microanalytical study of a number of gold-bearing samples where the gold was always associated with the pyrite, it was found that at least one level of the ore had a gold content which was approximately 65 times as high as the average for the deposit (1.30 oz/ton or 45 ppm, against 0.02 oz/ton or 0.7 ppm average, figure 3A). The pyrite content of such level was not significantly different from that of the other levels, thus misleading the less accurate researchers relating the gold to the pyrite abundance. The study also found that the gold in the high grade level leached four times as well as the gold in the other levels, with recoveries higher than 80 per cent against values of 5 to 30 per cent obtained from the rest of the ore (figure 3B). This resulted in extractions from the gold rich level which were between 250 and 300 times as high as those from the rest of the deposit.

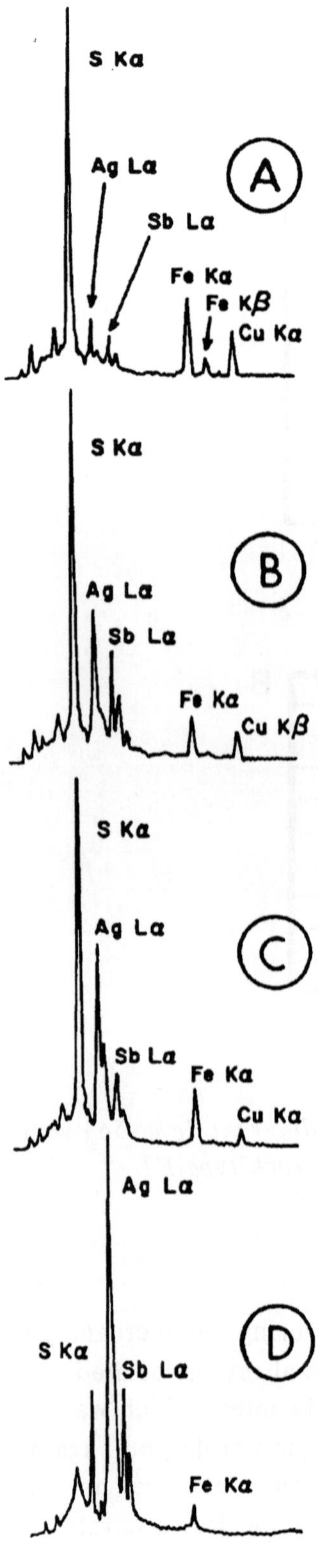

Detailed microanalytical studies to define the exact size of the high grade gold level, as well as to locate any other high grade levels, were recommended as the logical development of the project, after the discovery of such an interesting anomaly. These studies would be followed by decisions concerning the portions of the ore to be exploited, as against exploitation of the entire ore, practiced prior to the microanalytical study.

ORE CHEMISTRY

Knowledge of the general **ore chemistry** may in some instances be critical for the understanding of variations in recovery values.

Relative amounts of the major elements often vary within the same deposit. Spatial variations in the general chemistry of the ore may reflect in the crystallization of different minerals bearing the metal of interest. Some of these minerals may not respond well to the recovery agent commonly used, thus resulting in good extractions from some portions of the ore and poor ones from others.

Case study 5. Variations in the ore chemistry reflecting in decreased silver recoveries. Silver in a copper-zinc deposit, originally all reporting with the copper concentrates, started gradually to report with the zinc and to be lost in the tails, as different portions of the ore were being exploited. The microanalytical study, carried out on polished mounts of copper concentrates from which good and bad silver recoveries were experienced, as well as of zinc concentrates and tails with which some of the silver was reporting, determined that responsible for the differential separations was the presence of two silver minerals: tetrahedrite and pyrargyrite. Observations of chemical analyses representative of the general composition of the ore and electron microprobe determinations of mineral compositions, revealed that decreasing copper contents across the deposit were resulting first in a

Fig. 4. *X-ray spectra of three grains of tetrahedrite (A, B and C) and one grain of pyrargyrite (D) identified in a deposit characterized by spatial chemical variations.*

decrease in the copper content of the tetrahedrite, and eventually in the crystallization of a new mineral, the copper-free pyrargyrite. As long as in the tetrahedrite, the silver was reporting with the copper concentrates. Once in the pyrargyrite, the silver was reporting with the zinc concentrates or was lost in the tails. Figures 4A, B and C are the x-ray spectra of the tetrahedrite grains identified in copper concentrates from three consecutive runs. The spectra show decreasing heights for the copper peaks and increasing heights for the silver peaks. Figure 4D is the spectrum representative of the composition of the pyrargyrite identified in the zinc concentrates and tailings and containing no appreciable amounts of copper.

ROCK TYPES - PHYSICAL CHARACTER OF HOST ROCK

There are more than 50 **rock types** on the surface of the Earth's crust, distributed among the three major classifications of igneous (intrusive-extrusive), sedimentary and metamorphic rocks. An ore deposit often consists of several rock types. If such is the case, it is very possible that the metal of interest occurs in a different form in each rock type. Such an occurrence may result in the necessity for more than one approach to extraction. For example, grain sizes may vary, thus resulting in one rock type requiring a finer grinding size than another rock type, for liberation or exposure of the mineral of interest. The mineral in which the metal of interest is concentrated may change, like in case study 5 discussed above, thus requiring different flotation or leaching techniques for its separation. The host mineral may vary (like in case study 10, pages 113-114), which again may require a different approach to recovery.

The **physical character of a rock**, for example its resistance to breakage, or its permeability to solutions, may vary even within the same rock type, depending on the degree of compaction or fracturing, the degree of oxidation, weathering, etc. The physical character of the rock/mineral in which the mineral of interest occurs or with which it is associated *(the host rock/mineral)*, affects its resistance to crushing and grinding. It may also reflect in different extraction values, when employing hydrometallurgical techniques such as those applied to the recovery of gold and silver: a compact impermeable rock or mineral needs fine grinding to liberate or expose the mineral(s) bearing the metal of interest before the actual leaching step; a porous or fractured rock, on the other hand, does not need as much grinding. In many instances a rough grind may be sufficient, because fracturing and porosity allow penetration and contact of the leaching solution with the particles of interest. Fracturing also promotes breakage and therefore, particle exposure, as discussed in the example of figure 1A.

Fire assay and metallurgical test work indicate that oxidized and weathered rocks are often more amenable to heap leaching than their unaltered counterparts. Figure 5 is an example of how mineral porosity affects recoveries more so than the actual gold mineral grain size. Figure 5A is of a relatively coarse grain of native gold totally enclosed in a compact pyrite, figure 5B shows very fine particles of native gold distributed within the hollow layering of a concretionary grain of secondary iron oxide. Although finer, the gold in figure 5B is recovered easily by heap leaching, due to the cavities and gaps and to the higher solubility of the secondary iron oxide. The gold in the pyrite of figure 5A requires fine grinding for its exposure and will not be recovered unless more accurate techniques are used. Gold in compact pyrite, if occurring with sizes similar to those in the iron oxide of figure 5B would require roasting before any attempt at recovery.

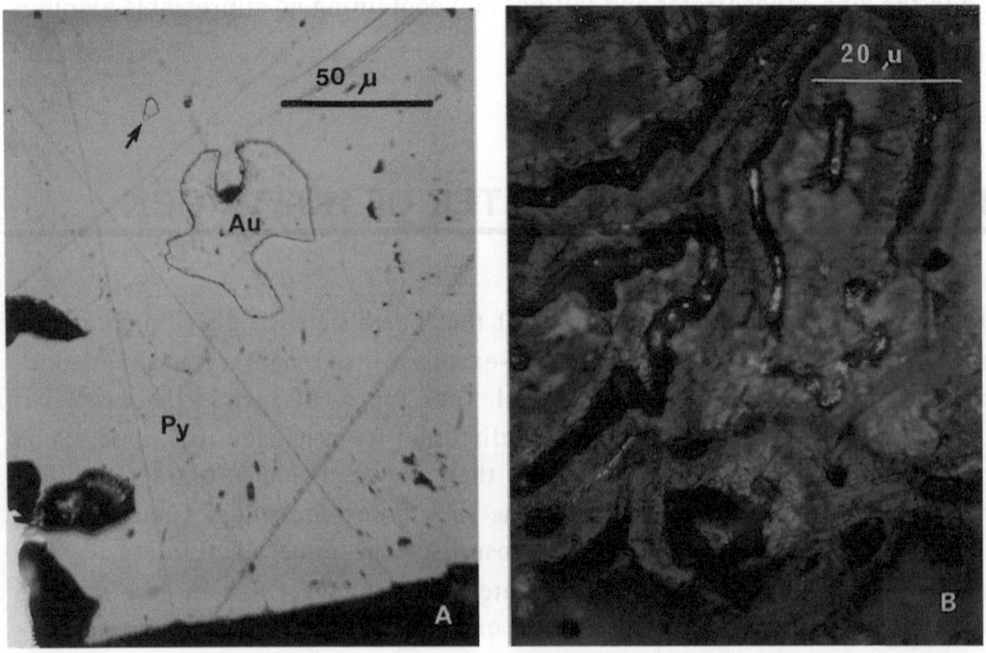

Fig. 5. *Native gold in pyrite (A) and secondary iron oxide (B) giving different types of recoveries, as a direct result of different host mineral permeability. Although much finer, the gold associated with the secondary iron oxide leaches freely, due to the ease of penetration of the cyanide into the hollow layering, thus only requiring a coarse grinding. The gold associated with the pyrite requires fine grinding for satisfactory recoveries, because it is compact and free of fractures. Host mineral permeability is the reason why unresponsive or impermeable pyrite becomes amenable to cyanidation, following the roasting process which converts it into iron oxide.*

The two case studies below are additional examples where the gold leachability is directly related to the rock or mineral porosity of the oxidized/weathered material, rather than to the grain size of the native gold.

Case study 6. Different leachability of oxidized and unoxidized gold ores. Some gold ores from a deposit in the Western United States consisted of unoxidized material where the native gold was coarse (20 to 50 microns in average size) and associated with the pyrite, and of oxidized material where the native gold was fine (1 to 5 microns in average size) and associated with the secondary iron oxides and various silicates. Despite the much finer grain sizes, heap leaching techniques were considerably more effective on the oxidized ore than they were on the unoxidized ore, the average cyanide soluble gold being 55 to 85 per cent in the oxidized zone and 5 to 30 per cent in the unoxidized zone (figure 6).

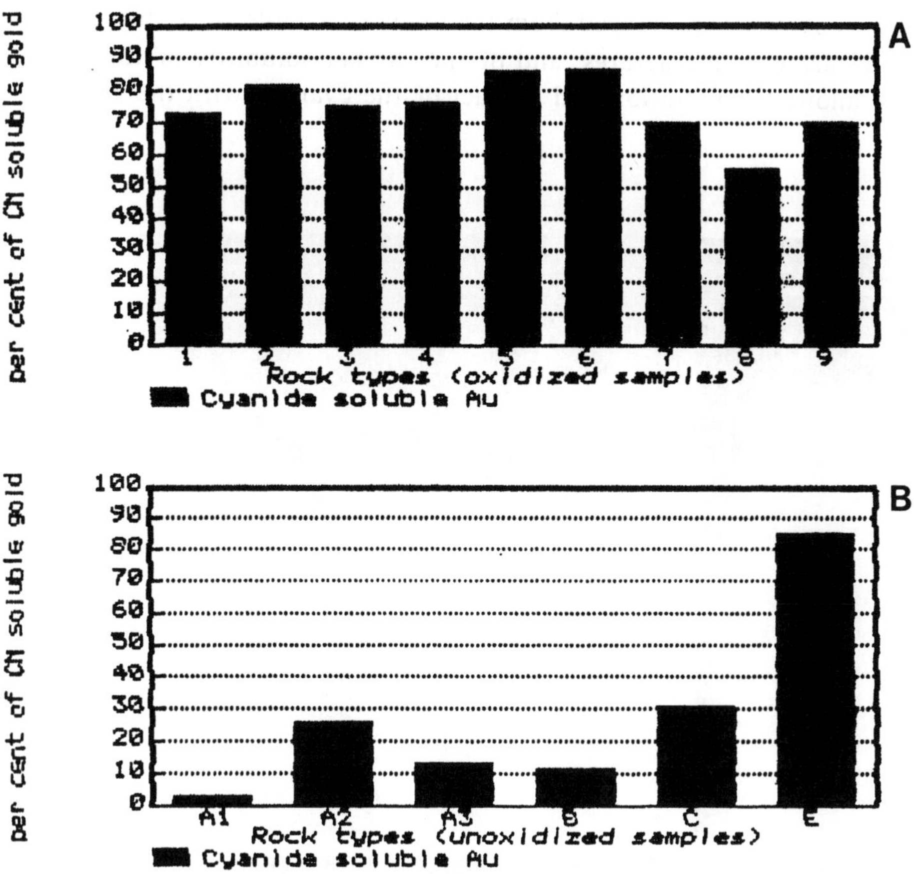

Fig. 6. *Solubility in cyanide of gold in different rock types from the oxidized (A) and unoxidized (B) zones of a deposit in the Western United States. Except for sample E (discussed in the case study of Fig. 3), the unoxidized ore is less amenable to cyanidation.*

Case study 7. Different leachability of oxidized and unoxidized gold-bearing pyrite crystals. Two samples, both consisting of pyrite crystals coated by secondary oxidation and therefore very similar when observed by the naked eye, gave significantly different recoveries, when treated by heap leaching techniques. The microanalytical study found that, although similar on a macroscopic level, the two samples were actually very different when observed under the microscope. The pyrite that did not leach well was almost completely unaltered (figure 7A). Much of its gold occurred as coarse totally enclosed particles (figure 7B) of native gold (with a few tellurides), while only a small fraction was in fractures and gaps. The pyrite that leached well was more than 50 per cent altered (figure 7C) and contained fine gold predominantly distributed in the fractures of the secondary iron oxide (figure 7D). Poor recoveries from the unoxidized sample were largely due to the totally enclosed state of the majority of the native gold particles within the pyrite. The coarser sizes did not contribute to improved recoveries, because of the impermeability of the host mineral. Good recoveries experienced from the oxidized sample were the result of the predominant gold occurrence in fractures, and of the higher solubility of the secondary iron oxide, which allowed penetration and contact of the cyanide with the gold.

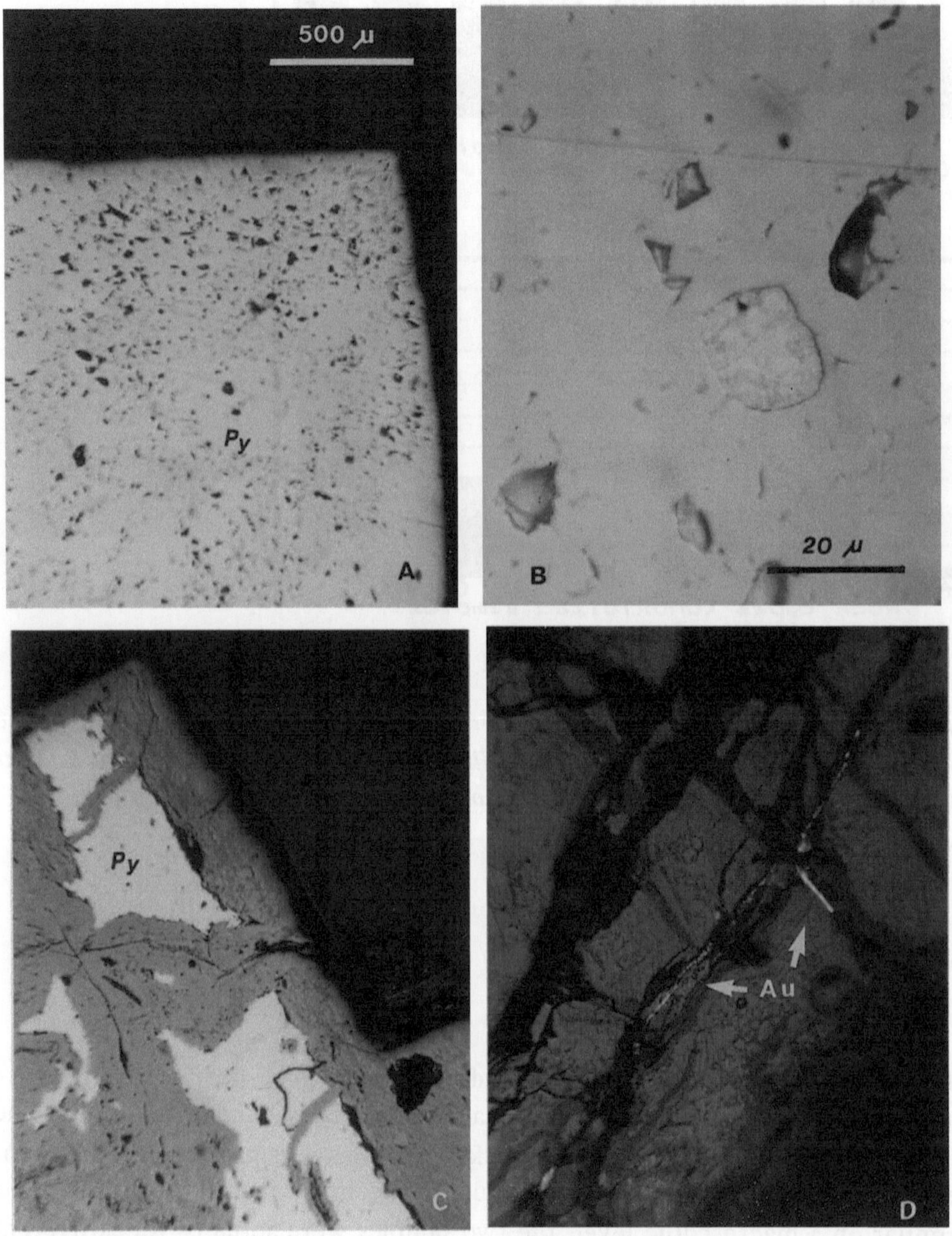

Fig. 7. *Unaltered and altered gold-bearing pyrite giving poor (top) and good recoveries (bottom), due to different host mineral permeability. A and C are low magnification photomicrographs showing a fresh pyrite crystal and a pyrite crystal replaced to a large extent by secondary iron oxide. B shows a grain of native gold totally enclosed in the fresh pyrite. D shows fine veinlets of native gold in relatively large gaps of the secondary iron oxide. Perhaps an even better example (when compared to figure 5) of how the roasting process renders the compact pyrite amenable.*

Best approach to situations such as those of the above two case studies, where the gold is totally enclosed in pyrite, but coarse enough for exposure by fine grinding, would involve the determination of the correct gold grain size, and then the grinding of the fresh pyrite to sizes similar to those of the native gold, followed by the cyanide treatment.

The two case studies prove that compact pyrite is not suitable to some of the less accurate extraction techniques, such as heap leaching, and that is independent of the particle size of the native gold. They also prove why the roasting process (the artificial equivalent of the natural oxidation) makes the ore amenable: it does so because the host becomes permeable. It does not do so because the gold is in solid solution in the pyrite and then crystallizes as native gold during the process of oxidation, as it is sometimes suggested.

MINERALOGY/ ORE MICROCOMPOSITION

Mineralogy and **ore microcomposition** were defined on page 7. Although the mineralogy defines the rock type, the microcomposition may vary within the same rock type. Mineralogy and microcomposition are extremely variable because they are the result of:

- The solidification of magmas of different chemical compositions.
- The solidification of magmas under different conditions of temperature and pressure.
- Physical and chemical processes affecting the solidified magma.

The above variable conditions result in rocks of different chemistry/mineralogy, on a macro, as well as on a microscopic level. In different textures for those rocks. And in minerals of different grain sizes and differently associated with one another.

Knowledge of the mineralogy/ore microcomposition is essential, when making decisions related to ore beneficiation and chemical separation by hydrometallurgical techniques. The mineralogy/ore microcomposition affect extractions to a much smaller extent, when using pyrometallurgical techniques. This is possibly one more reason why extractions of many of the base metals (those extracted by pyrometallurgical techniques) require less detailed mineralogical/microanalytical studies when compared to the precious metals.

Components of the mineralogy/ore microcomposition can be classified as follows:

Metal-bearing mineral(s)

Metals do not occur evenly distributed within their ores, rather they are concentrated in one or more minerals of which they may be essential or minor constituents. Different minerals bearing the same metal may respond differently, if using one recovery agent only, hence the importance of their correct identification.

The number of metal-bearing minerals is directly related to the ease with which our metal forms compounds with other elements. Gold, for example, forms few minerals and is concentrated predominantly in only one or two. Uranium, on the other hand, commonly occurs in a much greater number of minerals.

If the metal is concentrated in several minerals, such as uranium and also silver in many deposits, more than one agent may be necessary for its full or optimum recovery, as demonstrated by case study 5 on page 30.

Besides silver and uranium, some of the metals which are usually concentrated in several minerals are nickel, cobalt, copper and the platinum group elements. Some of the metals which, although known to occur in more than one mineral, tend to be concentrated predominantly in only one or very few minerals, are gold occurring mostly in native gold with or without silver, zinc occurring in sphalerite, less often in smithsonite, lead occurring in galena, less often anglesite, tin occurring in cassiterite, less often stannite, chromium occurring in chromite, molybdenum occurring in molybdenite.

Chemical composition of the metal-bearing mineral(s)

Knowledge of the chemical composition of the metal-bearing mineral provides two types of information:

- The exact amount of the metal of interest in the mineral. This value may vary for some metals. For example silver may occur in tetrahedrite and other copper sulfide/sulfosalts in variable amounts between 0 and several per cent (34.8 per cent for the tetrahedrite, Boorman et al 1982). Knowledge of the correct silver content of all the sulfide/sulfosalts in the ore, leads to (i) decisions concerning the economic minerals and the minerals that can be left in the tailings; (ii) the choice of agents for the separation of the economic minerals from the ore and of agents for the extraction of the metal from the economic minerals.
- The contaminant elements to be expected in the final concentrate of the mineral of interest. If separating zinc as a sphalerite concentrate, it may be possible, by use of the appropriate techniques and grinding sizes, to eliminate most of the iron carried by minerals other than the sphalerite. This means that any iron in pyrite, chalcopyrite, marcasite, pyrrhotite, etc., if in grains of adequate size, can be separated from the final sphalerite concentrate. Any iron in solid solution in the lattice of the sphalerite, cannot be separated. Iron values, when in solid solution in the sphalerite (in amounts up to about 20 per cent, S.D.Scott 1978, personal communication), are estimated more or less accurately by a microprobe or SEM determination (figure 8). Optimum grade for a zinc concentrate consisting of sphalerite bearing 15 per cent of iron sulfide, will not be higher than 85 per cent ZnS, the remaining 15 per cent being FeS_2 in solid solution.

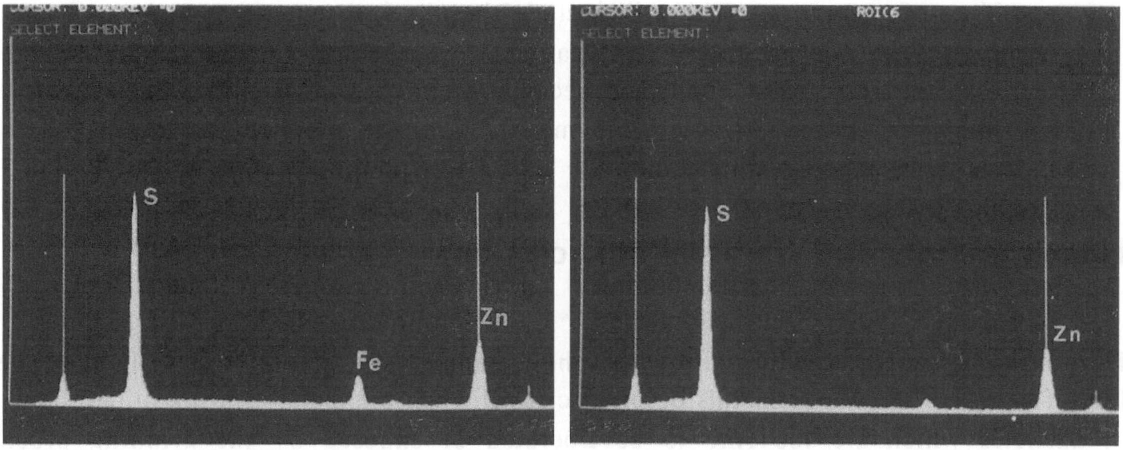

Fig. 8. *EDS-SEM x-ray spectra showing the presence of different amounts of iron in solid solution in two grains of sphalerite. Contaminant iron, when in solid solution in the lattice of the sphalerite, cannot be eliminated from the final zinc concentrates.*

Another situation where iron contributes to the final concentrate contamination, would be the one related to the concentration of the tin-bearing mineral cassiterite containing appreciable amounts of iron in solid solution.

Grain size of the metal-bearing mineral(s)

Where dealing with metals which are concentrated predominantly in one mineral only, losses during beneficiation or leaching are often due to their variable grain sizes. The coarser ones of these become liberated at a certain grinding size, the finer ones do not.

Grain sizes vary between a fraction of a micron and sizes which are visible to the naked eye. Minerals bearing the metals of lower market value, occur in higher concentrations which are usually coupled with coarser grain sizes. When in very fine sizes and if accounting for a minor portion of the total only, these minerals are left in the tailings, because their contribution to the final concentrates is negligible. Minerals bearing the more expensive of the precious metals, occur in lower concentrations which are usually coupled with finer grain sizes of 25 microns and less. These fine sizes account for a very major portion of the total, therefore the need for their correct determination before any attempt at recovery.

Knowledge of the correct grain sizes leads to decisions concerning:

- Correct grinding sizes. Grinding sizes which are too coarse prevent total particle liberation or exposure, resulting in poor recoveries. Grinding sizes which are much finer than those of the mineral of interest may be unnecessarily costly and time consuming.

- Type of separation. This will include the physical separation such, as flotation or gravity concentration, when dealing with grain sizes for the mineral of interest which are coarse enough to become liberated after grinding (down to 10-15 microns approximately or -200 mesh), or direct chemical separation, such as smelting, when dealing with finer sizes (for example those of the laterites).

Host minerals and mineral associations

Host minerals are the minerals with which the metal-bearing mineral is directly associated. Identification of the host minerals becomes critical when the metal-bearing mineral occurs in particles which are too fine to be liberated or exposed by the grinding process. Knowledge of the host mineral leads to the choice of the appropriate agent for its dissolution or other chemical breakdown.

Associations of the metal-bearing mineral with its host are very varied and largely dependent on their relative abundance. Attached particles and coarse and fine *intergrowths* (interlocked particles of two different minerals which have crystallized simultaneously) are frequent in those ores where host and metal-bearing mineral occur in similar amounts. These associations are typical of the base metals and of some of the silver minerals. Totally enclosed particles, particles distributed in fractures and at the border between grains are observed when the host mineral is predominant. These associations are common for the precious metals. Knowledge of mineral associations assist in decisions whether grinding sizes will lead to liberation or exposure, or chemical breakdown of the host is necessary. For example, a particle of gold which is only 5 microns in size, is likely to be exposed by a relatively coarse grinding, if occurring in a fracture of the host mineral, but will probably remain unliberated even after a very fine grinding, if totally enclosed in the same host.

Amstutz (1961) gave a graphic classification of different types of mineral associations which can be used in the understanding and planning of their separation by physical means. Such associations, which include free and poorly contaminated particles, as well as a number of basic intergrowth patterns, can be applied to minerals occurring in major amounts in their host rocks, such as those of the base metals (figure 9). Figure 10 shows associations of minor and trace minerals such as those of the precious metals. Figures 11 and 12 are actual occurrences representative of the mineral associations described schematically in figures 9 and 10.

Figures 11 and 12 show natural occurrences in which the metal-bearing minerals are associated with a host in the form of discrete particles of a separate mineral the grain size of which is variable and may be as small as a few microns. A competent microanalyst, if using the adequate ore microscope and magnifications, is always able to locate and observe such particles as distinct from the host mineral, no matter how small their size and concentration.

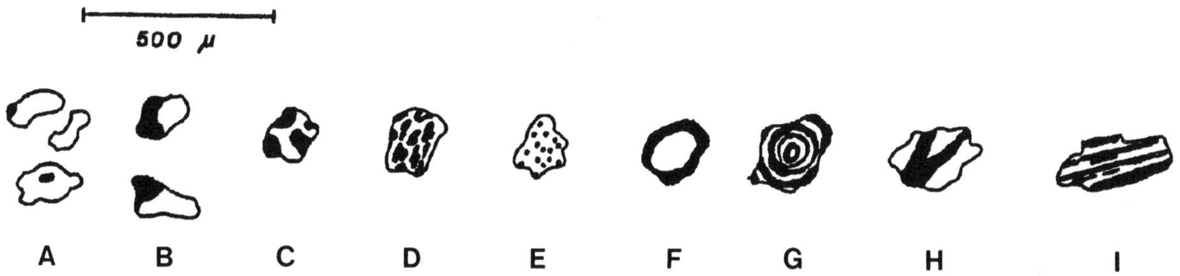

Fig. 9. *Associations between two major minerals or two minerals occurring in the ore in similar amounts (modified after Amstutz 1961). These associations are common among the base metals. A: liberated particles and particles associated with very minor contaminant. B: attached particles of similar size (common occurrence which can be corrected by fine grinding). C: coarse intergrowth (similar to the previous one, but finer particle sizes). D: medium size or eutectic type intergrowth (for example, chalcopyrite-stannite) and coarse exsolutions (for example, galena in sphalerite). E: fine or disseminated, emulsion-like, drop-like type intergrowth (for example, chalcopyrite-stannite) and fine exsolutions (for example, chalcopyrite in sphalerite). F: mineral coatings (for example, covellite, chalcocite or bornite over sphalerite or galena, galena over sphalerite, clay over cerargyrite, secondary silver sulfide over acanthite or electrum). G: concentric, spherulitic or multiple shell type association (for example, uranium/nickel minerals, manganese and iron oxide minerals). H: vein-like, stringer-like or sandwich-type association (for example, molybdenite/pyrite). I: lamellae, layered polysynthetic type association (for example, pyrrhotite/pentlandite, linnaeite/pyrite).*

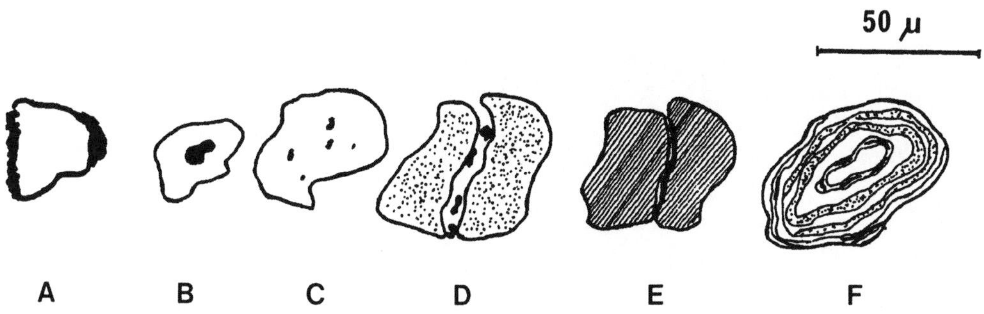

Fig. 10. *Associations between a major and a trace mineral (represented as fine black particles). These associations are common among the precious metals. A: particles distributed on the outside of the grain. B: relatively coarse, totally enclosed particle. C: fine, totally enclosed particles. D: particles distributed in large fractures, or at the border between grains. E: Particles and veinlets distributed in small fractures, or at the border between grains. F: particles distributed in concentric structures.*

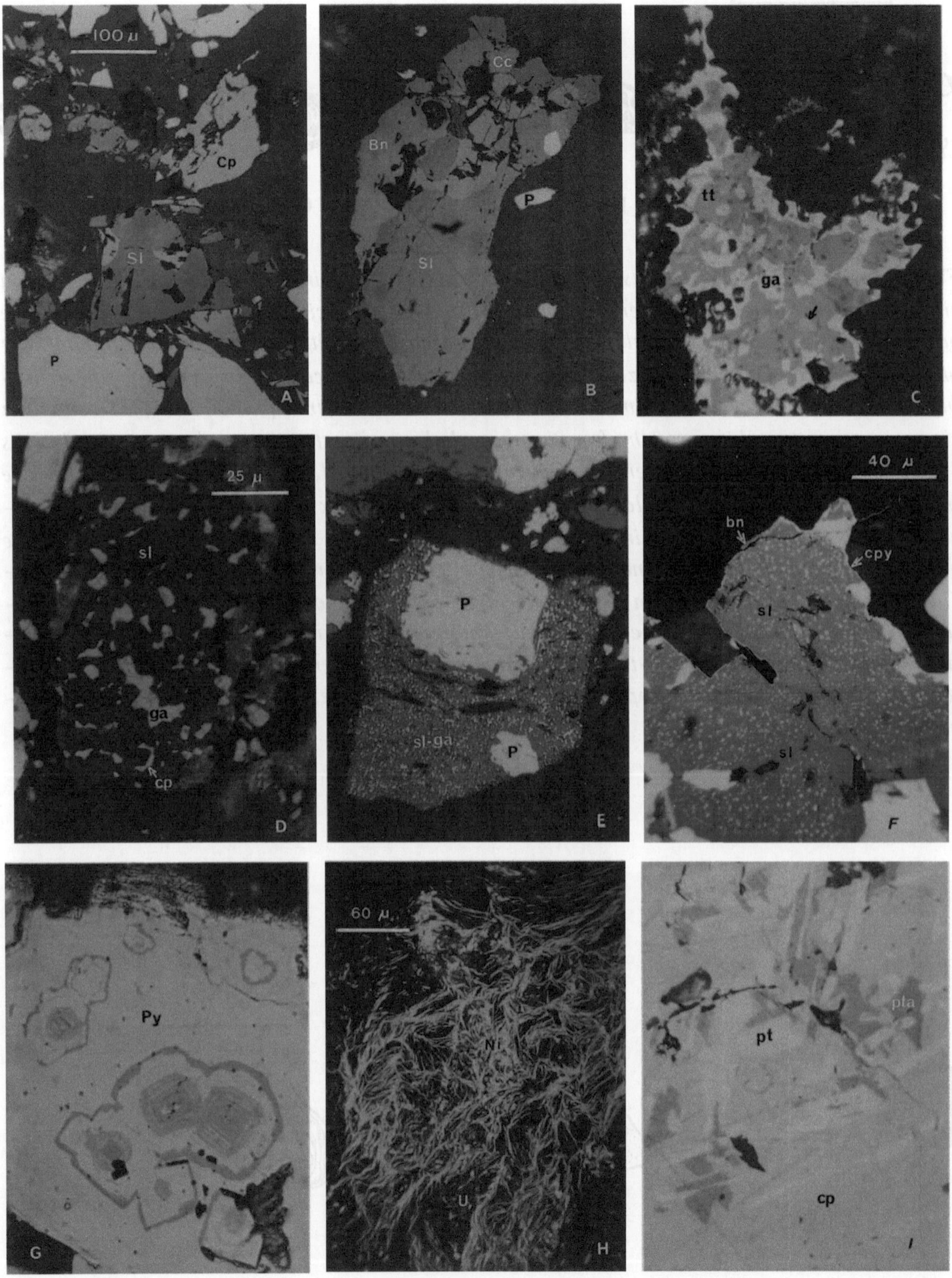

There are other situations where the metal of interest is not concentrated in a separate phase, rather it is distributed within the lattice of the host mineral, contributing to its crystal structure, in which it replaces some of the atoms of one of the major constituents. This type of occurrence is known as a *solid solution.* When dealing with solid solutions, direct

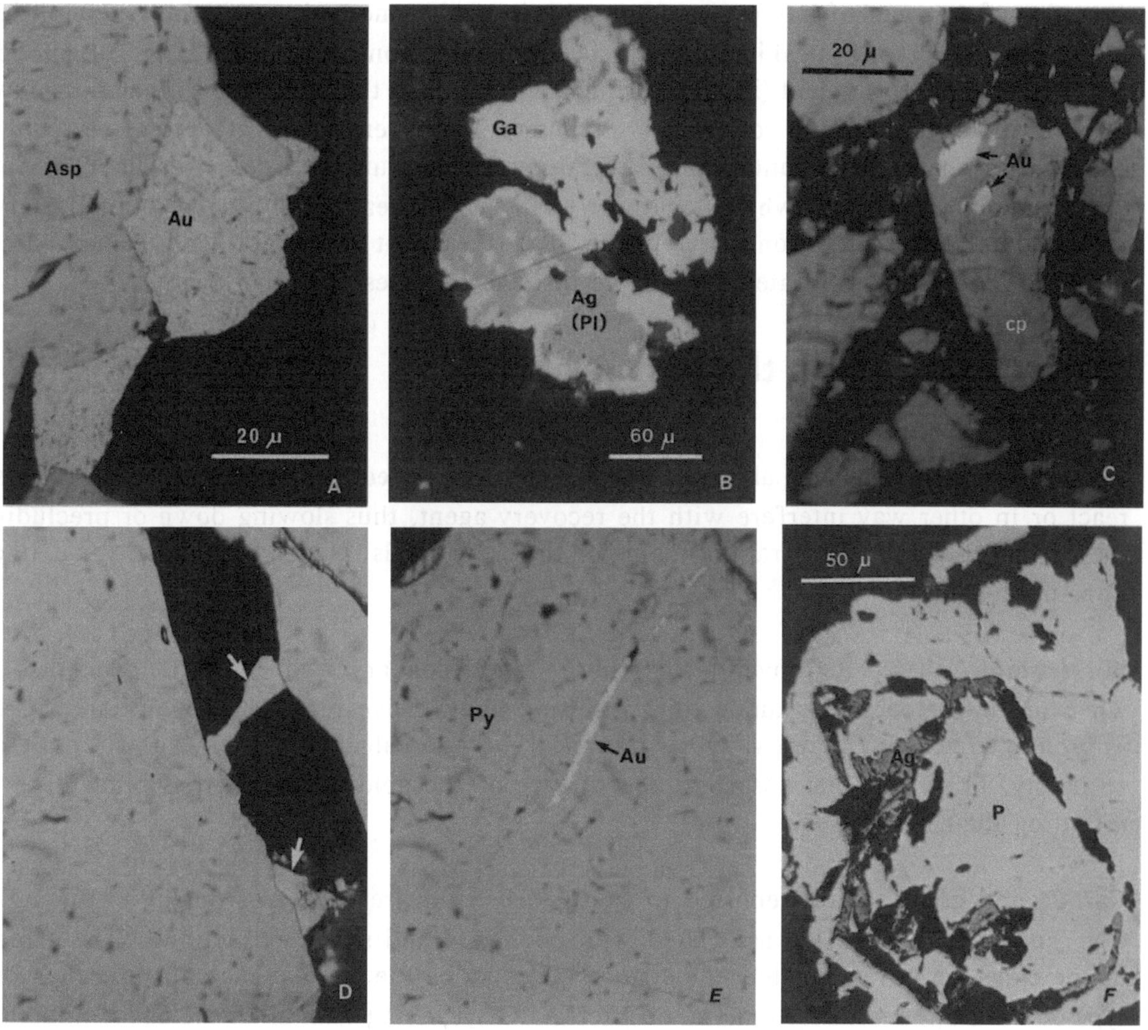

Fig. 12. *Occurrences representative of the mineral associations suggested and exemplified in figure 10. A: native gold on the outside of a grain of arsenopyrite. B: large polybasite enclosed in galena. C: fine native gold enclosed in chalcopyrite. D: native gold in a large gap between two grains of pyrite. E: fine veinlet of native gold in a thin fracture in pyrite. F: acanthite distributed between the core and rim of a grain of pyrite. An example of this last type of association could also be the gold distributed within the layering of the secondary iron oxide of figure 5B.*

Fig. 11. *Occurrences representative of the mineral associations suggested by Amstutz and exemplified in Fig. 9. A: grains of chalcopyrite and sphalerite liberated or attached to very minor contaminant. B: large composite grain of bornite-sphalerite-chalcocite. C: coarse intergrowth of galena and tetrahedrite. D: coarse exsolutions of galena in sphalerite. E: fine exsolutions of galena in sphalerite. The large light yellow mineral is pyrite. F: coating of bornite and fine exsolutions of chalcopyrite in sphalerite. G: concentric-type occurrences of cobalt minerals in pyrite. H: stringers of niccolite in pitchblende. I: lamellae type pentlandite and argentian pentlandite in chalcopyrite.*

detection of the metal of interest is not possible by optical methods and is only accomplished by the use of microanalytical instruments such as the electron microprobe/SEM, for amounts not lower than 0.01 per cent (if using the microprobe), and the ion probe for amounts down to parts per million. Solid solutions are common for silver, which often replaces copper and other elements in amounts up to several percent in many sulfide/sulfosalts. They are not as common for gold, which, contrary to the widespread belief in the United States, very seldom replaces the iron in the pyrite, if it replaces it at all. Rather it occurs within the pyrite grains, concentrated in separate discrete particles of native gold.

Minerals other than the host mineral

These are **minerals** which, although **not directly associated** with the mineral of interest, react or in other way interfere with the recovery agent, thus slowing down or precluding the normal process of separation. Some of these minerals affect gold and silver extractions by cyanidation.

For example, chalcocite, covellite, pyrrhotite and secondary iron oxides tend to dissolve when in the presence of cyanide, thus resulting in excessive reagent consumption, as well as prolonged leaching periods. Other minerals, such as realgar, orpiment, and sulfosalts of antimony, form ions that adsorb at the surface of the gold or silver, thus preventing or retarding their dissolution.

Carbon and graphite, are reported to cause early gold precipitations during cyanidation. Clay and sericite are often associated with poor gold and silver extractions. The effects of the clays are not completely understood and may be the consequence of more than one type of association with the mineral of interest. Clays, along with all of the above minerals, are discussed in more detail in the chapter on gold (pages 134-139).

Some minerals may affect the metal recoveries by forming thin coatings over the mineral of interest, thus altering its surface properties and reaction to the leaching or flotation agent. Examples of these coatings are galena, chalcocite and bornite over sphalerite (figure 11F), galena over chalcopyrite, pyrite over sphalerite, and clay minerals over cerargyrite (figure 44, page 170) or gold. Some of these coatings may be secondary and developed after the grinding process such as those observed in silver ores and in gold ores rich in silver. These secondary coatings are discussed in detail, when describing the gold and silver metallurgy (pages 122-124 and 171-172).

Awareness of the presence and control or elimination of all of these minerals may save recovery agent and time.

Mineral stability

Besides being related to the number of minerals in which an element is concentrated, the nature or electronic configuration of the atom of some elements affects the **stability**

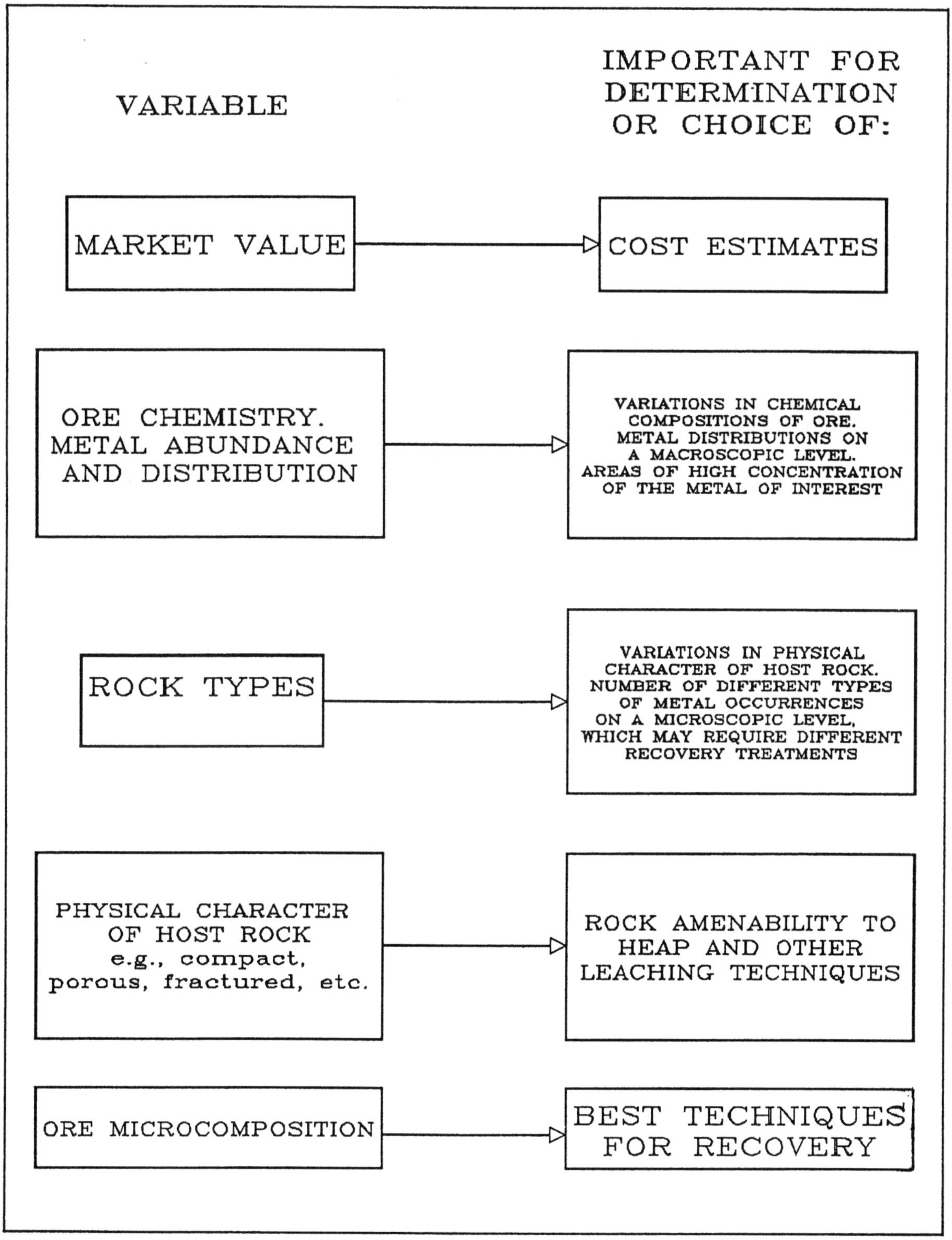

Chart 3. *Variables affecting decisions concerning metal recoveries.*

of their minerals Some such minerals can become so unstable that they decompose even at normal temperature-pressure conditions: exposure to light, small variations in temperature, atmospheric pressure and humidity are all factors which may alter their chemical structure. Such character, particularly common among the silver minerals, leads to chemical changes during and after the grinding process, causing decreased recovery rates which would not occur in the presence of the unaltered compound.

Again, awareness of the presence of unstable minerals is the first step in the control and prevention of recovery problems of this nature.

* * * * * *

Chart 3 is a schematic representation of the different variables described and of the way in which they affect the process of extraction.

CHAPTER 4

METHODS AND TECHNIQUES FOR THE STUDY OF ORES

INTRODUCTION

The variables discussed in the previous chapter are found in the literature or are determined experimentally.

Market values, which vary daily, are obtained for many metals, by consulting the financial newspapers and other publications which list the prices quoted on exchanges such as the London Metal Exchange (LME). North American papers providing the market prices of metals are **American Metal Market**, published daily in New York and **The Northern Miner** published weekly in Toronto. Price of metals is sometimes agreed upon through negotiations between buyer and seller, without use of the terminal markets.

The **ore chemistry** and correct **metal abundance** and **distribution** are determined by wet chemical analysis. Fire assay (Bugbee 1981, Smith 1987, Shephard and Dietrich 1989) is most commonly used for the determination of gold and other precious metals. Atomic absorption and neutron activation (Beamish and Van Loon 1972 and 1977, Van Loon 1980) are used as alternative methods to confirm and verify fire assay data. Atomic absorption is sometimes used for accurate determinations of silver when in the presence of high amounts of gold (page 92).

The **rock types** are defined by petrographic studies of thin sections, performed by use of a petrographic microscope. The **physical character of the ore**, focussing on rock permeability to solutions, is usually determined experimentally before the leaching process, if extracting gold or silver by hydrometallurgical techniques. To the author's knowledge, no systematic studies have been done nor tables are available showing the relation between rock type and permeability to solutions.

The **mineralogy** is determined by the use of an average ore microscope with or without the microprobe or SEM. The **ore microcomposition** is determined by the use of the microanalytical techniques, the main subject of this chapter.

THE MICROANALYTICAL TECHNIQUES

The microanalytical techniques give accurate information on the optical and chemical properties of the mineral particles in a rock down to sizes in the range of 1 to 2 microns. Although applicable to the study of any metal, microanalytical techniques are essential for the study of the precious metals because:

- Precious metals occur in their host rocks in very low quantities which are often difficult to detect by other methods of mineral identification.
- Their high market value requires as high an extraction as possible, therefore a thorough knowledge of their occurrence is necessary for the appropriate planning of efficient recovery strategies.

Microanalytical techniques require a high magnification/high resolution optical microscope (for the direct observation and for the determination of the optical properties of the particles under study) used in combination with an electron microprobe and/or a scanning electron microscope (SEM) equipped with an energy-dispersive spectrometer (for the determination of their chemistry). The ion probe is used for very specialized determinations such as the analysis of trace elements (in the ppm range) in particles down to a few microns in size. This instrument however, does not appear to be as widely employed for **routine** studies as the electron microprobe or the scanning electron microscope.

Although precious metals are the main subject, base metals are discussed in this chapter (like in the rest of the book) to highlight their differences from the precious metals and therefore to emphasize the need for a differential approach to their studies.

THE ORE MICROSCOPE

Also referred to as the optical microscope, this is a microscope with reflected light attachments (figure 13). Best ore microscopes, among the several makes available today remain, in the author's experience, those manufactured by Leitz and Zeiss. When observing and studying fine particles down to 1-2 microns in diameter, such as those of gold and platinum group elements, final magnifications of more than 1000x (provided by the objective and ocular lenses) and good resolutions at those magnifications, are essential. Microscope objectives can be purchased separately, but a good range of magnifications when

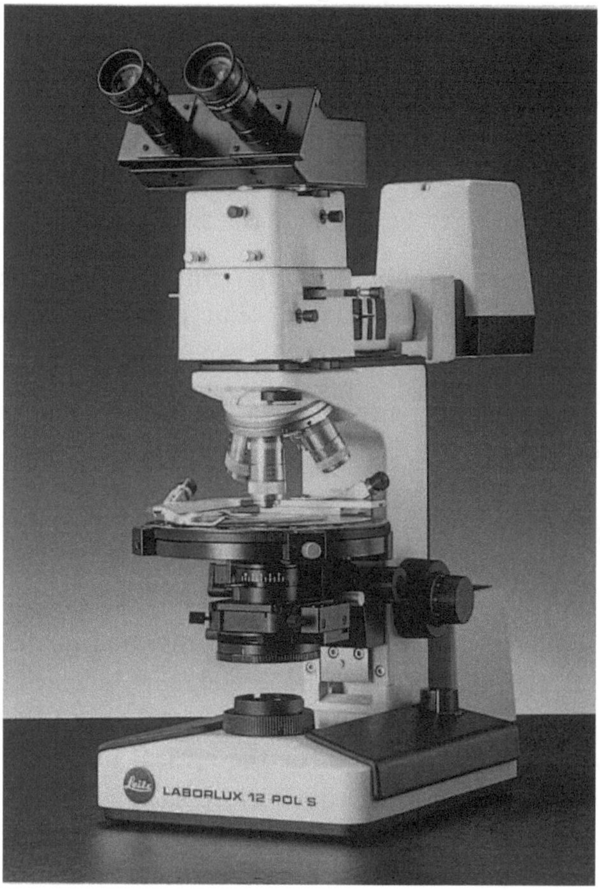

Fig. 13. *The ore microscope.* (*Courtesy of Leitz*)

making the initial purchase, could include 5x, 10x, 20x, 50x, and 125x. Oil objectives for higher magnifications give better resolutions and are recommended, despite their messier handling when compared to their equivalent dry forms.

Observation and study of fine particles by the ore microscope are conducted on *polished mounts*, also called *briquettes*, or *polished sections*. These are cylindrical molds, usually 1", 1.2" or 1.5" in diameter and between 1/2" and 3/4" high, made of bakelite or other plastic material, in which the sample is mounted. Polishing of the mount is obtained by grinding its surface down until sufficient sample is exposed, and then by further eliminating the finer scratches by use of diamond paste of different sizes, for example, 6 microns, 1 micron and 1/4 micron. Aluminum powder is sometimes used to complete the process. Polishing is usually done on appropriate machinery which process 6 or more mounts simultaneously by holding them against a rotating lap coated by a lubricant and smeared with the diamond paste. Best polishing laps are made of alloys the main component of which is lead, because the metal is sufficiently soft to handle most minerals (it is softer than most minerals), while still remaining flat, thus avoiding relief effects on the polished surface. Relief, due to differences in mineral hardness, is usually produced when using polishing laps made of cloth material, popular in some laboratories, because easier to handle. Polishing of ore and metallurgical samples for accurate optical microscope study and then for microprobe or SEM analysis, requires a certain amount of practice and skill. The availability of an experienced technician, as well as of the proper equipment, are therefore essential for reliable results. Polished mounts can be prepared of ore samples as well as of metallurgical products consisting of particles of any size down to -200 mesh.

Polished-thin sections, prepared by grinding the other side of the polished mount down to a sufficient thinness for transparent mineralogy studies, or by polishing the surface of a thin section (Amstutz 1960), allow the observation of both silicate/carbonate/sulfate and metallic minerals, thus providing a more complete sample description. Polished-thin sections are not usually recommended for the study of fine metallurgical products (for example those which have passed a -200 mesh sieve), because of the difficulty of determination of the optical properties of transparent minerals smaller than 50 microns in diameter.

The particles observed on the surface of the polished mounts are two-dimensional. Yet polished mounts represent a better and easier alternative to tri-dimensional observations where the particles are taken out of the ore in order to be studied, for two reasons:

- Separation of the particles would be considerably more difficult and time consuming than the polishing of the surface of the sample. For particles smaller than 10-15 microns, separation would be impossible.
- If we were to separate the particle from its ore, any information on host minerals and mineral associations would be lost.

THE ELECTRON MICROPROBE

The electron microprobe is an instrument for x-ray analysis capable of providing qualitative and quantitative compositions and of detecting minor and trace elements in particles down to a few microns in size (1 or 2 microns when using some of the newer instruments). These determinations are conducted on the same samples used for the optical microscope study, that is polished mounts or polished-thin sections, appropriately coated by carbon, gold or other metal, to enhance surface conductivity. They are obtained by analyzing the wavelengths or energies and by measuring the intensities of the x-rays emitted by the particle under study, after its surface has been excited with a beam of electrons. Electron microprobe determinations do not destroy the sample which remains suitable for further ore microscope or additional microanalytical determinations.

The basic design of an electron microprobe includes:

- An electron gun consisting of a tungsten filament to which high voltage is applied.
- A set of magnetic lenses focussing the beam of electrons emitted by the filament to approximately 1 micron in size (in the newer instruments) onto the surface of the particle to analyze.
- Detecting, amplifying and measuring units to convert the primary x-rays emitted by the particle under study into graphic or digitized signals.

An optical microscope, generally equipped with two magnifications (for example 10x and 200x) is attached to the instrument and is used to locate the particles to be analyzed. Figure 14 is the schematic representation of an electron microprobe.

The wavelengths and energies of the x-rays are used to identify the elements, their intensities are used to determine the elements' contents in the particles (Gasparrini, 1976).

Qualitative compositions of any element with atomic number above 10, that is sodium and up, and of some lighter elements such as C, O and B, are obtained by identifying inten-

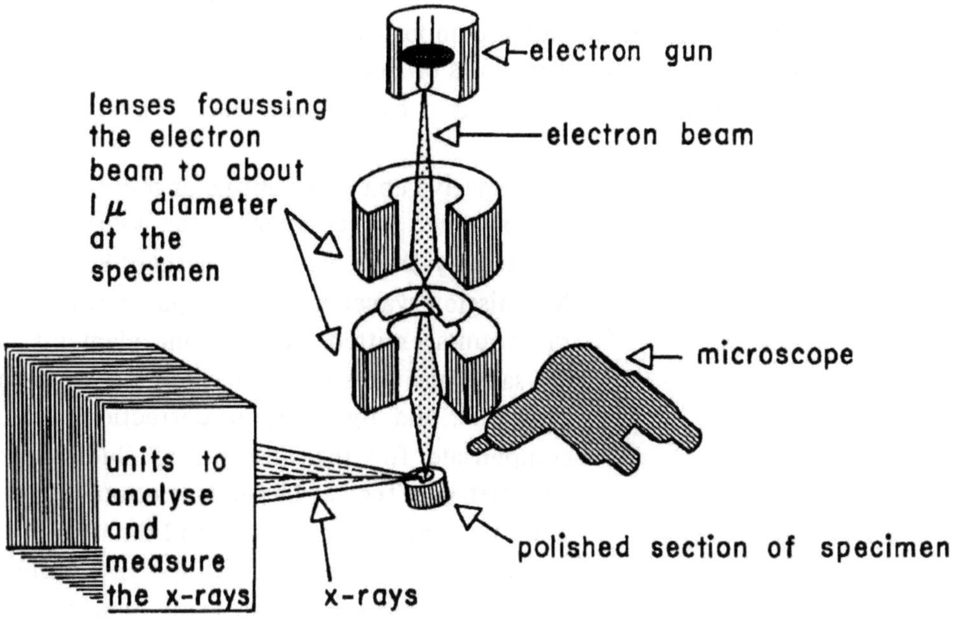

Fig. 14. *Schematic representation of an electron microprobe. (Modified after Birks 1963, p. 3).*

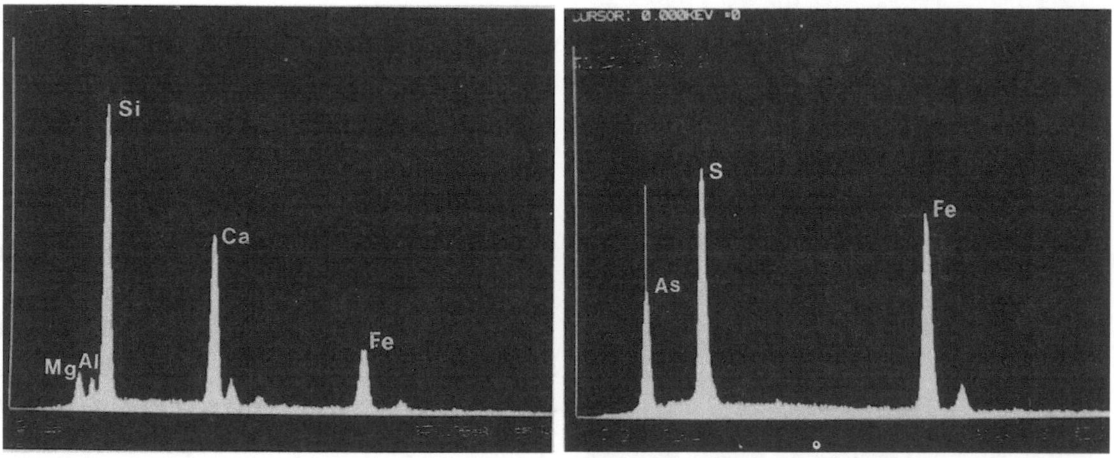

Fig. 15. *X-ray spectra showing the qualitative chemical composition of grains of augite (left) and arsenopyrite (right).*

sity peaks, based on their positions in the x-ray spectrum (figure 15). The limits of detection vary with the element analyzed, the chemical composition of the particle and the KV used during the analysis. The general range for limits of detection is between 0.1 and 0.01 per cent when some of the newer microprobes are used by an experienced operator.

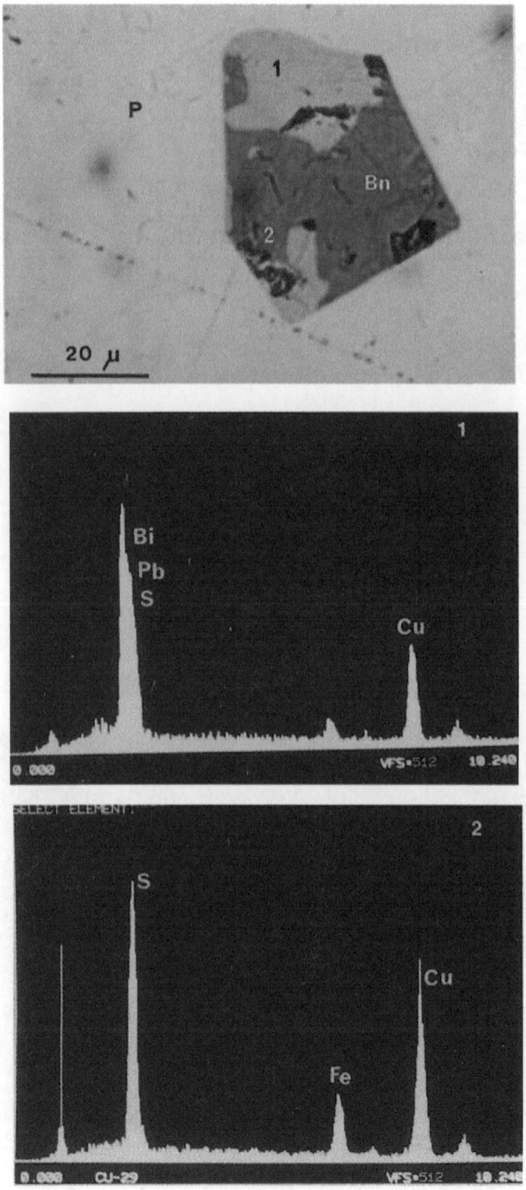

Quantitative analyses for one or more elements are obtained by comparing the x-ray counts emitted by these elements at their K, L, and M emission wavelengths/energies with the x-ray counts emitted by the same elements at the same positions in standards of known compositions, and by applying corrections which compensate for the absorption, fluorescence and matrix effects. These corrections are applied by computer programs and the results are usually obtained on line within a few minutes from the collection of the x-ray counts.

Mineral identifications are in most situations obtained based on their optical properties and qualitative compositions (figure 16). In those instances where we are dealing with unusual minerals or compositions, quantitative analyses to recalculate the formula may be necessary for positive identifications. When dealing with very fine metallurgical products, where optical properties cannot be determined accurately, identifications of the more common silicate minerals, such as quartz, feldspar pyroxene, etc., are obtained based on the position and relative heights of the x-ray peaks.

Spectra of mineral particles can be collected by analyzing the x-rays according to their wavelengths by use of the crystal spectrometers. This is the *wavelength-dispersive system* or *wavelength-dispersive spectrometry*, or *WDS*. The spectra can also be collected by analyzing the x-rays according to their energies by use of the solid state detec-

Fig. 16. *A composite grain of lindstromite-bornite in pyrite (top), and x-ray spectra of the lindstromite (1) and bornite (2). The two minerals were identified based on their optical properties and qualitative chemistry together.*

tor for energy-dispersive analysis. This is the *energy-dispersive system*, or *energy-dispersive spectrometry*, or *EDS*. The EDS was developed approximately 10 years after WDS microprobes became common instruments for analysis and at the beginning was considered not as accurate a technique as the WDS. Although perhaps not as accurate for **some** determinations, the EDS offers the advantage of giving significantly faster results because 30 seconds (against several minutes of the average WDS) are generally sufficient for a good indication of the chemistry of the mineral particle under study. Limits of detection of the EDS are poorer by a factor of at least 10 when compared to those of the WDS.

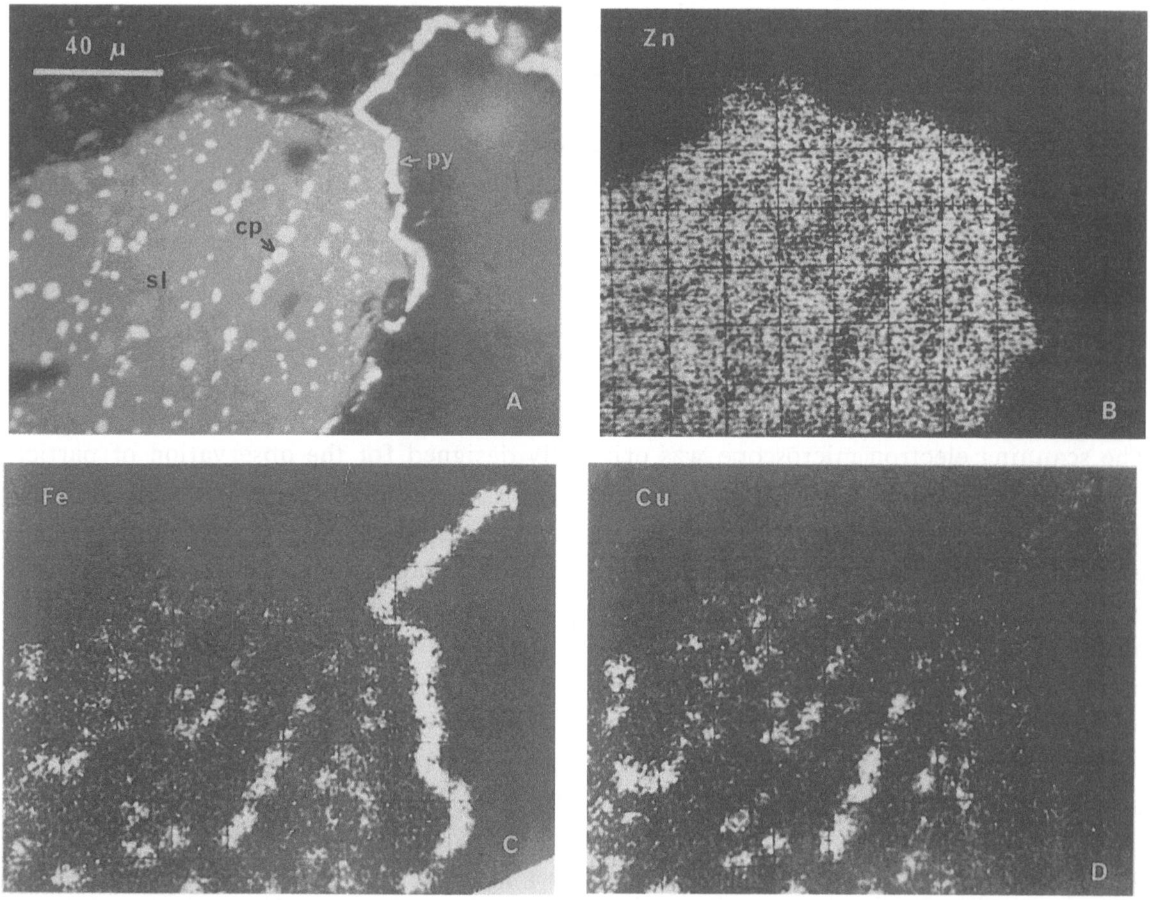

Fig. 17. *A grain of sphalerite showing exsolved chalcopyrite and coated by a fine rim of pyrite (A), and the x-ray distribution images for Zn, Fe and Cu (B, C and D).*

Another determination which is done by the electron microprobe is the one that gives a visual representation of the elements' distribution in the particle under study. When an element is in a mineral which forms a fine vein or a rim around a grain, or which occurs in lamellae or exsolution blebs, or is part of an intergrowth, it may be more informative to look directly at the distribution of the various elements in the particle, rather than to determine their average amounts. Information on element distributions is obtained by sweeping the surface of the particle in question with the electron beam and by observing the x-rays emitted by the same surface as they are collected on an oscilloscope screen. The image can be photographed for a permanent record of the elements' distribution. Figure 17 shows the photomicrograph of a grain of sphalerite with exsolved chalcopyrite and partly coated by a rim of pyrite (taken in an optical microscope, A) and the x-ray distribution photographs of the zinc, iron and copper in the same grain (taken on the oscilloscope screen, B, C, and D).

X-ray distribution images along with backscattered electron images (differentiating particles based on their mean atomic number) became very popular when microprobes were first introduced, because of the very effective visual representation of the element distributions. Because they are very slow to produce, these images are probably not as common in mineralogical or microanalytical reports today.

For additional information on the electron microprobe and on the energy-dispersive techniques, see Gasparrini (1976, 1980 and 1985) and the other references listed in those three papers.

THE SCANNING ELECTRON MICROSCOPE (SEM)

The scanning electron microscope was originally designed for the observation of particles the size of which was too fine to be magnified by an optical microscope (less than 1 micron). The instrument uses a beam of electrons, instead of a beam of light on the sample, because their smaller wavelengths allow for a better particle resolution and therefore for a higher magnification. The general design is similar to that of the electron microprobe and includes an electron gun with a tungsten filament to which high voltage is applied, and magnetic lenses to focus the electron beam onto the surface of the particle under study. When using the SEM as a microscope, the particle or area to be magnified is swept by the beam, thereby the smaller the particle or area swept, the higher the magnification. Particle resolution, for the newer instruments, is down to 70-80 angstroms.

When using the SEM as a microanalytical instrument, it is possible to focus the electron beam down to a similar size to that of the electron microprobe (1 to 3 microns) and then to collect and analyze the primary x-rays emitted by the particle. For such purpose, energy-dispersive spectrometers similar to those attached to electron microprobes, have been adapted for use with SEM's with comparable, although not usually as accurate results in the quality of analysis. Crystal spectrometers for accurate quantitative analyses and lower limits of detection, have also been designed to be attached to SEM's. Corrections compensating for fluorescence, absorption and matrix effects are again applied by computer programs for the recalculation of quantitative analyses within short periods of time. Element distribution images within the magnified particles are easily collected, provided that the proper attachments are available.

Advantages derived from the use of an energy-dispersive spectrometer attached to the SEM rather than to the microprobe, are in the capability of the SEM to easily locate the grains to analyze, particularly those of high mean atomic number enclosed in a matrix of lower mean atomic number. This is important when studying mineral particles in the 5 micron diameter range and finer, such as those of native gold. Particles of that size, when carbon coated, are difficult to see when using the optics of the *average* electron microprobe. Of course the above discussion, and all other discussions in the remainder of the book, refer to the *average* electron microprobe popular in the 1980's. Most of the recent models are equipped with scanning facilities which give similar results to those of a good SEM.

Because of the lower initial and running costs, the ease of operation, the often more modern designs and the needs of the mining industry for only relatively accurate quantitative data (in many instances), but in large volumes, the SEM with energy-dispersive spectrometer (EDS-SEM) is probably a more widely used instrument in North America than the microprobe.

Additional information on the SEM is in Goldstein et al (1984).

* * * * *

The importance of the combined use of the ore microscope/microprobe or ore microscope/SEM for precious metal mineral identifications, becomes apparent when compared to the more traditional methods based on optical or physical properties, chemical composition and atomic structure, and developed for the study and identification of minerals of coarser sizes such as those of the base metals. Precious metal minerals are of microscopic size, in the 50 to 100 micron diameter range, often much smaller. It may be difficult to recognize them by methods other than the microanalytical ones.

Some of the drawbacks of the more traditional methods of mineral identification, when used for the study of the precious metals, are:

- **Optical methods.** The hand lens, a very popular instrument among field geologists, permits identification (or at least observation) of the more common minerals, such as pyrite, chalcopyrite, galena, sphalerite, but is effective only for the observation of particles several 100 microns in diameter. The ore microscope alone, even when capable of very high magnifications, is not always sufficient for positive identifications, because many minerals have similar optical properties, but different chemical compositions.
- **Physical methods**, such as streak, luster, etc. These methods, often quoted in metallurgy texts, are not accurate and require large amounts of the sample.
- **Chemical methods**. Chemical methods alone, such as wet chemistry or x-ray fluorescence techniques, or even the microprobe and SEM, are not always capable of giving positive identifications, particularly when dealing with *polymorphs* (two or more mineral species with identical chemical compositions, for example, pyrite-marcasite) or with minerals having overlapping chemical ranges. Moreover, chemical methods other than the microanalytical ones require a large amount of sample and are not applicable to the analysis of fine particles. They

do not give information on minor element distributions, for example, whether they occur in separate very fine inclusions/exsolutions, or they are evenly distributed in solid solution in the lattice of the host mineral.

- **X-ray diffraction techniques.** Another method of identification is the one based on the measurement of atomic structures by the study of their x-ray diffraction (XRD) patterns. This method applies to fine grains, down to less than 100 microns in size (when used by a competent scientist) and, because the XRD patterns have been defined as *as distinctive of a mineral as fingerprints are of a human being*, they have for a long time been regarded as the most reliable means of mineral identification. XRD patterns become confusing or difficult to obtain when the mineral studied is part of an intergrowth or forms grains which are much finer than 100 microns. Because they are based on atomic structures, XRD methods do not identify *amorphous* minerals (or minerals without a crystal structure), nor do they give information on minor elements, unless these are present in the mineral in amounts sufficiently high to modify its structure. XRD methods do not give information on *metamict* minerals, that is minerals which behave like amorphous phases when examined by x-ray diffraction techniques, unless heat treated to induce recrystallization or to nucleate crystallites of oxide-breakdown products (for example, ZrO_2 and SiO_2 from metamict zircon $ZrSiO_4$, Morton 1978). More information on metamict minerals is in Lima de Faria (1964), Mitchel (1973) and Vance and Routcliffe (1976).

As opposed to the above techniques, the ore microscope and electron microprobe/SEM, when used together, provide identifications of minerals of any sizes down to a range below the minimum required by the other methods. These identifications, based on the optical properties and chemistry of the grain, are obtained rapidly and, in most cases, are as accurate as those obtained by the XRD techniques. Techniques involving the use of the ore microscope and electron microprobe/SEM require small amounts of the sample even when searching for gold and other precious metal contents which are in the range of a few ppm, an additional advantage, when the laboratory is at a distance from the location of the mine.

Capacity to produce results such as accurate analyses of particles the size range of which is not covered by other methods, capacity to determine element distributions in veinlets, exsolutions, intergrowths and rims as fine as 0.5 microns across, capacity to detect minor and trace elements in fine particles, make the microanalytical techniques particularly suited for the study of the precious metals and their very fine particle sizes and low concentrations. Microanalytical techniques are also important for the study of uranium occurrences, due to the metamict state of several of the radioactive minerals (see chapter on uranium).

CHOICE AND PREPARATION OF SAMPLES

When characterizing an ore, the first and most important decision is the one related to the choice of the samples on which the microanalytical studies will be conducted. Unless the samples examined are representative of the ore, no matter how accurate the microanalytical work, the resulting data will not help in the planning of the following recovery steps. Detailed microanalytical studies of metal distributions in samples which are only vaguely known to be associated with an economic mineralization can be worthless or even misleading. So can studies of one or two head samples only, when hundreds of such samples are being run routinely through the mill. Yet, this is unfortunately the type of information which many companies consider sufficient for the planning or understanding of their recoveries. Some of the more common errors in selecting samples for microanalytical studies include the use of:

- High grade material, particularly when dealing with gold, upon the assumption that the particles of interest will be easy to find (discussed in case study 1 on page 18). The particles may be easy to find, but they are not representative of the average gold occurrence in the deposit.
- Material originated from the general mineralized area, but not analyzed for precious metals. These samples may be barren.
- A very small number of samples. Sometimes only 2 or 3 samples are chosen at random from the available cores, upon the assumption that they will be sufficient to give a good indication of the overall occurrence of the precious metals.

How can representative material be chosen? A reliable way of doing so could include the examination of a number of thin sections of samples selected from various portions of the ore (in order to define the number of existing rock types) and the preparation of fire assay analyses for precious metals of the **same** samples. The number of samples analyzed and examined in thin section could be in the range of 100 or more. Let us assume that the thin section study indicates that these samples belong to a certain number of rock types, for example 10, and let us also assume that the fire assay shows that these 10 rock types all contain the precious metal of interest in economic amounts. Then one sample could be selected for each identified rock type (for a total of 10) and examined in detail by the microanalytical techniques. It is very possible that the manner of occurrence of the precious metals is different in each rock type (page 31). It is important during the petrographic study that quantitative data are obtained for the major minerals, because some of these minerals, such as sericite/clay, graphite, etc., when accounting for more than a few per-cent of the total, may affect the precious metal recoveries (pages 134-138).

Once the representative samples are selected, the next decision relates to the number of polished mounts to be prepared for each sample, for a statistical determination of the manner of occurrence of the metal in question. The lower the amount of the metal, the higher the number of polished mounts. For example, two or three polished mounts 1" in diameter are sufficient for the study of the base metals, six polished mounts may be necessary for the study of precious metals occurring in the ore in amounts smaller than one ounce per ton (34.2 ppm) and down to 0.1 ounce (3.42 ppm) per ton. The preparation of two or three rather than one polished mount, is advisable when dealing with amounts of the metal of interest which are much greater than 1 per cent, because one mount may only be representative of the occurrence of a portion of the metal. On the other hand, preparation of several polished mounts is necessary when dealing with precious metals, because the mineral(s) bearing these metals are often not present in sufficiently large quantities to be representative of the occurrence or even to be located on the surface of one or two polished mounts only. In general, when examining polished mounts of ores with similar grades, the metal is found more readily when in finer sizes, because of the larger number of particles present (McLean 1982). However, because the grain sizes are not known before starting the study, it is always safer to have a larger number of mounts available than what is actually needed.

When studying gold, determination of the *cyanide soluble gold* at the same time as and on the same samples on which the microanalytical work is being conducted, provides a quick and easy preliminary indication of the amount of gold extractable. These values can be tied with the microanalytical data for a better understanding of the ore, and for a more accurate prediction of the type of extraction to be expected during the actual separation.

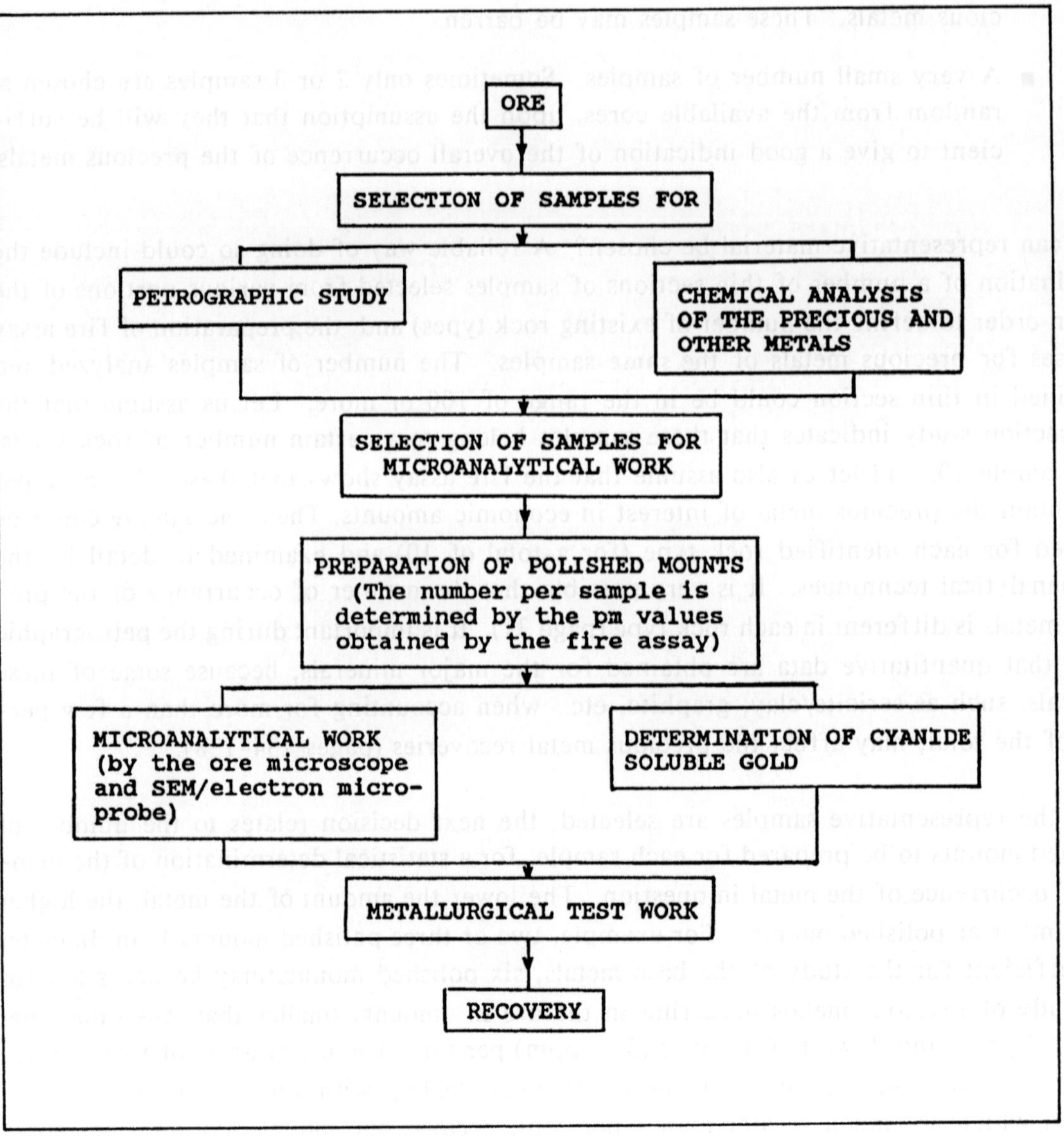

Chart 4. *Steps that may be followed during the planning of gold and other precious metal extractions.*

Cyanide soluble gold is determined on the rejects from the fire assay by adding cyanide to the solution containing the pulp and by calculating the amount of dissolved gold.

Chart 4 summarizes the different steps for the planning of gold and other precious metal extractions.

When conducting a microanalytical study on samples of which chemical analyses are available, use of the ground rejects from the analyses, may represent a good approach, if we want to examine identical samples to those analyzed by the fire assay. In such case, the grinding size of the rejects should be fine enough to make the samples homogeneous, but not too fine for the mineral particles bearing the metal of interest to become liberated, thus depriving the microanalyst of information on original particle sizes, host minerals and mineral associations. This size is in the range of a few 100 micron or -35 mesh. If such an alternative is chosen, preparation at the same time of a number of polished mounts and polished-thin sections of the original uncrushed hand samples is advisable, for the determination of the distributions and associations of the major minerals.

Concentrations below 0.1 ounce per ton or 3.42 ppm, require the preparation of a very large number of polished mounts, unless the metal of interest occurs in particles which are 1-2 microns in diameter and finer, an information not available beforehand. In such a situation the microanalytical study becomes too slow and the examination of heavy mineral concentrates, prepared by flotation, gravity separation, super-panning or other technique for physical separation, may represent an easier, faster alternative. If such an approach is chosen, fire assay analyses of both the sulfide or heavy fraction and the tail or light fraction should be obtained before proceeding with the microanalytical study, for an indication of the gold distribution in the two products. A common error in the study of precious metal ores is the one neglecting such step, and thus leading to time wastes searching for gold in the wrong product. Presence of pyrite and other sulfides does not imply a gold association with the minerals in the heavy fraction.(for example, chart 16, page 148).

Heavy mineral concentrates have the advantage of providing a fast means of locating the precious metals, but present two drawbacks:

- Any fine precious metal mineral associated with the light fraction will not be observed during the study of the heavy fraction.
- Unless still associated with the sulfides and other minerals separated with the heavy fraction, precious metal particles will provide no information on original grain sizes and host minerals.

Incorrect approach

Charts 1 and 2 of case study 3 on pages 22 and 23, already discussed in chapter 2, provide a good practical example of the manner in which a project, although including microanalytical studies, produced no positive results, only more confusion. The charts are schematic representations of how the study should have been conducted and how it was actually conducted during the course of a project related to a gold-bearing deposit in Nevada, and containing variable amounts of clay.

The project involved the recovery of gold from a matrix showing different degrees of alteration. The gold was partly coarse and associated with the sulfide or liberated, and partly of submicroscopic size, possibly colloidal, and associated with the clay. During the course of the project, a sufficient number of samples was selected and submitted for petrographic work. Although the ore contained substantial amounts of clay, the accurate content and distribution of which should have been determined, no quantitative data were obtained of the major minerals. The microanalytical study was then arranged, but the samples submitted were not the ones of which petrographic descriptions had been prepared. When separation of the heavy minerals was recommended, this was done on samples other than those on which the microanalytical study and petrographic descriptions had been conducted. Needless to say, the project was costly, but provided very few answers to the poor recoveries.

BASIC STEPS FOR THE STUDY OF METAL DISTRIBUTIONS

The study of polished mounts of an ore containing base metals is very quick, because minerals bearing elements in amounts above 1 per cent are readily located by use of an average microscope with magnifications of 100x-200x. Identification of many of the more common base metal minerals, such as chalcopyrite, galena, sphalerite, cassiterite, is easy and can be performed by an experienced mineralogist without the aid of an SEM or electron microprobe. An exception can be the determination of minor amounts of cobalt and nickel in pyrite or pyrrhotite, or the analysis of contaminant elements, such as iron in sphalerite and cassiterite.

When dealing with precious metals, the study is slow and time consuming. The smaller quantities of these metals by factors of a million or more has already been mentioned (pages 3-4, and 14). A plan is therefore necessary which may include:

- Preparation of polished mounts of the ore or metallurgical product, for examination in the ore microscope and electron microprobe or EDS-SEM. For soft and incoherent rocks which break and crumble during the sawing, mounting and polishing processes, impregnation may be necessary to hold the sample together. If impregnation is not sufficient, mounting and polishing of the crushed material may represent a good alternative. The technique may not provide as comprehensive information as when looking at the uncrushed material, for example major mineral assemblages and distributions will not be observed, but will give all the mineral identifications and all the correct mineral abundances. Given the fine size and scarcity of the precious metal particles, the variety in the polishing properties of the host and other minerals, as well as of the different rock types, the importance of the use of a professional person whose skills are used exclusively for sample preparation, should be obvious.

- An optical microscope examination of the entire surface of the polished mounts to locate the minerals bearing the metal of interest. During this study, the particles are photographed and marked for fast location in the microprobe or SEM. Besides locating the metal(s) bearing particles, the ore microscope study leads to their tentative identification, the tentative identification of the host minerals, the determination of the grain sizes, the determination of the type of association with the host minerals, for example, totally enclosed particles, intergrowths, distributions in fractures, etc., and the determination of the general opaque mineralogy of the sample. If instead of polished mounts, polished-thin sections are prepared, the transparent mineralogy can also be determined, as well as the proportions of the major silicate/carbonate/sulphate minerals. An efficient approach for a complete microanalytical description might include the preparation of several polished mounts per sample for the detailed search of the precious metal(s), and of one or two polished-thin sections cut from the same sample, for the determination of the general transparent mineralogy.

- An electron microprobe or SEM study of the particles located and photographed during the microscope examination, to determine their qualitative and/or quantitative compositions. An electron microprobe or SEM study of some of the major sulfide/sulfosalt minerals, to detect any solid solutions of minor amounts of the metal of interest.

- Mineral identifications, based on optical properties (such as color, reflectivity, hardness, iso/anisotropism), determined during the microscope study, and on the qualitative/quantitative compositions, determined during the microprobe or SEM study.

Some laboratories prefer to use the SEM or microprobe alone without the aid of the optical microscope for the location of the precious metals. There are two drawbacks to such method:

- The method does not give information on optical properties and therefore does not always provide positive mineral identifications.

- The method is expensive, because it uses a much greater amount of microprobe or SEM time.

Chart 5 compares microanalytical studies of gold, silver, platinum group metals, uranium and base metals, and the amounts of time required by each metal or group of metals for satisfactory results.

Additional information on microanalytical methods has been given in this same book with the descriptions of the individual metals.

METAL(S)	COMMON GRADE	NO. OF POLISHED MOUNTS	MICROSCOPE QUALITY	TIME SPENT ON MICROSCOPE	TIME SPENT ON SEM	TIME SPENT ON MICROPROBE	TOTAL TIME/ SAMPLE	TOTAL TIME/25 SAMPLES
GOLD	Less than 1 oz/ton	Six	High mag/ high resol	3 days	Half day or less	Not usually necessary	3 and 1/2 days	87 and 1/2 days
SILVER	5 oz/ton or more	Two to four	High mag/ high resol	1 to 2 days	Half day or more	Often necessary 1 day	2 and 1/2 to 4 days	62 and 1/2 to 100 days
PGM's	Less than 1 to several oz/ton	Six (for the lower grades)	High mag/ high resol	3 days	Half day or more	Usually necessary 1 day	4 and 1/2 day or more	112 and 1/2 days or more
URANIUM	1 per cent ore more	Two to three	Average	1 hour	Half day	Not usually necessary	Less than 1 day	Two weeks plus time for autoradiograph
BASE METALS	More than 1 per cent	Two to three	Average	1/2 hour	Less than half day or none	less than half day or none	Less than 1 day	One week

Chart 5. *Schematic comparison of the study of the precious and base metals.*

Misconception

The most common misconception associated with the study of low quantities of metals in their ores has already been mentioned in several occasions (for example, on pages 15 and 42) and is related to the occurrence of gold in pyrite and other sulfides. According to this misconception, if gold (or any other precious metal) is not readily located during the ore microscope study, it is assumed that it can be detected by the microprobe or by the SEM, as long as it occurs in particles which are resolved by those instruments' beams (1 micron for the microprobe and 70-80 angstroms for the SEM). If these instruments fail to locate any gold bearing particles after a search which is not any more detailed than the ore microscope search, it is then concluded that the gold is in solid solution in one of the major minerals. The assumption is incorrect in that it does not take into account that:

- The chances that an SEM or microprobe will find the precious metal particles are not any better than those of an ore microscope. The only difference is in the means by which the three instruments single out the particles. The ore microscope does so based on optical properties, the SEM and microprobe based on chemical compositions and/or mean atomic numbers.

- The optics (magnifications and resolution) of a good ore microscope are superior to those of the average microprobe. This makes the ore microscope a more likely instrument to locate the same particles. This statement especially applies to native gold which in most instances can be identified by the ore microscope alone based on its very distinctive yellow-orange color and high reflectivity. A gold particle 5 microns in diameter or smaller, if carbon coated, is much less likely to be found when using the *commonly* poor optics of a microprobe, and therefore to be analyzed by the spectrometer.

- The assumption that the SEM can analyze particles down to its limits of resolution when used as a microscope, is also an illusion. Perhaps many of the more recent instruments can magnify and resolve particles as small as 70 angstroms, very few however can provide their accurate chemical composition. The electron beam used for the microanalytical determinations has a diameter of 1 to 2 microns, very similar to the beam of an electron microprobe. This means that, although finer particles can be magnified, only those the size of which is in the range of 1 or 2 microns can be accurately analyzed for identification. The x-rays emitted by anything finer will have too small an intensity and will be mixed with the x-rays emitted by the surrounding area. Therefore they may not be recognized by the detector as emitted by a separate phase.

In the light of the above considerations, the ore microscope remains the best instrument for the location of minerals occurring in trace amounts before any microprobe or SEM work can be performed.

APPLICATIONS

The ore microscope is used in combination with both the electron microprobe and the EDS-SEM for the study of precious metal distributions in their ores. Although both instruments give similar results in many situations, there are applications which are best handled by the electron microprobe and applications which are best handled by the EDS-SEM. Examples of some such applications are:

Use of the microprobe preferable to that of the EDS-SEM

- For the detection of low quantities of an element when below the limits of detection of the EDS-SEM (between a few and 0.01 per cent). This is important for determinations of elements such as silver when occurring in minor amounts in one or more of the major sulfide/sulfosalt or other minerals.
- For the quantitative analysis of particles for which a high accuracy of analysis is needed, for example for the analysis of a new mineral, or of minerals with unusual chemical compositions. These analyses may be necessary for the recalculation of mineral formulae before positive identifications can be made.

Use of the EDS-SEM preferable to that of the microprobe

- For the qualitative analysis of particles containing elements in amounts above a few per cent (value varies with instrument and mineral analyzed). For the qualitative analysis of particles which are very fine or which form very fine intergrowths. When carbon coated, these particles are unresolved by the electron microprobe optics, but are clearly visible when using the higher SEM magnifications. Use of the SEM is generally preferable to that of the microprobe for the analysis of any grains less than 10 microns in size, because these are easier to observe on the display screen. The SEM is particularly well suited for the study of gold ores because gold often occurs in particles of fine sizes and for the preliminary study of complex ores of silver when occurring in fine intergrowths.

CHAPTER 5

GENERAL PRINCIPLES OF MINERAL PROCESSING

INTRODUCTION

Metallurgy, the study of the physical and chemical properties of metals, is a relatively new science. Many of the elements in the periodic table were discovered during the past 250 years (see chapter on base metals and table 17). Their systematic study has been carried out for less than a century.

Mineral processing/extractive metallurgy (Gilchrist 1980, Hayes 1985) is the branch of metallurgy which studies the separation or **recovery** of the economic metals from their ores. During recovery, a certain metal which is chemically bound with one or more elements to form a mineral, which is in turn associated with other non-economic minerals (the *gangue)* in a rock, is taken out of such rock and mineral and converted, through a number of physical and chemical processes, into the pure or almost pure state, to be used for practical or other applications. In the more common situations, the basic steps in the process of recovery can be summarized as follows:

- **Comminution**, or **crushing and grinding**, to prepare the ore for the following separation steps.
- **Physical separation**, or **mineral dressing**, or **beneficiation**, to separate the mineral(s) bearing the metal of interest from the other non-economic minerals in the ore, or gangue. This stage may or may not precede the chemical separation step.

- **Roasting**, applied to many sulfides, some oxides and other minerals occurring in forms which do not lend themselves readily to the conventional chemical processes.

- **Chemical separation**, including hydrometallurgy and pyrometallurgy, to extract the metal of interest from its mineral(s).

- **Recovery from solutions and refining**, to remove impurities and produce a high grade metal product.

Chart 6 is a schematic representation of the sequence of events taking place during the process of metal extraction.

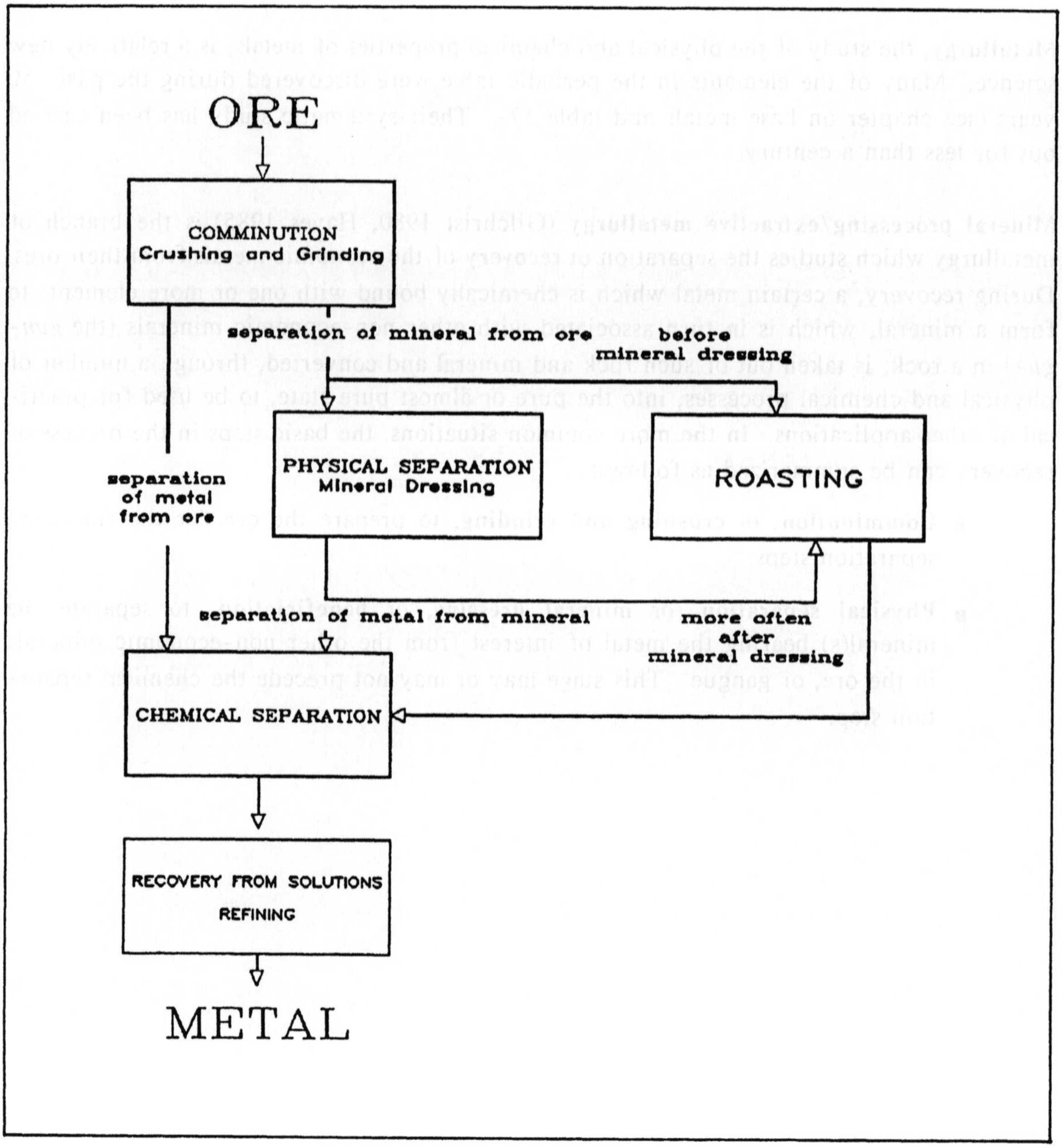

Chart 6. *The process of metal extraction.*

COMMINUTION

Comminution, the first step in the extraction of a metal, involves the breakage and pulverization of the ore until reduced to particles which are fine enough for the mineral of interest to become available for further processing. Comminution is necessary where the metal of interest occurs in a *lode* or a *vein deposit*, also defined as a *massive ore*, because in such an ore the particles of interest are not accessible. Comminution is not necessary or is reduced where dealing with *placer* or *alluvial deposits*, due to the disaggregated state of the rock. In this latter type of deposits, the less stable minerals in the original formation have been eroded away, leaving the minerals of interest unaltered and liberated, therefore ready for the separation process. As one would expect, based on the definition in the introduction, pages 3-4, corrosion resistance coupled with the higher density, allow gold and the platinum group elements to concentrate in placer deposits, but not silver. More common occurrences of silver in placer deposits are represented by minor amounts of the metal distributed in the lattice of the native gold. Besides the precious metals, economic minerals found in placer deposits are diamonds, monazite (for rare earths), ilmenite, rutile and leucoxene (for titanium), cassiterite (for tin), zircon (for zirconium), pyrochlore/native tantalum (for tantalum and niobium), scheelite and wolframite (for tungsten), cinnabar (for mercury). Fine quartz (sand) and sometimes magnetite (black sand) are the major components of the placer product. Another mineral commonly found in placer deposits is staurolite. When the original massive ore from the erosion of which the alluvial deposit originated, is traced back, this is known as the *mother lode*.

"Comminution is one of the basic industrial processes involving large expenditures for heavy equipment, energy, operation and maintenance. Its mechanical efficiency is extremely low, possibly close to 1 per cent and almost all of the energy required to break the rock appears as heat. The cost of the metal worn away is great, particularly in wet treatment. The size reduction of the rock has been a long continuing economic challenge to science and industry." (Bond 1985, p. 3A-3).

"Crushing may be viewed as a mechanical operation where a sufficient force is applied to relatively brittle solid particles in such directions that failure of the bonding forces in the particles occurs. Gaudin (1939) defines crushing as that operation or group of operations in a mineral processing plant, the objective of which is to reduce large lumps to fragments or smaller particles, the largest particles in the crushed product being 1/20 in. or greater in size. Generally the objective of crushing is to accomplish size reduction, and occasionally the additional requirement to generate a minimum amount of fine particles is imposed. One distinction between crushing and grinding equipment is that the working surfaces of crushing equipment do not make contact with each other during size reduction. *Crushability* may be defined as the ease of crushing a sample under standard conditions. Various testing procedures may be used to measure crushability." (Horst 1985, p. 3A-5).

"The reasons for comminution of raw materials and mineral products are numerous, however the major purposes may be summarized in the following four categories:

- The liberation of one or more economically important minerals from the gangue components in an ore matrix.
- The exposure of a large surface area per unit mass of material to facilitate some specific chemical reaction.
- The reduction of the raw material to the desired size for subsequent processing or handling.
- The satisfaction of market requirements concerning particle size specifications.

Generally, the objective of comminution, at least in the mineral industry, is to achieve liberation of the mineral species so that separation of the desired minerals can be attained..... When *crushed* ore particles consist of a single mineral, they are called free (*or liberated*) particles; when *the particles of the mineral of interest are still associated with the host in a manner which prevents any contact with the flotation or leaching agents* (words and sentence in italics added or altered by the author), they are called locked particles. The degree of liberation of a particular mineral species is the percentage of that mineral occurring as free particles in relation to the total quantity of that mineral occurring in both the free and locked forms. Liberation means severance of the mineral components in the ore. Severance and fracturing in most raw materials can be traced to one of the following causes:

- Macrostructural weaknesses, as bedding planes in coal.
- Microstructural weaknesses, as schistosity of some ores.
- Microstructural differences in physical properties of adjacent minerals, as hardness, brittleness and cleavage.

Examples of the latter case would be experienced with a mixture of chalcopyrite, pyrite, galena and quartz." (Horst 1985, p. 3A-5). Average grinding sizes for the ore can be brought down to 10-15 microns at the very most, if using long grinding periods on a product which has already passed a -200 mesh sieve. Some specific mills are reported to be available, which are capable of grinding the ore down to sizes of 5 microns or less. (Report published by Kilborn Limited in 1988, entitled A **generic study of the capital and operating costs for the recovery of precious metals from refractory ore**).

In this book the term *totally enclosed particle* is used to indicate the most difficult type of occurrence of the mineral of interest in its host, in relation to the extraction process. Such occurrence is represented by a mineral enclosed within a host which is compact and/or consists of a single particle or crystal, like the one represented in figure 1B on page 26. The term totally enclosed applies to occurrences of particles which have been **directly** observed in the ore. The terms *locked* or *encapsulated*, preferred by the mining industry, are not considered sufficiently representative of the occurrence, because they are the result of a conclusion drawn when extractions are poor and **not** of direct observations. If that is the case, the particle may be *locked* in a host which is compact, but may also be *locked* in a host which is fractured, like the one represented in figure 1A. We have seen that fine grinding is more likely to liberate or at least expose the locked mineral whose host is fractured, than it is to liberate or expose the locked mineral whose host is compact. The need for a terminology which is descriptive not only of the actual occurrence of the mineral in its host, after the ore has been crushed or ground to a certain size, but of its potential to become liberated if further ground, becomes obvious: such terminology will assist in the decision concerning whether liberation or exposure can be achieved by grinding only, or by chemical techniques in addition to grinding.

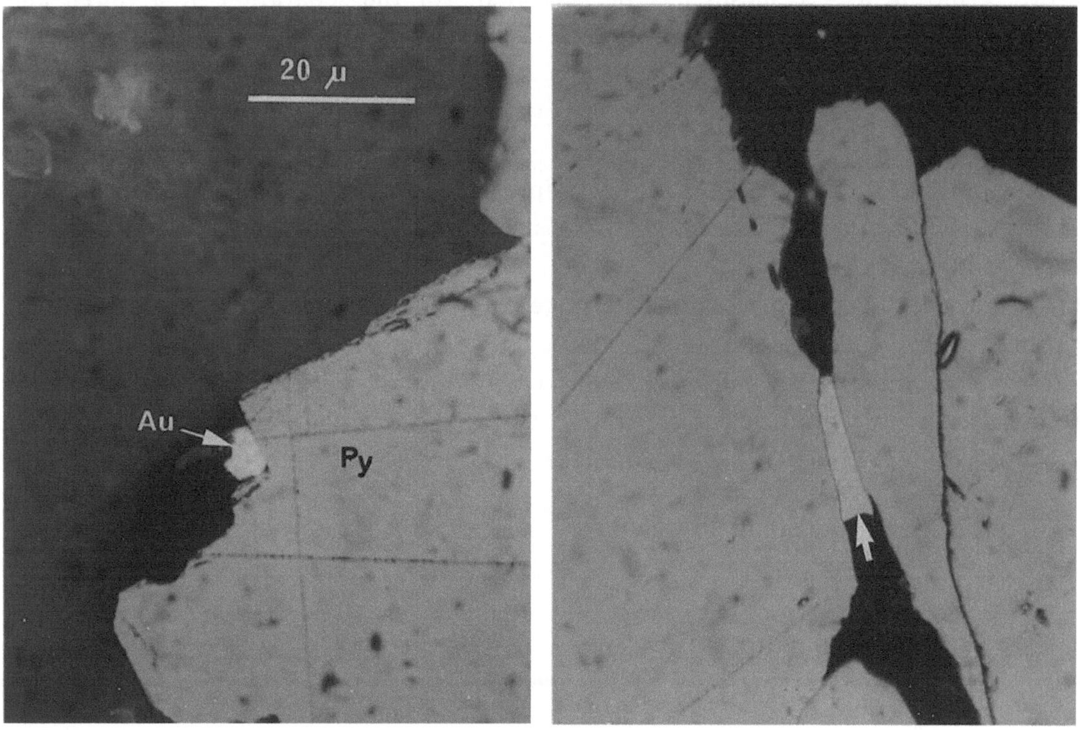

Fig. 18. *Particle (left) and veinlet (right) of native gold which, due to their occurrence in relation to the host pyrite (along the outside and in a large gap), do not require fine grinding for successful extractions by cyanidation.*

Complete particle liberation is necessary during beneficiation, but not during the chemical separation. When smelting the product, neither liberation nor exposure of **all** the particles of interest are necessary. When leaching the product, exposure of the particles (which may still be attached to the host) rather than liberation, is sufficient for the dissolution process to take place. For example, if leaching a gold-bearing pyrite concentrate where the native gold occurs along the outside or in fractures of the host, grinding sizes will not need to be as fine as those of the gold (see discussion on page 27 and figure 1), as long as the gold particles occur in a manner which can be reached by the cyanide solution (figure 18). Grinding sizes are decided upon, based on the average size of the mineral of interest and on whether liberation, exposure, or a rough mineral concentrate are needed. A chalcopyrite concentrate prepared for smelting of the copper, will be much coarser than a **compact** ore to be leached for gold because:

- The chalcopyrite occurs in particles the average size of which is considerably larger than that of the native gold.
- Neither liberation, nor exposure are essential, for the preparation of a chalcopyrite concentrate to be smelted, the main requirement being the elimination of a large amount (but not all) of the gangue.

Therefore, grinding to -10 mesh may be sufficient for the separation of a rough chalcopyrite concentrate, but the finest grinding of -200 mesh will be necessary for the preparation of an ore for leaching of gold which is 10-15 microns in average size and **totally enclosed** in its host.

Direct observations of the mineral of interest represent the most reliable means for decisions concerning the choice of the correct grinding sizes.

Crushing and grinding are often done in stages, and may include some beneficiation or other processes applied between. For example, if dealing with a gold-bearing ore where the native gold occurs with particles showing a wide range of grain sizes, an initial rough grind may prepare the product for separation of the coarser gold by amalgamation. The same product may then be further ground for exposure of the finer particles, and treatment by cyanide.

PHYSICAL SEPARATION

Physical separation, or **mineral dressing,** or **beneficiation** represents the stage of metal extraction during which the appropriately ground or pulverized minerals bearing the metal of interest are separated from the worthless particles, or gangue.

During the physical separation, advantage is taken of the more common optical, physical, chemical, magnetic, and electrical properties of the mineral of interest. According to Lorig (1988), the basis of an economically practical process is that the particles containing the metal of interest and those of gangue must not respond alike in any important degree to the specific separating force employed. Some of the more common processes of separation are:

HAND SORTING. This is the simplest form of concentration. The Chinese are reported to have used hand sorting in 500 BC. Minerals to be sorted must have sufficient distinguishing optical features to be singled out. The method is practiced largely where dealing with precious or semiprecious stones, with contaminants and with any minerals the specific gravity of which is such that they do not lend themselves to easy separation by other techniques. Some of the minerals which are hand sorted are those of antimony, beryllium, bismuth, niobium and tantalum.

GRAVITY CONCENTRATION "may be defined as that process where particles of mixed sizes, shapes and specific gravities are separated from each other by the force of gravity or by centrifugal force. The nature of the process is such that size and shape classification are an inherent part of the process in addition to separation on the basis of specific gravity from whence the process got its name" (Aplan 1985, p. 4-1)

"When applicable, gravity concentration provides a simple, high capacity, low cost, reasonably effective means of concentrating ores. It works best in the following instances:

- For rich ores.
- For those ores showing a coarse size of liberation.
- For those ores where, e.g. flotation, low intensity magnetic separation are not effective.
- For large volume and/or low value material.
- For placer deposits.
- For preconcentration.
- For processing in remote areas.
- For economic reasons where the situation dictates the least expenditure of money". (Aplan 1985, p. 4-2).

Gravity concentration methods are applied to the sulfides of the major metals, when in coarser grain sizes, and to most of the placer minerals including native gold.

Techniques used for the concentration of gold include **jigs** (for the separation of coarse and medium size particles), **shaking tables, strakes, blankets** and **corduroy tables** (for the separation of fine particles). " **Panning** is a type of gravity concentration used by prospectors for the recovery of gold from river beds. It concentrates the heavy gold particles at the bottom of the pan while the light gangue is washed off on top." (Lorig 1988, p. 352).

"**Jigging** is one of the oldest methods for concentrating ores and is still widely used in some areas. In jigging, water is pulsed strongly upwards and downwards through a bed of particles. After sufficient pulsation, the top portion of the bed becomes an accumulation of the lighter gangue, which can be rejected, while the lower part is the heavier enriched concentrate" (Lorig 1988, p. 352). " The **corduroy table** consists of wide sloping plates with shallow sides which hold a coarse corduroy cloth. Periodically the corduroy is removed and washed by hand in boxes partly filled by water to recover the gold which concentrates." (Wise 1988, p. 411).

MAGNETIC SEPARATION takes advantage of the magnetic properties shown by some minerals such as magnetite, chromite, some pyrrhotite, etc., while the other particles in the ore remain inert. It is practiced by use of magnetic separators equipped with either electron or permanent magnets. More common use of magnetic separation is in the beneficiation of iron ores.

FLOTATION. Fine particles of certain minerals such as graphite, talc and sulfur, float on water because they resist wetting. Most other minerals do not and require a water repellent coating for their flotation. The mineral particles are coated by agitating the mixture of ore, water and suitable chemicals. The chemicals become bound to the surface of certain particles thus making them water repellent, but not to others. These water repellent particles then adhere to air bubbles and rise with them to the surface where they are removed in the froth. Minerals that naturally resist wetting may be treated so that their surfaces will be wetted and sink. Flotation is widely used to concentrate copper, lead and zinc minerals when together in the same ore and for the beneficiation of many more minerals of economic metals. (Rephrased from **Encyclopaedia Britannica-Micropaedia).**

Flotation "involves the aggregation of air bubbles and mineral particles in aqueous medium with subsequent levitation of the aggregates to the surface and transfer to a froth phase..... Whether or not bubble attachment and aggregation occur is determined by the degree to which a particle surface is wetted in water. When a solid surface shows little affinity for water the surface is said to be hydrophobic, and an air bubble will attach to the surface. The stability of this attachment is measured by the contact angle developed between the three phases: liquid, solid and gas.... When the air bubble does not displace the aqueous phase, the contact angle is zero. On the other hand, complete displacement of the water represents a contact angle of 180 degrees. Values of contact angle between these two extremes provide an indication of the degree of surface hydration, or conversely, the hydrophobic character of the surface." (Arbiter 1985, p. 5-40).

"From its beginning in the first decade of this century, flotation has gradually moved to a predominant role in mineral separation. There are several reasons for this, the continuing trend toward treatment of lower grade and more finely disseminated ores being the major one. Not only has this been true for sulfide ores to which the process was first applied commercially, but also for most non-sulfide ores as well. A second factor has been the broad applicability of the process with respect to particle size; it is effective from 8 to 10 mesh to below 10 microns. Finally, more so than for any other separation process, flotation has almost no limitations in separating minerals. Gravity, magnetic or electrostatic processes are either applicable to particular minerals or to a restricted number of mineral combinations. Flotation, on the other hand, can utilize the wide range of surface chemical differences among minerals, and an equally wide range of reagents. The recent availability of selective flocculation capabilities on a commercial scale as an adjunct to flotation should remove one of the limitations to the process in the very fine particle range." (Cooper et al. 1985, p.5-1).

OTHER TECHNIQUES. Some of these are separation in magnetic fluids, sorting by electric selection, air scrubbing, grease tables, thermoadhesive methods, etc.

Chart 7 is a schematic representation of the various processes within the physical separation.

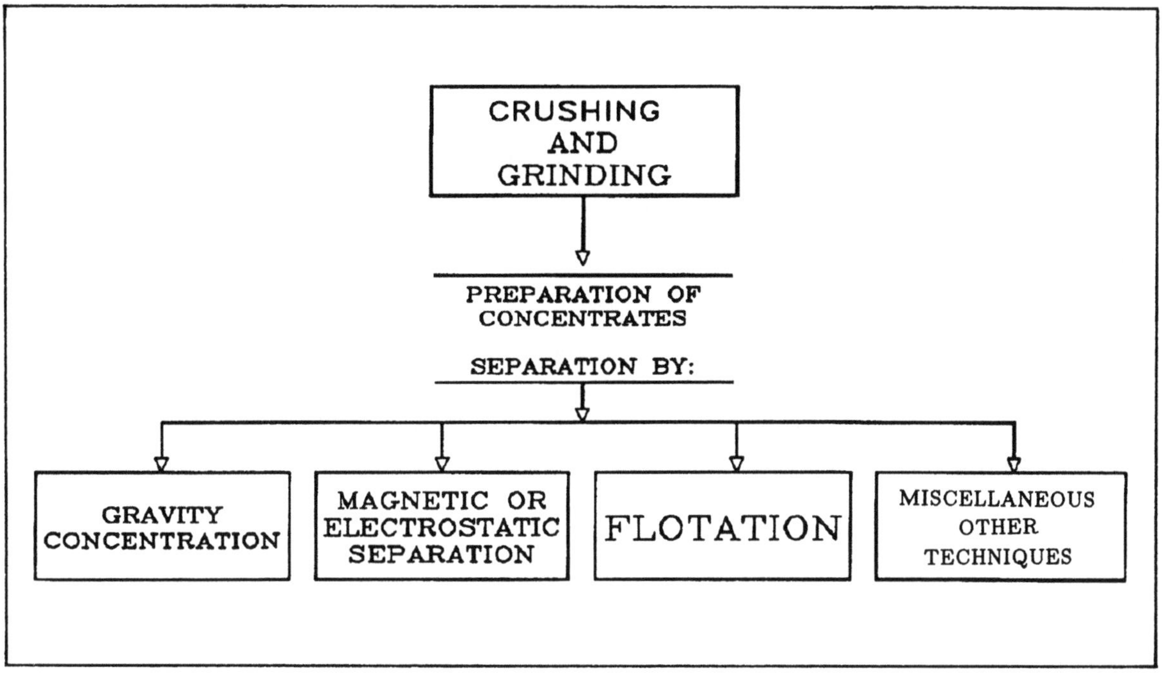

Chart 7. *The physical separation (mineral dressing).*

ORES TO WHICH LITTLE BENEFICIATION IS APPLIED. A type of ore to which beneficiation is not usually applied, is the *laterite or lateritic ore,* because, due to their finely divided state, separation of the economic minerals from the gangue is not always possible. Laterite ores are the result of the gradual decomposition of rocks which have been exposed to the atmosphere at the earth's surface, for prolonged periods of time. During the process leading to the formation of this type of ore, the continued chemical and mechanical action of air, water and temperature changes breaks the original minerals down to soil and clay. If the rocks contain economic metals in appreciable amounts, the weathering process may act to concentrate these metals to a degree that the resulting deposits can be mined. Essentially the metal is taken into solution in the ground water and redeposited at a greater depth, producing a zone where its content is abnormally high. The process is called lateritic weathering, hence the generic term laterite ores. (paragraph modified after Boldt and Queneau 1985, p. 17-4). Laterites form in humid environments, such as those of the tropical and subtropical regions, and under strongly oxidizing conditions. Laterite soils may contain clay minerals, these however represent an intermediate stage of alteration because the final product tends to be poor in silicon. Laterites are not uniquely identified with any particular parent rock, geologic age, or single method of formation. They are products which form in response to a set of physicochemical conditions including a metal containing parent rock, a well drained terrain, abundant moisture for hydrolysis during weathering, relatively high oxidation potential and persistence of these conditions over thousands of years. Common laterite deposits are those rich in iron, nickel, and manganese, although the manganese is usually found in the form of nodules and surface rubble distributed within the clay. Bauxite is the aluminum rich representative of iron and nickel laterites. Laterite ores, if consisting of much fine disaggregated soil, need a limited amount of comminution. Some other laterites are so compact that they need explosives for their breakage.

In addition to the laterites, little beneficiation is usually required by ores, when using some of the hydrometallurgical techniques, such as heap leaching for the separation of gold, copper or uranium.

SINTERING AND PELLETIZING are processes used for the agglomeration of the ore fines and concentrates, for their subsequent transportation to smelting and refining plants. Sintering is produced "through partial fusion for production of indured agglomerates which are irregularly shaped" (Ban 1985, p. 11-8) and is used during the processing of base metals such as iron, nickel and copper.

ROASTING

The purpose of the **roasting process** is to convert the mineral of interest into a soluble phase or into a phase with lower melting point, in preparation for the leaching or smelting processes. Roasting is also practiced to liberate very fine particles of precious metals from otherwise insoluble hosts, such as pyrite and arsenopyrite. Because the roasting reaction is in many instances exothermic (it gives off heat), no fuel is required, once initiated.

"............ roasting can be considered a preparation process because it has the following effects:

- Volatilization of impurities that would interfere with the chemistry of the metal dissolution or recovery process, *for example, arsenic and antimony from copper concentrates* (words in italics added by the author).
- Alteration of associated minerals so that they do not enter into the chemical dissolution process.
- Opening of the mineral structure so that the leach solution can penetrate, for example, roasting of gold-bearing arsenopyrite ores.
- As a source of leaching reagent, e.g. sulfation to produce sulfuric acid in copper ore leaching.
- Assistance to grinding, as roasted ores are usually easier to grind.

A disadvantage in roasting sulfide ores or concentrates is the production of SO_2 requiring gas treatment facilities to protect the environment. This has resulted in recent research to develop aqueous oxidation and reduction processes as substitute for roasting.....

In principle, roasting systems can be described as gas-solid systems in which the reaction time and temperature can be controlled, so that only the desired changes take place. The common classifications with some examples of the most commonly used roasters are:

- Static bed - batch furnace, heap roasting.
- Moving bed - rotary kiln, shaft furnace, single hearth furnace, multiple hearth furnace, spouted bed.
- Fluidized bed.
- Pneumatic flash roaster.

These systems and the same equipment can also be used for drying and for roasting either wet or dry feed with suitable changes in operation, mechanism, and temperature within the system. Two or more chemical or physical changes can be carried out in sequence by suitable profiling of the temperature, charge and gas composition within one roaster. A common example is the low temperature oxidation and volatilization of arsenic as As_2O_3 from arsenopyrite, followed by high temperature oxidation of iron sulfide in gold re-roasting....

Although there are many types of roasters and roasting systems in use, one of the most important at the present time is the fluid-bed system." (Picket 1985, pp. 13-5 and 13-6).

Roasting processes, carried out after the physical separation, occasionally before, include **oxidation, chlorination, reduction, sulfation** and **sulfidation**.

OXIDATION of sulfides by roasting is often necessary to lower the sulfur content, because the average furnace does not reach sufficiently high temperatures for their reduction. The process involves heating in air without fusion, and has the purpose of transforming the sulfides into oxides, a more suitable state for conversion to the metal by carbon. More common sulfide minerals roasted before smelting are those of copper, lead, zinc, nickel and molybdenum. Mineral oxidized before leaching is sphalerite (electrolytic process). Gold-bearing pyrite and arsenopyrite are reduced to the less stable oxide phases, in order to make their fine gold particles accessible to the cyanide solution (page 35).

In **CHLORINATION**, the oxides of some metals are converted to chlorides by a reaction with chlorine, which takes place with or without carbon. Chlorination is especially applicable in the processing of complex ores where the conventional metallurgical methods are inadequate, for example in the processing of titanium ores, and of the copper oxide CuO.

REDUCTION. As opposed to the sulfides, the more common of which are roasted before smelting, the oxides of many metals may be treated directly and do not need reduction.

Some of the minerals or ores which are reduced to the metallic state before leaching, are chrómite, ilmenite, cassiterite and nickel laterites. Reduction agents used are carbon (coke, coal, carbon in solution), various metals (sodium, magnesium, aluminum, potassium, calcium, silicon, iron), gases (hydrogen, carbon monoxide, methane, sulfur dioxide, sulfur), metal hydrides (calcium hydride) and electricity.

In **SULFATION AND SULFIDATION** the minerals of interest are converted into soluble sulfates and sulfides, before the leaching process. Some of these minerals or ores are copper oxides and nickel laterites (converted to sulfates) and chalcopyrite (converted to the sulfide CuS). (Information derived and interpreted from Picket 1985).

CHEMICAL SEPARATION

Chemical separation is conducted at low temperatures (ambient or slightly above), and at high temperatures (500 to 1600°C). **Hydrometallurgy** is the study of the techniques for the separation of metals at the lower temperatures. **Pyrometallurgy** is the study of the techniques for the separation of metals at the higher temperatures. **Electrowinning** and **electrorefining,** which use electricity applied to the metal while dissolved in a liquid, are sometimes classified as belonging to hydrometallurgy, and sometimes as belonging to a third branch of chemical separation, **electrometallurgy**.

Once the ore has been prepared, hydrometallurgical processes include "the **leaching** itself, which generally applies to the selective dissolution of metal compounds and minerals from ore mixtures; the **solution purification**, which depends largely on the chemistry of the positive and negative ions in the solution; the **precipitation** of the metal to be extracted by electrolysis with direct current or by treating the solution with a reducing gas such as hydrogen; and the **regeneration** of the solution for its reuse" (Lorig 1985, p. 357). The **SME Mineral Processing Handbook** (section 13) summarizes the hydrometallurgical processes as follows:

- Leaching, the actual separation of the metal from its minerals, following liberation or exposure by comminution with or without physical separation.
- Solid-liquid separation, immediately after leaching.
- Solution purification and concentration, which include ion exchange, solvent extraction and charcoal adsorption.
- Recovery from solutions, which include precipitation, electrolysis and crystallization.

LEACHING. Leaching is conducted through the use of aqueous solutions containing chemicals capable of dissolving the mineral(s) in which the economic metal is concentrated, while leaving the associated gangue undissolved. There are three basic requirements for a successful leaching process. These are that:

- The economic mineral(s) is (are) soluble in the dissolving agent.
- The gangue is not soluble in the dissolving agent.
- The economic minerals are exposed to the dissolving agent. Particle exposure, rather than total liberation is usually sufficient. (Gasparrini 1992-A).

Many metals are extracted by leaching techniques, from at least some of their ores, for example, gold, silver, copper, uranium, nickel, iron, zinc, manganese, potassium, chromium, rare earths, tin (see descriptive sections for details). The dissolving agent varies of course with the nature of the mineral(s) from which the economic metal is being extracted, but it is often represented by hydrochloric acid, sulfuric acid, ammonia, or water. Most commonly used agent for gold and silver ores is presently cyanide (pages 119 and 165). Thiourea leaching (page 120) is being applied commercially to gold ores, although still in a very limited manner. Other processes, such as those using iodide (Hiskey and Peihao Qi 1991), thiosulfate (Langhans et al 1992), or lime-sulfur synthetic solutions (LSSS - Zhang Jian et al 1992) are still in an experimental stage. If the metal is concentrated in more than one mineral, more than one dissolving agent may be necessary. There are different means of applying the leaching technique, and they largely depend on the nature of the ore, the accuracy of recovery required and the available budget. These are:

In-situ leaching "refers to the leaching of ore left in a mine after it has been worked out, or the application of leaching solutions directly to the body in the ground." (Wadsworth 1979). "The leach solutions are introduced above the level of the ore body and are allowed to percolate by gravity through the ore and are collected at a lower level by a network of sumps. The pregnant leach liquor is then pumped directly to the surface. The technique relies on the permeability of the ore body. If the natural porosity of the rock is low, then it may be necessary to induce fracture and shatter the rock by the use of explosives. One of the difficulties of in-situ methods is that extreme care is to be taken to avoid loss of the leach liquor to natural ground water. This is undesirable on two counts, firstly it represents the contamination of the water table in the surrounding area, secondly, it represents a loss of metal, reducing the efficiency of recovery of the extraction process. For these reasons, it is necessary to obtain a complete knowledge of the hydrology of the region before such a technique is attempted. The extraction of metals by in-situ leaching appears to be an extractive technique because of potential savings in the costs of mining the ore. However, these savings may be offset by the low extraction efficiencies, low extraction rates and low metal concentrations in the final liquor." (Hayes 1985, p. 186)

Dump and heap leaching. The terms "refer to the leaching of mined material which has received little or no size reduction following removal from the ore body. The ore, which may have a maximum size of between 1.0 and 0.1 m diameter, is piled in dumps or heaps on a pre-prepared impervious base." (Hayes 1985. p. 186).

"The solutions are distributed over the surface of the dumps via a variety of systems, which may be ponds, small holes drilled at intervals in 2 in. bore plastic pipes, or similar pipes fitted at intervals with plastic needle valve for individual flow control. Some plants use a sprinkler distribution system.

In operation, the pH of the liquor applied to the dumps may be adjusted to an optimum value, determined by the chemical and bacterial reactions involved, by the addition of sulfuric acid; nutrients may also be required. The overall recovery of metal in this type of leaching operation may not exceed 60% and the leaching period may extend over months or years to achieve this recovery; however this method requires the lowest capital cost of all alternative methods. If reagent additions of the leach circuit are required, their consumption per pound of metal recovered over the total life of the operation may be high." (Forward and Warren 1985, p. 13-12).

"Leaching rates can be improved with only modest increases in processing costs by reducing the mean size of the ore particles and by the screening out of fine clayey material. These treatments have the effect of substantially improving the exposure of the minerals to the leach solution and improving the permeability of the dump. The typical extraction efficiency of this type of process is of the order of 60%." (Hayes 1985, p. 186).

Heap leaching techniques are used for many gold deposits where the rock porosity allows for easy penetration of the cyanide solution (pages 120-121). Heap leaching is also applied to some low grade uranium deposits.

Percolation leaching. "In this technique, which is operated on the batch principle, ore, which is crushed and sized to +3/8-3/4 in. and importantly containing no fines or with the fines agglomerated by moistening and rolling the ore, is leached in large vats or tanks. A typical plant using this technique will consist of a number of vats, from 60 to 175 ft on a side and 16 to 20 ft deep, arranged side by side and spanned by loading and excavating bridges of the gantry type. The vats are constructed of reinforced concrete with lining such as wire-reinforced mastic, lead, etc. The bottoms are covered in turn with timbers, cocoa matting, saran-screen filter media, or other suitable material and additional timbers to prevent their damage during the charging and excavation of the ore.

The leaching operation is carried out by charging an empty vat with ore by means of the loading bridge. The ore is then hand leveled and the metal values extracted by submersion in a series of leaching and washing solutions. The ore is drained and finally removed from the vat by the bridge, sluicing and other means. The total process cycle may be of the order of 100 to 200 hours. In this leaching technique recoveries of up to 95% of the readily soluble metal values, such as in oxide copper and uranium minerals, may be obtained." (Forward and Warren 1985, p. 13-12).

Agitation leaching. "Agitation leaching involves the leaching of fine (approximately 200 microns) high grade concentrates in which the slurry or pulp is continuously mixed and stirred to improve the rates of reaction." (Hayes 1985, p. 187).

"Agitation leaching yields the highest metal recoveries in the shortest time, but requires the highest capital investment per annual ton of material processed. To achieve these results, finer comminution of the ore is necessary to ensure suspension and maximum exposure of the value-containing minerals to the leaching reagents. The leaching operation may be performed in the batch, cocurrent or countercurrent nodes at slightly subatmospheric, atmospheric or superatmospheric pressures, and at ambient or elevated temperatures with slurry densities of approximately 75% solids.

Batch leaching is applied particularly where small quantities of material are to be treated; it is also employed in cases where close control of process conditions is necessary. For continuous cocurrent operation at least three leaching vessels are employed in series to narrow the dispersion of the residence time of the mineral particles in the leaching system about the mean residence time. Since the concentration of soluble material in the ore and that of the reagent in solution both decrease as leaching proceeds, the reaction rate decreases. To offset the effect of this event on efficiency of metal extraction a countercurrent leaching system may be employed. In such a system, fresh ore in a first stage is leached with depleted solution from a second stage; this partially leaches the ore. The liquid and solids are then finally separated, the strong (pregnant) solution proceeding to metal recovery and the partially leached solids being mixed with fresh leaching solution and placed in a second stage unit. When leaching in the second stage is complete, the liquid and solids are again separated, the liquid returned to mix with fresh ore in a succeeding cycle, and the solids are sent to waste or such subsequent treatment as may be required. A countercurrent leaching system may involve as many as five individual stages. The capital cost of a plant employing countercurrent leaching is higher than one employing cocurrent leaching because of the requirement for a solids/liquid separation device (usually a thickener) between each stage." (Forward and Warren 1985, p. 13-12).

Pressure leaching (Berezowsky and Weir 1983, and Weir and Berezowsky 1984), **pressure cyanidation** (Girardi, Alruiz and Anfruns 1989), and **oxidative acid leaching** (Lakshmanan and Childs 1984, Scheiner, Lindstrom and Henrie 1971, and Guay 1980), promote penetration sometimes at high temperatures of the leaching solutions. They are often used for gold extractions, after oxidation of impermeable host minerals. They are some of the most expensive of the leaching techniques (Mason and O'Kane 1985). Sherritt Gordon uses pressure leaching processes for the treatment of nickel and zinc concentrates.

Bacterial leaching. "In certain cases, microbial leaching can be used for the profitable recovery of metal values. *Most commonly treated metals recovered on a commercial basis are* (sentence altered by the author) copper and uranium. Microbial leaching is aimed currently at low-grade ore in heaps, dumps, or in situ in mines or deposits. The process is attractive, because relatively little labor and capital investment are required compared with conventional techniques. However, it is a much slower process than other commercial leaching systems. The organism involved is called *Thiobacillus ferroxidans*, a small rod shaped cell about 1 micron long. It uses atmospheric CO_2 as its carbon source for cell growth and oxidizes ferrous iron and sulfides to obtain energy for growth. Sulfuric acid is a product of the metabolism of the organism. Sufficient dissolved oxygen must be available during these oxidation reactions. The other main growth requirements are: ammonium, nitrogen, phosphate, suitable temperature (about 30°C), and acidity (pH about 2.0). In general, other temperatures (5 to 50°C) will not kill the organism, but its activity can be severely curtailed. T. ferroxidans is usually present in a natural, acidic, sulfide environment and only needs encouragement to accelerate the leaching process. Some metals, e.g., mercury, silver and possibly molybdenum, could retard or stop a leaching process by inhibiting or killing the bacteria. Also, strong sunlight shining on shallow ponds containing bacteria could inhibit their activity." (Fletcher, Finkelstein and Derry 1985, pp. 13-21 and 13-22).

During the late 80's, bacterial leaching was being used by at least one South African gold mine (Gencor, see section on gold metallurgy, page 128), and was being tried on an experimental basis in North America for low grade gold deposits in which the native gold is fine and totally enclosed in pyrite.

Metals more commonly recovered by hydrometallurgical techniques are gold, silver, uranium, the oxidized ores of copper and nickel, some unoxidized ores of copper, and many ores of zinc (electrolytic process).

Following the leaching process, the pregnant solution is separated from its undissolved gangue material by:

SOLID-LIQUID SEPARATION. This "is an important part of essentially all hydrometallurgical operations. This is caused by such factors as the relatively large amount of gangue solids that must be separated from the resulting pregnant liquor after leaching; high clarity requirements for liquors going to ion exchange, solvent extraction, precipitation or other processing; and the recovery of valuable solids after precipitation or crystallization while simultaneously demanding a high purity for these products." (Dahlstrom and Emmett 1985, p. 13-26). Washing of the gangue and/or discard residue is particularly important to maintain acceptable values of recovery.

SOLUTION PURIFICATION AND CONCENTRATION. **Ion exchange.** This is "defined as a reversible process in which ions attached to or incorporated in the molecular structure of a solid, exchange ions with a surrounding solution" (Porter 1985, p. 13-33) and are then separated to isolate the metal of interest. "The application of ion exchange systems to hydrometallurgy has not been great, except for the recovery of uranium from both acid and carbonate solutions. Except for uranium, hydrometallurgy ion exchange so far has been limited to special applications, such as rare earth separations, and recovery of metallic ions from plating, pickling liquors, and other industrial wastes. However, it will increase in its applications, as more emphasis is centered on control of mine and plant effluents." (Porter 1985, p. 13-35).

Solvent extraction. "Solvent extraction, as used in extractive metallurgy, consists of contacting an organic solvent with an aqueous solution containing a desired constituent. During the contact, for chemical reasons, the desired metal is transferred preferentially from solution in the aqueous phase to solution in the organic phase. Because the solvent and aqueous liquids are immiscible, they separate when the mixing ceases. In this way the desired constituent can be separated from other undesirable materials in the aqueous solution. After separating the organic and aqueous phases, the organic can be stripped of its contained metal and returned to the extraction circuit for reuse." (Hazen 1985, p. 13-38). Metals separated by solvent extraction are tantalum, niobium and the platinum group elements.

Charcoal adsorption. "Active carbon is unique and one of the most powerful agents employed for recovery, isolation and purification of various materials. Although the effectiveness of charcoal as a purifying agent for water and spirits was recognized centuries ago, it has been only in the last four or five decades that its usefulness through activation has been recognized and applied successfully on a scientific and engineering basis. The first use of charcoal as a metallurgical reagent was in connection with the extraction of gold by the chlorination process" (Bhappu 1985, p. 13-44). "In the earlier applications of activated carbon in hydrometallurgical processes, gold and silver were the only metals recovered from leach solutions on a large scale. Today (1973), however, the use of the activated carbon has been extended to the recovery of platinum, palladium and osmium from dilute chloride solutions. Active carbon is also used for adsorbing neodymium, samarium, and gadolinium.... Of course, the most interesting and economic viable use of activated carbon is in the treatment of low grade gold and silver ores." (Bhappu 1985, p. 13-50).

Besides **roasting**, already discussed (pages 68-70), pyrometallurgical techniques include:

SMELTING. "The term smelting is a corruption of the German *schmeltzen*, meaning melting, and is used to describe the process whereby an ore or concentrate is melted and chemical reactions take place

Modern non-ferrous smelters normally use oxygen enriched air or commercial oxygen (approximately 95 per cent O_2) to utilize fuel value contained in the feed materials. In sulfides, it is the oxidation of iron and sulfur which provides a large proportion of the heat required. Where there is little or no fuel value in the feed, as is the case for oxides, auxiliary fuel and a reductant, frequently in the form of coal or coke, are used.

Fluxes, usually silica or limestone, are added to combine with oxidized impurities to form a fluid *slag* and concentrate the values in a *matte* or crude metal product. Any nonferrous values, such as gold, silver, etc. contained in the flux, will be collected in the matte or crude metal, thereby improving recovery. Differences in specific gravity produce separate molten layers, with the slag remaining on top and the matte or crude metal settling slowly to the bottom of the furnace. This provides a physical separation of values from the impurities. The top layer, slag, is *skimmed* off and discarded directly or charged to a slag cleaning circuit, depending on economic considerations. The values are *tapped* from the furnace hearth for further processing.

When a matte is produced, this will normally involve a converting step where the remainder of the iron, and in some cases the sulfur, are removed. Crude metals such as blister copper, ferronickel, etc., usually require fire refining.

The process gases from a smelting operation frequently contain substantial quantities of SO_2, as well as a high proportion of the volatile impurities, for example arsenic and mercury. In the majority of modern smelters these gases are cleaned to remove the solid particles and the sulfur is fixed in a sulfur fixation plant, in most cases as sulfuric acid." (J.C. Taylor 1988, personal communication by letter).

Smelting techniques are used for the treatment of ores of the major base metals such as iron, cobalt, nickel, copper, lead, zinc, and tin. Smelting may also be used to prepare some ores of copper and nickel for the leaching process.

Processes of smelting to upgrade the quality of the metal, include **in-vacuum induction, melting** and **electrolysis and electroslag refining** (Lorig 1988). **Fire refining** is used for the refining of *blister copper* (see section on copper, page 256), due to environmental regulations limiting the use of hydrometallurgical techniques. The process is conducted in an anode furnace where oxidation of the sulfur is obtained with air.

CASTING is a process by which the molten metals from the smelter are molded into shapes to be transported or stored. Most castings are obtained by pouring the metal into sand molds prepared from mixtures of moistened sand and clay. In recent years more accurate molds have been prepared such as plastic mold, composite mold, investment, permanent mold and die casting.

RECOVERY FROM SOLUTIONS - REFINING

PRECIPITATION. "The term precipitation has been used in the metallurgical and mineral industries to describe a number of processes. For example, this term has been used to denote:

- The deposition of an insoluble compound from solution by the action of a chemical agent or heat.
- The deposition of a metal or metallic oxide when an electric current is passed through a solution.
- The collection of dust particles or suspended droplets from a gas stream by passing the mixture through an electrostatic field.
- The formation of a new solid phase within an alloy by solid-solution reaction.

In a hydrometallurgical approach to processing of ores and concentrates, precipitation reactions may be employed both in solution purification and in the final production of a metal or compound." (Merritt et al. 1985, p. 13-51).

ELECTROLYSIS. "When a direct electrical current is passed through an aqueous solution of a metallic salt, metal ions (*cations*) migrate towards the cathode, and ions of the acid radical (*anions*) migrate towards the anode. Chemical changes occur at the surface of both electrodes. At the cathode cations are discharged, with acceptance of electrons, in a process of reduction. The principal cathode product is a metal deposit, but in some cases there is also a tendency toward reduction of hydrogen ions to form hydrogen gas as a secondary product. At the anode there is a loss of electrons in a process of oxidation. If the anode is of the same material as the metal being deposited at the cathode, the anode material is oxidized and dissolves. However, if the anode surface is of a nature which withstands attack or oxidation, the principal anodic product is gaseous oxygen from oxidation of the hydroxyl ion derived from ionization of water. The anion of the salt remains in the solution (electrolyte) associated with the hydrogen originally associated with hydroxyl ion which has been discharged; therefore, free acid is the second anodic product." (Chapman 1985, p. 13-59).

In metallurgical processes, the substance to be transformed may be the electrode, it may constitute the solution, or it may be dissolved in the solution. Electric current (e.i. electrons) enters through the negatively charged electrode (cathode). Positively charged components of the solution travel to this electrode, combine with the electrons and are transformed into neutral elements or molecules. The negatively charged components of the solution travel to the other electrode (anode), give up their electrons and are transformed into neutral elements or molecules. If the substance to be transformed is the electrode, then the reaction is generally one in which the electrode dissolves by giving up electrons. (Rephrased from the **Encyclopaedia Britannica-Micropaedia).**

The process of electrolysis is applied for the refining of many metals which have been separated by both the smelting and the leaching processes. Electrolysis is used to separate gold from silver in gold-silver bullions.

Electrowinning uses the process of electrolysis to extract the metal in the place of smelting, for example, zinc and cadmium from aqueous sulfate solutions, magnesium, aluminum and calcium from molten halide baths. **Electrorefining,** or **electrolytic refining,** uses electrolysis to refine the metal after smelting or leaching, for example, gold and silver.

EVAPORATION AND CRYSTALLIZATION. "**Crystallization** is the process in which molecular compounds in solid form originate and grow because of solubility changes of the compounds in solution. The process produces a crystalline rather than an amorphous, colloidal or gelatinous material. In industrial crystallization, the physical size, shape, rate of growth, and purity of the product are controlled by reproducible conditions of operation to give a product with economic value. The temperature, ionic components and amount of water constituting solutions are described in solubility tables. The data of these tables are plotted in phase diagrams...from which the quantity of crystals formed at specific temperatures can be predicted.

The process of concentration starts with a solution having molecular concentration below the saturation point required for crystallization. Solubility changes are made by addition of other ions, by removal of water by freezing or evaporation, or by changes of temperature that increase the concentration first to the saturation point in the solubility tables and then to a higher concentration called supersaturation. The amount of supersaturation that can be obtained varies with the solution components and is the driving power that causes crystallization to progress as a continuous process....**Solar evaporation** uses ambient atmospheric conditions and direct heat from the sun in the production of chlorides, sulfates, carbonates, the borates of sodium, potassium, magnesium and lithium salts from natural brines and from solutions produced by solution mining." (Wilson 1985, p. 13-67 and 13-70).

REFINING. "Metals produced in a high temperature smelting operation commonly contain in solution impurities derived from the original ore, reducing agents, fuels, fluxes and furnace materials. Refining processes must be employed to reduce these impurities to below the limits permitted by the specifications for marketable metals: these processes include precipitation of impurities, chemical reactions, distillation and electrolysis. Occasionally a reverse refining operation is performed in which the main metal is removed from a residue containing an accumulation of impurity." (Davey 1985, p. 12-2).

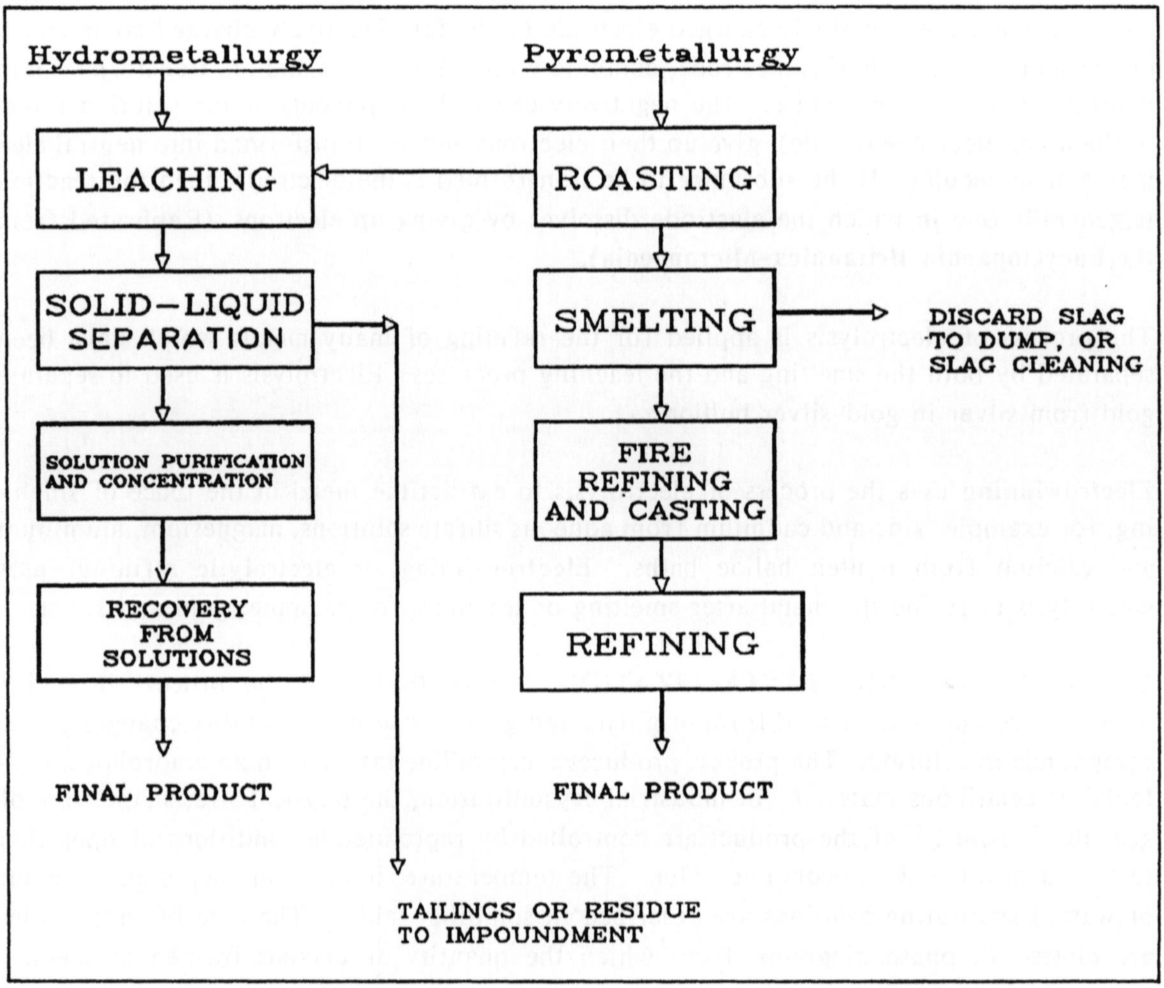

Chart 8. *The chemical separation.*

"While several metals are actually produced by an electrolytic process, **electrorefining** is commonly employed for copper (the recovery of nobler metals would otherwise be impossible or at least extremely difficult), and sometimes for nickel, lead (when there is a high bismuth content) and tin." (Davey 1985, p. 12-2).

In addition to the **SME Mineral Processing Handbook** (1985), theoretical descriptions of the metallurgical processes are in **Process Selection in Extractive Metallurgy** (Hayes 1985) and in **Principles of Extractive Metallurgy** (Habashi 1986), both of which provide further and sometimes updated details on extraction processes and practices.

Chart 8 shows the various phases or branches of the process of chemical separation.

PART II

CHAPTER 6

GOLD

INTRODUCTION

"Gold is fascinating. It is the most romantic of all metals. It receives affection and attention to a degree that borders on worship. Great is the mystique of gold. A yearning for gold appears to be ingrained in the human psyche." (Mohide 1981, p. 3). There is something magic about gold that has attracted human kind since the beginning of time. Ancient populations in Europe, Northern Africa, the Middle East, Mexico and South America have left their jewelry and ornaments as testimony to their fascination for gold. In medieval times, alchemists racked their brains trying to find the philosopher's stone, a substance which supposedly converted lead into gold. In more recent times, the gold rushes in rich territories such as California, Colorado, Alaska, British Columbia, the Yukon, attest to the mysterious attraction exercised by gold. What is amazing is the hardship which people have endured in order to find gold, a hardship not justified by the dream of becoming rich alone. It is a strange drive which cannot be explained on a rational level and which compels the individual to search and discover the most precious of all metals. How a prospector is capable of finding those minute practically invisible flakes in a river bed, the patience he exerts in separating such minute flakes, are also part of the mystique of gold. Just as much as the extraction of the invisible micron size gold from a compact ore. Most of the gold produced today does not become visible until converted into bullion.

Gold exercises the same mysterious power when observed under the microscope. No matter how many thousands of polished mounts are examined, studies of gold ores are never tedious. The finding of a shiny particle, even one as small as a few microns, is one of the most exciting moments in the study of a gold ore. The longer it takes to locate one, the more exciting the moment. The shine, the color, the luster, everything contributes to a feeling of marvel unequalled by most other visual experiences. It is experiences such as these that make one wonder how much in this world is beyond the rational and in the realm of magic. But perhaps these thoughts do not belong in a scientific treatise. They must be the result of the unexplained effects of gold on the author's mind.

Needless to say, the most common use for privately owned gold is and always has been for jewelry and personal ornaments. In recent years, uses of gold have extended to practical applications such as electronics (electroplating, electrodes, aerospace applications, computer science applications, etc.), dentistry and medicine. Gold coins, minted by many nations, are an attractive solution to the problem of the small investor. (Mohide 1981).

* * * * * *

Gold, together with silver and copper was the first metal to be known and used during the latter part of the stone age. In those early days, it was probably recognized when occurring in clearly visible form, such as large nuggets dispersed in river beds.

A report entitled World Mine Production of Gold, 1986-1990 (1987), published by the Gold Institute in Washington DC, states that largest gold producers in the world during 1986 were South Africa (38.5 per cent) and the USSR (18.8 per cent), followed by the United States (which replaced Canada in 1986 with 7.0 per cent), Canada (6.4 per cent), Australia (5.0 per cent), China (4.1 per cent) and Brazil (4.1 per cent). Largest gold producer in the non-communist world remained South Africa by a very wide margin (51.6 per cent), followed by the United States (9.4 per cent) Canada (8.5 per cent per cent) and Australia (6.2 per cent). South African gold contribution to the free world has decreased since the beginning of the 1970's when the total amount produced was 60 per cent. More recently, the April 1991 issue of **Precious Metals News and Review** published that "preliminary data from the US Bureau of Mines have shown that 1990 gold production in the United States exceeded output in the Soviet Union for the first time in nearly 50 years, making the United States the second largest gold producer after South Africa."

"More than 95 per cent of all the gold ever produced in South Africa has its source in the conglomerate beds of the Witwatersrand Triad (the Dominion group and Witwatersrand and Ventersdorp Supergroups which are closely associated, both areally and structurally). Current production is confined to reefs of the Witwatersrand Supergroup and the basal formation of the overlying Ventersdorp Supergroup. Gold production from the underlying Dominion Group in the Klarksdorp area has been discontinued more than two decades ago." (Engelbrecht 1984, pp. ii-iii). Gold in the Soviet Union has been produced since ancient times by territories such as the Urals, the Caucasus, Kazakhstan and Central Asia. 60 per cent of the current Soviet Union production derives from placer deposits (Mohide 1981). Gold in the United States and Canada comes from a very large variety of mines. States producing most of the United States gold are Nevada, South Dakota, California, Montana and Utah (Lucas 1985). Provinces and territories producing most of the Canadian gold are Quebec, Ontario, Northwest Territories and British Columbia (Law-West 1986).

Average grades and actual sizes of the producing mines are seldom published or made available by the mining industry. Even without those values, when taking into consideration that:

- The size of South Africa is smaller than that of Alaska;
- Most of the gold in South Africa is produced from one specific locality only;
- Recovery plans and microanalytical studies are more systematic in South Africa than they are in North America;

one cannot help wondering how much of the total gold contributed by that country comes from natural resources and how much is the direct result of the proper planning and application of the appropriate recovery strategies.

MARKET VALUE

According to Wise (1988), the development of complex economic systems has given gold a major function as a high denomination currency and as a backing for paper currency systems, a function which it retains in many nations. Approximately 60 per cent of all the gold mined today is held by governments and central banks. Gold is presently the most significant means of international payment.

The price of gold at the time the last draft of this book was being prepared (October 1988) was below US$ 400 per ounce, down by about US$ 50.00 from the previous year, and remained within a similar range up until the time when its second edition went to the presses (September 1991). During the past ten to fifteen years, the price of gold has been similar, although usually lower by about US$ 100, to the price of platinum. Along with rhodium, trading at US$ 1,185-1,200 per troy ounce on September 29 1988, gold and platinum trade at the highest metal values on the market.

Over the past 100 years the price of gold has moved from about US$ 20, in 1873, to US$ 35, set in 1934, to over US$ 100 in 1973. Once past the psychological barrier of US$ 100, it rose steadily to US$ 200 (in July 1978), US$ 300 (in July 1979), US$ 400, then US$ 500 and US$ 600 during the next few months. 1980 was a most remarkable year in the history of gold when its price rose to a record high of US$ 850 on January 21, then came down to US$ 474 towards the end of March. It broke through the US$ 700 mark again on September 23 when it reached another record high of $ 750.50, then fell to US$ 526.75 by the end of the year (Mohide 1981). By that time many predictions were made regarding an increase above the US$ 1,000 mark, but the price went down instead, although never below US$ 300. Considering the steady price always below US$ 50.00 which gold has kept during more than half of the 20th century, the strong moves upwards during the 70' and 80's are difficult to explain in rational terms. At the cost of contradicting the general opinion that the price of gold, in the range of US$ 350, is very low today (September 1991), such price would rather seems high instead. Particularly so when compared to the price of many other commodities. The high gold to silver price ratio during the 80's (discussed on page 151) would also confirm such an observation.

Fluctuations in the price of gold **are** in fact irrational. The consistently inaccurate predictions, so frequent during the past two decades, more than prove such a conclusion. According to Mohide (1981), the price of gold is, at least in part, the product of emotions such as: **fear,** of inflation or perhaps even of a major war involving a superpower; **anxiety,** about the uncertainty of the future: gold, in the form of coins and bullion is a more secure investment *psychologically* than paper money; **speculation** or **greed,** where one hopes to buy low and to sell high; **resentment,** when the gold price goes down; **hope of a bonanza,** which may result from investments in mineral exploration. Chart 9 shows the fluctuations in the price of gold during the 70's and 80's, and some important factors which *may* be responsible for its variations (from Mohide 1981).

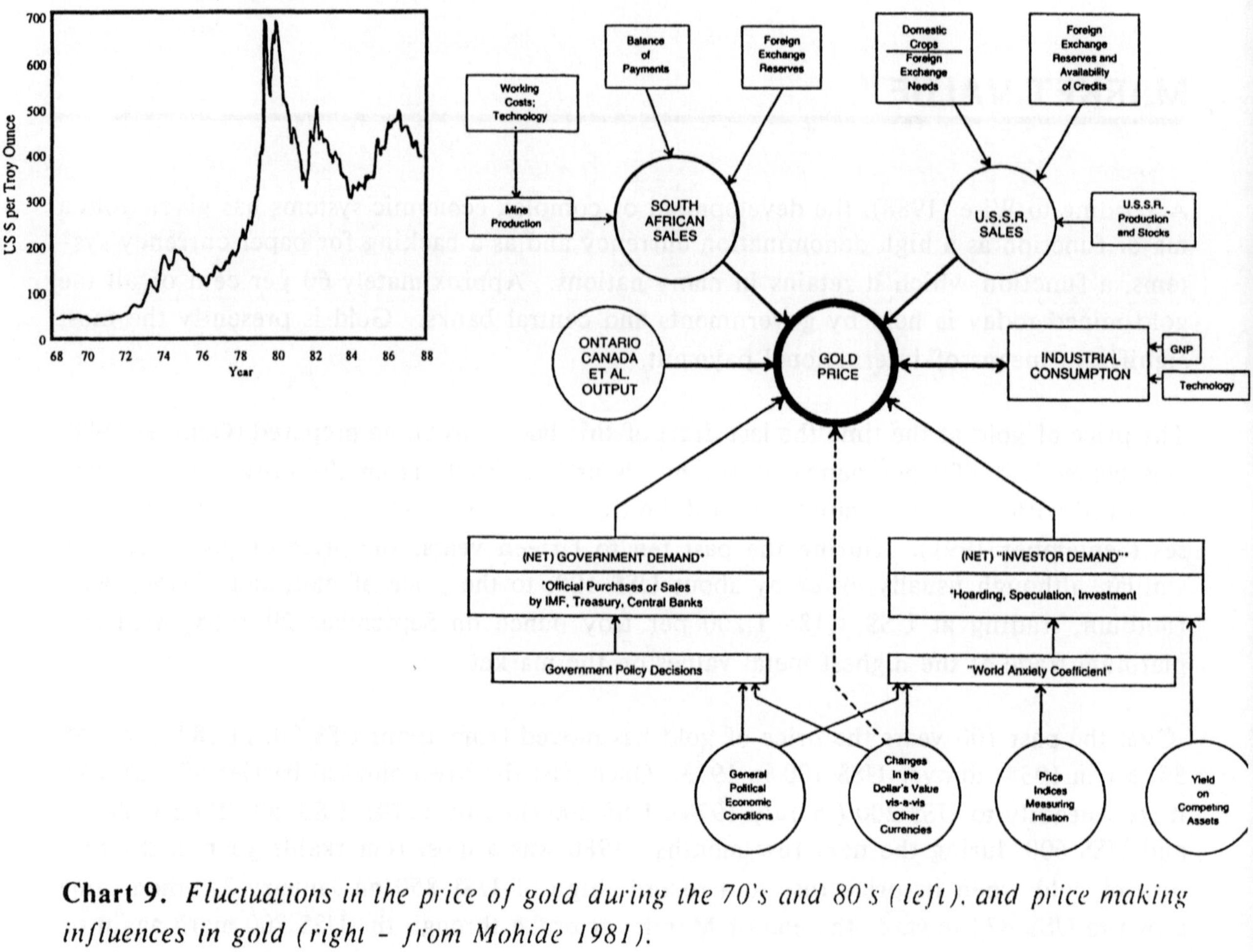

Chart 9. *Fluctuations in the price of gold during the 70's and 80's (left), and price making influences in gold (right - from Mohide 1981).*

GEOCHEMISTRY SYMBOL: Au ATOMIC #: 79

"Gold is a member of group IB of the periodic system which includes copper (Cu), silver (Ag), and gold (Au).....As might be expected from electron orbit considerations, there is some similarity between Ni, Pd and Pt (Group VIII) and the corresponding elements of group IB, namely Cu, Ag and Au. Thus, platinum has only one electron less than gold, and since the energy differences are small in the electronic constitution of the two elements, the differences in the chemical behavior of the two elements in a few respects are slight. In natural environments, however, there is no close relationship between gold and platinum except that they tend to occur in the native state, and some platinum may be present in certain varieties of native gold and vice versa. There are no particular similarities between gold and mercury, the corresponding element in group IIB of the periodic system. As is well known however, gold and mercury form an amalgam, and well defined compounds occur in the system Au-Hg; natural amalgam is known, but formal Au-Hg compounds do not appear to occur in nature, or if they do, they have not been adequately characterized. *(More recent data on mercury are discussed on page 101* - sentence in italics added by the author*)*. There are marked differences in the chemistry of silver and gold, a consequence that is attributable to the diminishing effects of the lanthanide contraction between atomic numbers 57 (La) and 79 (Au).

Gold is considered by some to be a transition element of the third series, although it does not fulfil the necessary electronic conditions that characterize transition elements. Gold is a soft yellow metal with a moderately high melting point (1063.0°C) and a high electrical and thermal conductivity exceeded only among the elements by silver and copper. It is the most malleable and ductile of metals; one ounce of gold can be beaten into foil some 300 square feet in area. Of all the elements, gold in the pure state is undoubtedly the most beautiful especially when it occurs in crystalline groupings or in dendritic and arborescent arrangements." (Boyle 1979, pp. 9-10).

Gold abundance in the Earth's crust is 0.002 per 10^6 atoms of silicon (Brownlow 1979). Gold average economic abundance in the ore is today in the range of 0.1 to 0.2 oz/ton or 3.42-6.84 ppm, but deposits yielding grades to the fire assay which are as low as a tenth of those values, are exploited.

As shown in figure 2, on page 28, gold distributions tend to be irregular. They appear to be so on a macro, as well as on a microscopic level. This means that although usually present throughout the ore, larger concentrations are often found within confined levels. And when conducting the microanalytical studies, we may see no gold in one or two polished mounts cut from a certain sample and then discover five or six particles all within short distances in the third and fourth mounts cut from the same sample.

The discovery of a high gold concentration coupled with high gold solubility in cyanide, is described in case study 4, on pages 29-30.

Fire assay

Fire assay is the most commonly used technique for the determination of gold in ores throughout the world. Other methods of analysis are by neutron activation (NA) and by atomic absorption analysis or spectrometry (AAA or AAS, references on page 45).

"Fire assay is a branch of quantitative chemical analysis in which metals are determined in ores and metallurgical products with the aid of heat and dry reagents. The method is applied today principally to the determination of silver, gold and some members of the platinum group although lead, tin, copper and mercury in ores are also in some instances determined by fire-assay methods.

The fire assay for silver and gold depends mainly upon:

- The very great degree of solubility of these metals in molten metallic lead and their almost complete insolubility in slags of suitable composition.
- The decided difference in specific gravity between the two liquids, lead and slag, which permits the completion of the separation of the precious metal-bearing alloy from the slag.
- The fact that lead may be practically completely removed from the precious metals by a carefully controlled oxidizing fusion, involving differential filtration, in a porous vessel known as cupel.
- The ready solubility of silver and the complete insolubility of gold in dilute nitric acid." (Bugbee 1981, p. 1).

"The assay of gold ores was probably practiced by the Romans, and was elaborated in the Middle Ages. The German assayers had already reached a fair degree of proficiency in the year 1500.

The assay of gold is generally conducted in the dry way, i.e. by furnace methods. The plan of operation is to concentrate the precious metal in a button of lead either (1) by fusion in a crucible; or, more rarely, (2) by scorification." (Rose 1986, p. 490). During fire assay, the ore is finely ground and mixed with a flux of soda, borax, sometimes silica, lead monoxide (litharge) and flour to reduce the lead oxide to metallic lead during fusion by heat. The reduced lead collects the precious metals and settles to the bottom of the crucible, while the silicate compound and remainder of the flux float to form a glass slag which is discarded. The button of lead obtained by this method is then subjected to cupellation, during which gold and/or silver are separated by melting the impure metal in a flat porous dish made of refractory or other high temperature material called cupel, and by directing a blast of hot air in it in a special furnace. During cupellation, any impurities, such as those of lead, copper, tin and other unwanted metals, are oxidized and partly vaporized/partly absorbed into the pores of the cupel. The resulting bead, consisting of precious metals only, is weighed. Silver, if present, is removed from the gold bead by a process known as *parting*, consisting of a treatment with nitric acid. For proper parting the silver content of the bead should be about three times that of the gold. The platinum group elements, if present, alter the appearance of the bead, thus giving an indication of their presence for further treatment.

The fire assay technique is reliable for the analysis of most ores (Bacon et al 1989, Nadkarni and Savolainen 1992). Losses to the cupel are generally minor, except for some specific cases. "The loss is greater in the presence of impurities. Copper carries gold into the cupel. One gram of copper in 25 grams of lead causes a loss of 10 per cent of the bead of gold, 1 milligram in weight, and even 0.05 grams of copper doubles the ordinary loss. Tellurium and selenium are still more dangerous in large amounts. S.W. Smith (1908) has shown that tellurium is slowly removed into the cupel during cupellation, and if there is not enough lead to carry it off, so that towards the end the tellurium becomes equal in amounts to the gold plus silver, then the surface tension of the globule breaks down completely and the alloy spreads over a wide area, wets the cupel and is completely absorbed. The lead should be 80 to 120 times as much as the tellurium to obtain good results in cupellation. The loss of gold is then small.

The elements copper, bismuth, sulfur and especially tellurium are oxidized with difficulty and tend to remain in the button, increasing its weight. Nickel is intractable in cupellation because its oxide is insoluble in litharge, and if copper is also present, the difficulty of removing nickel is greatly increased.

The formation of scoria in the cupel, owing to the presence of certain base metals, is usually accompanied by low results. Minute beads of fine metal may be entangled in the scoria.

Indications of the presence of some base metals in cupellation are as follows: copper stains the cupel green if a small quantity is present and dark brown or black if the amount is large. If the amount is large enough, the button spreads out and adheres to the cupel. The green stain is used by assayers, who add copper to some of the assay pieces in a large muffle charge so that the stains produced will indicate the correct order of the assays. Tellurium sometimes gives a pink stain and sometimes merely a brown color in a ring round the litharge stain. Bismuth gives an orange-yellow ring. Nickel and cobalt form black or dark green scoriae with a green stain." (Rose 1986, pp. 500-501).

According to K.J. Henley (1988, personal communication by letter), when dealing with ores "where sulfides are abundant, it may be necessary to roast before fire assay, and some variation in flux composition to cope with variations in rock matrix may be appropriate. Extraction by aqua regia followed by AAS generally gives similar results to fire assay except in the case of sulfide calcines, where the iron oxides appear to protect some of the gold from the aqua regia, and the AAS results are lower than the fire assay results."

In the author's experience, the data obtained through microanalytical studies are usually in good agreement with those of the fire assay. But there are exceptions.

Case study 8. Fine gold in oxidized ore. Samples from the oxidized zone of a deposit in California contain, according to the fire assay, between 0.01 and 0.05 oz per ton or 0.34 and 1.7 ppm of gold. When searching for gold during microanalytical studies of samples characterized by such low grades, the chances of locating a sufficiently large number of particles to be representative of the general occurrence, are small even if 6 polished mounts per sample are examined. Yet when looking at samples from this deposit, gold is **easily** found in 1, 2 or 3 polished mounts per sample (depending on the grade). The gold is silver-free, thus reducing the amount of particle surface necessary to account for the values given by the fire assay. It is fine, less than 10 microns in diameter, with many veinlets, but not sufficiently fine and too frequent to be representative of grades as low as 0.01 to 0.05 oz/ton. Finer material and just as frequent, may be observed in samples of much higher grades.

Case study 9. Coarse gold in oxidized ore. Oxidized and unoxidized samples of metallurgical products derived from a deposit in Central America contain very similar amounts of gold, according to the fire assay. Yet the results of the microanalytical studies disagree: many more particles are found in the oxidized samples than in the unoxidized samples. A comparison of the fire assay values and microanalytical data in two representative samples of oxidized and unoxidized products is as follows:

	Fire assay		Native gold found
	Au oz/ton	ppm	(in six polished mounts)
Unoxidized sample:	0.231	7.9	3 particles
Oxidized sample:	0.220	7.52	17 particles

The gold is between 20 and 30 microns in size in both samples and contains similar amounts of silver (in the range of 30 to 35 per cent). It is unlikely that additional gold occurs in solid solution in some of the sulfides or other minerals of the unoxidized sample, or is in other way invisible: the ore gives recoveries which are in the range of 80 per cent by the standard techniques and such high values would not be obtained from an ore containing high amounts of invisible gold.

In the two case studies, the gold shows different grain sizes and silver contents. The only similarity between the two ores is in their oxidized state.

Although aware that fire assay is the best, if not the only technique for accurate determinations of precious metals in ores, Sawyer (1992, p. 37) recognizes that "there are some naturally occurring materials which cannot be successfully determined" by such a technique. "Some of those same materials can only be determined by instrumental techniques, if their chemistry is understood."

Problems associated with reliability of the fire assay data were discussed with many metallurgists, chemists, and earth scientists. Some of the more interesting responses were:

- **S.A. Hiemstra, Mintek, Johannesburg - South Africa.** Although cautious, recognized the difficulty of finding gold in amounts of 0.01 to 0.05 oz/ton or 0.34 to 1.7 ppm, like the one described for the ore of case study 8. He recommended the use of a technique for quantitative determinations of gold contents in polished mounts, by use of a computerized image analyzer developed by Oosthuyzen also at Mintek, before coming to any final conclusions. Advantages of the technique are discussed further in this chapter, in the section describing the method of study for gold ores (page 118).

- **K.J. Henley, Amdel, Frewville - Australia.** Also cautious, but admitted to the difficulty of seeing gold when in amounts of less than 0.1 oz/ton or 3.42 ppm. He suggested the use of aqua regia/AAS and neutron activation in addition to fire assay. If the values obtained by the two techniques are similar to those of the fire assay, then the conclusions can be drawn that the analyses are accurate.

- **D.A. Pretorius, Witwatersrand University, Johannesburg - South Africa.** Stated that the problem about reliability of gold assays is one almost as old as gold mining itself. Over and over again, there have been records of more gold appearing in mineralogical examinations than in fire assay and of more gold being recovered in actual mining operations than was anticipated from assaying. He admitted of having been involved, on several occasions, in the prospecting of tracts of well developed and well mineralized conglomerates, which consistently assayed less than 0.05 oz/ton or 1.7 ppm Au, but which in the simple mills of small workings, yielded in excess of 0.20 oz/ton or 6.84 ppm Au. He added that the only suspicion, not confirmed by experimental data, was that the gold might contain up to 5 per cent mercury which would promote its volatilization during the fire assay.

- **J.C. Taylor and W.R. Snelgrove, Jan H. Reimers and Associates, Oakville - Canada.** Snelgrove agreed with the possibility of gold volatilization during fire assay, due to the presence of mercury in the native gold or in the ore. He suggested that arsenic might also contribute to gold losses. Taylor recommended verification of such theory by collecting and analyzing for gold the fumes liberated during the fusion of the samples being prepared for fire assay.

The ore from central America associated with the coarse gold and described in case study 9, does contain mercury. Chemical analyses indicate a definite increase in the element with the degree of oxidation. Electron microprobe analyses for mercury of native gold particles from the ores of case studies 8 and 9, gave values in the range of 0.2 to 0.3 per cent. The ore described in case study 8 contains arsenic values above background, and because no discrete minerals of arsenic were identified during the microanalytical studies, it would be possible for this element to be concentrated in minor amounts in the native gold. Accurate determinations for arsenic by the electron microprobe on selected particles of native gold from the deposits of the two case studies, were never obtained. The techniques recommended by Hiemstra and Henley were not tried, nor were the experiments suggested by Taylor ever conducted.

Obviously problems associated with the production of correct fire assay analyses are complex and far from resolved. Having the microanalytical studies, and sometimes even the recovery rates, uncovered the presence of possible inaccuracies, perhaps some systematic research on the subject will clarify the contradictions above.

TABLE 1 - GOLD DEPOSITS

DEPOSIT	STRUCTURAL CHARACTER	AGE	TYPE LOCALITY	MAIN MINERALS	ASSOC. MINERALS	TYPICAL ELEMENTS	METALS MINED	% OF WORLD PROD.
Gold-quartz lodes	Hydrothermal veins	Mostly Precambrian	To be expected in Precambrian rocks, such as the Canadian, Australian, Brazil-Guayana and African-Arabian shields	Low-silver native Au	Quartz	Si	Au	20-25
Epithermal deposits	Hydrothermal veins	Tertiary	Emperor deposit (Fiji)	High silver native Au, tellurides	Quartz, carbonate, barite, fluorite	Te, Bi, Pb, As, Sb, Hg, etc.	Au, Ag, some base metals	very small
Young placers	Unconsolidated or semiconsolidated sands and gravels	Recent	River beds that assay for gold	Native Au	Heavy minerals such as monazite, ilmenite, zircon, apatite, pyrochlore, etc.	Ti, RE's, Nb	Au, Ti, RE's, Nb	5-10; 25 to 35 in the past
Fossil placers	Conglomerates consisting of quartz pebbles in matrix of pyrite and mica	Precambrian	Witwatersrand. To be expected in the Precambrian shields	Native Au	Heavy minerals including uraninite and pge's	U, pge's, etc	U, Au, Pge's	More than 60
Disseminated gold	Silty carbonaceous dolomitic limestone	Mesozoic-Tertiary	Carlin, Cortez, Getchel, Gold Acres (Nevada type ores); Bald Mountain (South Dakota)	Very fine native Au	Silica, barite, pyrite and other sulfides	Fe, As, Sb, Hg, C Very little Ag	Au	Has been increasing. Brought into prod. in 1935
Greenstones	Metamorphosed volcanics	Precambrian	Kalgoorlie, Hemlo, Timmins, Val D'Or Kirkland Lake	Native Au, calaverite	tellurides of Au, Hg, Pb, Cu, Bi, compounds of gold with Hg, Sb, and other elements, native tellurium	Hg, Pb, Bi, Sb	Au	Has been increasing

TABLE 1 - CONT.

DEPOSIT	STRUCTURAL CHARACTER	AGE	TYPE LOCALITY	MAIN MINERALS	ASSOC. MINERALS	TYPICAL ELEMENTS	METALS MINED	% OF WORLD PROD
Young greenstones ?	Metamorphosed volcanics	Mesozoic-Tertiary	Deposits in South-eastern California such as American Girl, Cargo Muchacho, Mesquite, etc.	Native Au	Pyrite, Fe oxides, various silicates	Fe, Si	Au	Has been increasing
Byproduct gold		-	Any gold-bearing base metal deposit	Native Au, other gold minerals	Base metal minerals	Cu, Pb, Zn, As, Sb, Bi	Au, base metals	5-10
Gold in sea water	0.0011-0.05 ppb in solution	-	-	-	-	-	-	-

ROCK TYPES HOSTING GOLD

According to Rose (1986, p. 83) economic amounts of gold are obtained from:

- "Quartz veins (also called lodes, reefs and leads) in rock formations. In this division may be included replacement deposits, disseminations in rocks, and also for example, the marine deposits accumulated in shallow water, such as the conglomerates of the Transvaal.
- Placers or alluvial deposits of ancient and modern streams. Modern beach deposits and loose sands or gravels generally, may be included in this section.
- Deposits worked chiefly for base metals and from which gold is obtained as a by-product" during the refining of the copper or lead.

In the author's experience, gold occurs in economic to sub-economic amounts in a large variety of geological environments including metamorphic, volcanic and sedimentary rocks, compact and incoherent, altered and unaltered, Archean and Recent. Gold tends to be in, but is not limited to the environments suggested by Rose, although perhaps, when in other environments, it does not occur in as high concentrations. These observations are confirmed by Clarke (1924) who reports a similar widely diffused occurrence of gold in the Earth's crust.

Gold is also detected in sea water, but in too low amounts to justify the development of recovery techniques.

Table 1 is an attempt at classifying the more common gold deposits. The deposits and other data were put together based on updated information from Henley (1975).

ORE MICROCOMPOSITION

Gold-bearing minerals

Gold occurs in a limited number of minerals. Although 40 are reported (Nekrasov in press), only **native gold** and **electrum** are common. The **tellurides** are uncommon. The others are rare. Table 2 lists 38 gold minerals available from the more recent Western literature. Of those, several of the alloys are poorly defined and, rather than actual minerals, may be varieties of native gold, or simply intergrowths of the two metals (Boyle 1979).

In addition to the minerals in table 2, a number of compounds of gold with tellurium, selenium, antimony, chlorine, iron and sulfur (probably of secondary origin and occurring in a barite environment), were discovered in the St. Joe Minerals property of Tambo in

TABLE 2 - GOLD-BEARING MINERALS

MINERAL	FORMULA	COLOR (WHEN AVAILABLE)
Native gold	**Au**	**Orange yellow**
Electrum	**(Au,Ag)**	**Bright to pale yellow**
Kustelite	(Ag,Au)	Pale yellow
Cuproauride	(Au,Cu)	Orange red
Cuprian gold	(Au,Cu)	
Cuprian electrum	(Au,Ag,Cu)	
Porpezite	(Au,Pd)	White
Palladian gold	(Au,Pd)	
Rhodite	(Au,Rh)	White
Rhodian gold	(Au,Rh)	
Iridic gold	(Au,Ir)	White
Platinum gold	(Au,Pt)	White
Bismuthian gold	(Au,Bi)	White
Auricupride	Cu_3Au_2	
Tetraauricupride	CuAu	
Rozhkovite	$(Cu,Pd)_3Au_2$	
Maldonite	Au_2Bi	White
Amalgam	Au_2Hg_3 (?)	White
Goldamalgam (gamma)	Au_2Hg_3	
Weishanite	$(Au,Ag)_3Hg_2$	
Calaverite	$AuTe_2$	White/creamy yellow
Muthmannite	(Ag,Au)Te	
Krennerite	$(Au,Ag)Te_2$	Creamy white
Sylvanite	$(Au,Ag)Te_4$	Creamy white (lr)
Petzite	Ag_3AuTe_2	Grayish white (lr)
Kostovite	$AuCuTe_4$	
Montbrayite	$(Au,Sb)_2Te_3$	Creamy white
Nagyagite	$Pb_5Au(Te,Sb)_4S_{5-8}$	Grayish white (lr)
Bilibinskite	$Au_3Cu_2PbTe_2$	
Bogdanovite	$Au_5(Cu,Fe)_3(Te,Pb)_2$	
Bezsmertnovite	$Au_4Cu(Te,Pb)$	
Uytenbogaardtite	Ag_3AuS_2	
Liujinynite	Ag_3AuS_2	
Aurostibite	$AuSb_2$	White
Criddleite	$Ag_2Au_3TlSb_{10}S_{10}$	
Fischesserite	Ag_3AuSe_2	
Penzhinite	$(Ag,Cu)_4Au(S,Se)_4$	
Petrovskaite	AuAg(S,Se)	

The bold characters are used to indicate the more common minerals. (lr) signifies low reflectivity

Chile (Gasparrini 1983-B and C). At the time the book was being written, no x-ray diffraction data or quantitative analyses had been obtained and therefore, although new, the minerals are not reported to the New Minerals Commission (R.A. Gonzales 1988, personal communication by letter). Figure 19 shows occurrences of native gold, of electrum and of one of the new minerals from Tambo.

Minor to trace amounts of gold may occur in solid solution replacing silver, the platinum group elements and copper in the crystal structure of their minerals, and in some native metals. Reported minerals in which small quantities of gold have been detected are **native tellurium**, Te, **atokite** $(Pt,Pd)_3Sn$ (Kuhnel, Prins and Roorda 1980), **atheneite**,

$(Pd,Hg)_3As$, **palladoarsenide** Pd_2As, and **zvyagintsevite,** Pd_3Pb, (Cabri 1981). Similar gold solid solutions are reported in the more common sulfides, but these are far less widespread. Confirmed solid solutions of gold in sulfides, detected by the ion probe, were found to be in the range of 0.2 ppm for pyrite and up to 158 ppm for arsenopyrite in gold-bearing samples from Golden Pond, Quebec. (Martha Sizgoric 1989, personal communication).

The many mineralogical reports describing small concentrations of gold in solid solution in pyrite, have already been mentioned (page 15 and 19). Unless supported by data collected by the ion probe, the conclusions of those reports have no experimental or scientific grounds.

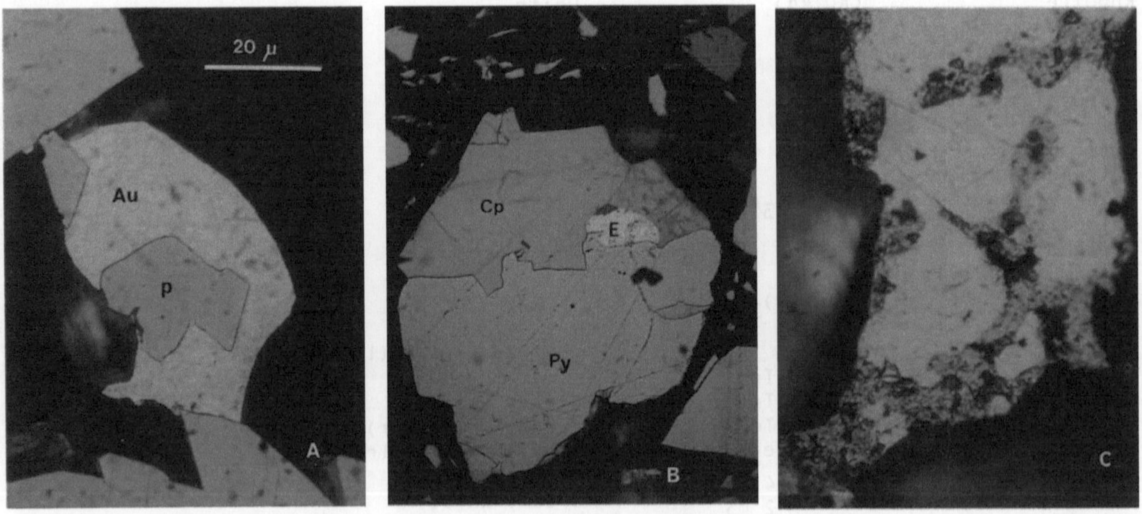

Fig. 19. *Gold-bearing minerals. A: native gold enclosing pyrite. B: electrum at the border between pyrite and chalcopyrite. C: native gold partly replaced by a secondary compound consisting of gold with minor amounts of antimony, chlorine and iron. This last compound is a new unnamed mineral from the Tambo property in Chile.*

NATIVE GOLD AND ELECTRUM account for the majority of the gold occurring in nature. There is no sharp distinction between the two minerals. In fact electrum (or argentian gold) is considered to be only a variety of native gold containing silver in amounts of 18 to 36 per cent, and representing gold to silver ratios of 2 1/2 to 1 and 1 to 1 (Dana 1963). According to Boyle (1979), although not observed in nature (or at least not reported), a complete substitutional series probably exists from gold through electrum, through **kustelite** (aurian silver), to native silver. Complete substitutions between elemental gold and silver are obtained in laboratory studies, and are compatible with the very similar crystal structures of the two native metals, both isometric with a face-centered cubic lattice. In ores in which the gold to silver ratio is variable, both gold and electrum may occur, although zoned or composite particles consisting of the two minerals together, are unusual. Native gold in most primary deposits usually contains variable amounts of silver above 5 per cent. Native gold in oxidized environments tends to become depleted in silver and may be totally silver-free (Desborough 1970, Santosh et al 1992).

Optical properties of native gold and electrum, such as reflectivity, iso-anisotropism, etc., are similar, although the color varies from bright orange for native gold, to yellow which becomes lighter with increasing silver content for electrum. Both minerals are very soft and do not polish well when preparing the mounts for optical microscope study. High reflectivity, lack of total extinction under crossed polars, bright orange and yellow colors, and abundant scratches are usually sufficiently distinctive properties for positive identifications under the microscope.

Copper is similar to silver in its association with native gold. It may occur (although it does not do so as commonly as silver) concentrated in its lattice in minor amounts as well as in larger quantities, thereby forming new varieties and separate minerals such as **cuproauride** (poorly defined but containing up to 20 per cent Cu), **tetraauricupride** and **auricupride** (40 per cent Cu). According to Boyle (1979), the complete artificial isomorphous Cu-Au series exists, but in nature the copper end members with high gold contents (more than 3 per cent Au) have not been recognized so far.

Other elements which are sometimes detected in minor amounts in native gold, are the platinum group elements, mercury, tellurium, antimony, arsenic, selenium, lead, bismuth. Elements also reported (Boyle 1979), such as silicon, magnesium, iron, zinc, chlorine and sulfur, are probably present in the form of fine impurities rather than in solid solution in the lattice of the mineral itself.

Mercury is an interesting element, because of the suggestion made earlier (page 95) that it may cause gold volatilization during the fire assay, thus resulting in low gold values for the samples being analyzed. In addition to the minerals **amalgam** (poorly defined), **goldamalgam** and **weishanite**, mercury may occur in native gold in amounts of up to 20 per cent, without modifying the crystal structure and without appreciable changes in the optical properties of the particle (D.C. Harris 1988, personal communication). The presence of mercury in native gold has seldom been reported in the past (see excerpt from Boyle 1979, on page 91), largely because of its difficulty of detection. Several of the mercury emission lines are masked, due to their vicinity to the gold energy/wavelength range, and this has made accurate analyses a very delicate operation by the microprobe and an impossible one by the EDS-SEM. Awareness of the possible presence of mercury has lately led to the detection of the element in the native gold of several deposits (the Hemlo deposit for example, in Northern Ontario). And several gold-mercury minerals have lately been submitted to the New Mineral Commission. (D.C. Harris 1988, personal communication).

THE OTHER GOLD-BEARING MINERALS have distinct optical properties and structures which are different from those of the native gold/electrum. When present, **tellurides**, usually account for not more than 20 per cent of the total gold in many deposits, the remainder being native gold or electrum. Only occasionally do they account for more than that amount or are they predominant. According to Henley (1975) there are two main geological environments in which telluride mineralizations predominate:

- Veins and fissures in Tertiary volcanic rocks. For example, the Emperor deposit in Fiji;
- Precambrian greenstones and metamorphosed volcanic lavas and flows. For example Kalgoorlie in Australia.

The other reported gold minerals are rare and, when present, account for a very small fraction of the total gold in most deposits.

The new gold minerals from the Tambo deposit in Chile are abundant and form particles of large irregular sizes (up to 100 microns) often composite, and consisting of two or more phases. Microanalytical studies conducted on a limited number of samples indicate that these minerals account for about 30 per cent of the total gold, the remainder being native gold. The gold-bearing minerals, together with the *exclusively* barite host, represent a very unusual occurrence.

Grain size

Sizes for gold-bearing minerals vary from grains and veins which are visible to the naked eye, without the aid of a hand lens, to particles which are only a fraction of a micron in diameter (often occurring in clay ores). Particles not visible even at the highest SEM magnifications (15,000x) are reported (Hausen 1981), but like in the case of the solid solutions, are not proven. These particles were assumed to be present because an SEM study of the surface of the pyrite had failed to locate any coarser gold occurrences.

There is often (but not always) a direct correlation between ore grade and gold particle size, with the lower grades coupled with average sizes in the range of a few microns (only visible when using the highest microscope magnifications). Irregular gold distributions, characterized by higher concentrations in restricted portions of the ore, are usually also coupled with coarser grain sizes for those higher concentrations.

Average grain sizes for native gold in most deposits today are between a few and 50 microns. Placer gold is coarser and usually visible to the naked eye or with the aid of a hand lens. "Crystals an inch or more across have been found in alluvial deposits in California. Boulders of pure gold were found in the early days of the Australian gold rush of 1851." (Wise 1988, p. 411). Some spectacular gold specimens, including nuggets, large crystals and dendritic incrustation, are on display at the Los Angeles Museum of Natural History where they stand testimony to the beauty of gold and to that magnetic attraction which gave a new direction to so many people's lives during the second half of the 19th and the beginning of the 20th century.

The native gold/electrum particles, when occurring in sulfides, usually show irregular rounded smooth contours (like the grain shown in figure 1B on page 26), rather than the crystalline sharp shapes which one would expect from a mineral belonging to the isometric system. When in fractures, native gold fills the space by developing into veinlets and

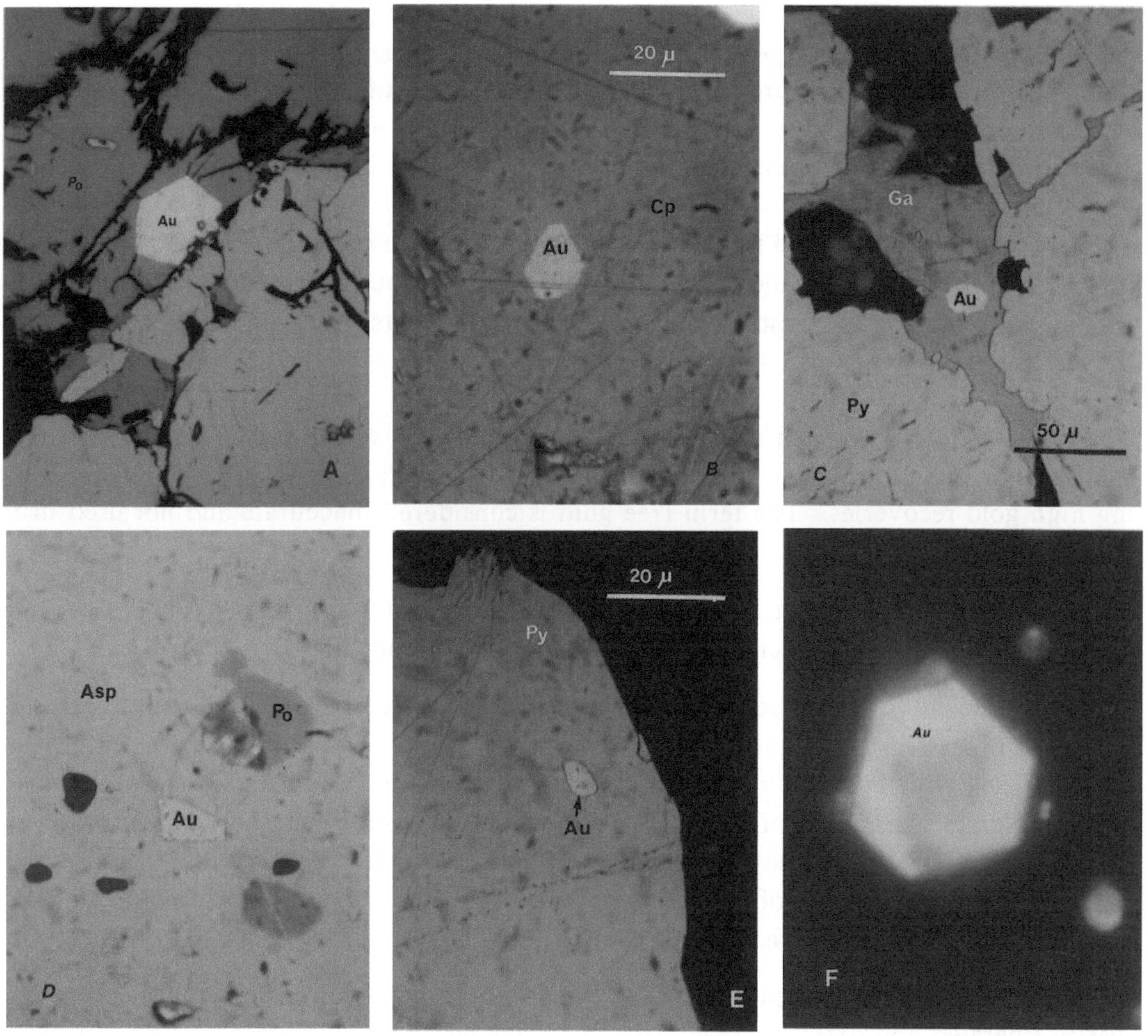

Fig. 20. *Native gold enclosed in pyrrhotite (A), chalcopyrite (B), galena (C), arsenopyrite (D) and pyrite (E). The particles show eu, sub and anhedral shapes, depending on their host mineral. F is the photomicrograph of an artificially developed crystal grown onto algae material and showing a perfectly euhedral shape . (Photomicrograph A provided by Hans Von Michaelis, Randol International. Photomicrograph F provided by Robert Gee, Imperial College). The magnification is 500x for A and C and 1250x for B, D and E.*

other elongated blebs (like those shown in figures 1A on page 26 and 18 on page 67). Irregular shapes for gold particles are the result of their low energy of crystallization when compared to that of most host minerals: the higher the energy of the host, the less regular the shape of the native gold. Figure 20 is an example of how gold crystals grow into eu, sub and anhedral shapes, in hosts such as pyrrhotite, chalcopyrite, galena, arsenopyrite and pyrite, the pyrite being the mineral with the highest energy of crystallization. An artificially developed gold crystal, grown onto algae material imposing no limitations to its expansion (P.M.J. Gray 1988, personal communication), and showing a sharp crystalline contour is also included in figure 20 for comparison. The data and photomicrograph of the artificial product were provided by Robert Gee (1988) of Imperial College in London, responsible for the development of the crystal.

The larger gold particles and nuggets with anhedral smooth contours are sometimes porous. Their irregular shapes and the porosity result in large amounts of exposed surface, an advantageous feature when dissolving the particle in cyanide (see section on metallurgy, pages 121-122 and case study 11).

The term *free gold* is often used by the mining industry to indicate a gold which is not enclosed or in solid solution in its host. The term is used in connection with ores yielding high gold recoveries. The term free gold is considered inaccurate and not used in this book because:

- It is not the result of direct observations, but of the type of recovery obtained. The actual occurrence of the gold therefore is not known.
- Except for the placer deposits, gold is never free before the crushing/grinding processes. Depending on its grain size, it may become free at a certain grinding size, but may remain locked at a coarser one. Should gold show a range of grain sizes, some of it may be free, while some other may not at a certain grinding size, but it may all become free, after the grinding size has been reduced. If it is a question of liberation only, any gold above 10 microns in diameter could be classified as free gold at the correct grinding sizes.

Better terms for the description of gold particles are considered here to be *coarse gold* (easily liberated and extracted) and *fine gold* (smaller than 15 microns, when the ore requires fine grinding sizes and/or special treatments before the leaching step). Where dealing with *fine gold*, knowledge of the host mineral and type of association becomes essential.

Host minerals

Predominant gold association, according to Boyle (1968), is with a quartz gangue in which the gold is either within or in other way associated with arsenopyrite, pyrite and chalcopyrite. Boyle interprets this association, known for centuries, as the most common, rather than the only one, because quartz, pyrite, arsenopyrite and chalcopyrite are common minerals, and also perhaps because gold sometimes occurs in these minerals with coarser grain sizes.

Although not as commonly as with the above, gold occurs directly associated with probably any mineral. The author has observed gold in sulfides, sulfosalts, oxides, silicates, carbonates, sulfates, carbon and uranium minerals. She has only looked at a very small portion of the gold deposits in the world.

When occurring in a silicate or carbonate ore in which sulfides are present, gold tends to be associated with the sulfides as totally enclosed particles, in gaps or fractures, or along their outside. If in the gangue, it remains in their close proximity. If secondary iron oxides are present instead of sulfides, the gold tends to be associated with the secondary iron oxides in a similar fashion, although the totally enclosed particles are not as frequent, due to the porosity and lesser coherence of these minerals. This is however not a rule and gold may be found associated with the silicate (for example quartz or clay) or with the carbonate gangue, even if such gangue is rich in sulfides and other metallic minerals.

In many metallurgy texts, gold is classified as *native* or *free gold* as opposed to *gold in sulfides*. The considerations regarding *free gold* on the previous page also apply here. In most deposits characterized by a gold/sulfide association, gold occurs within the sulfide in the form of native gold, just like it does when in other hosts, although the average grain size is perhaps smaller. The following is an attempt at as clear an illustration as possible of the actual gold occurrences within the sulfide and other minerals, in order to replace the simplistic view which today predominates in the gold mining community.

PYRITE. Pyrite is the most common host sulfide mineral to gold, because pyrite is the most common sulfide mineral. Gold occurs in pyrite most of the time as native gold or electrum in all sizes and in a large variety of forms which depend on the grade and general textural features of the ore. Figure 21 summarizes the *type* associations, where the gold occurs with particles of different sizes, in fractures and gaps of variable thickness and totally enclosed. What is sometimes called a *refractory pyrite* is a pyrite in which most of the gold is very fine (less than 10 microns in diameter) and totally enclosed, like the one in figure 21E. *Refractoriness* is caused by the unliberated state of the gold, even after the finest grinding, because grinding can only reduce the ore down to sizes in the range of 10 to 15 microns at the most (by the average mill), thereby never allowing for total liberation or exposure of the finer particles. What the mining industry calls a *refractory ore* is an ore in which the gold occurs and remains totally enclosed in pyrite, arsenopyrite or other mineral even after fine grinding. A *refractory ore* may also be one where the gold-bearing mineral is represented by an insoluble telluride; or it may be one which is rich in clay or in graphite, because these minerals affect the gold dissolution (see section on gold metallurgy, pages 134-138). In the light of the detailed classification presented in this book, as well as of the clear representation of the different reasons for unsatisfactory recoveries, the term *refractory* has too broad a meaning. Therefore, as discussed in the introduction (page 8), it is not used for the description of any type of ore.

As already mentioned (pages 15, 19 and 100), gold occurrences, although unsupported by scientific evidence, are often reported as solid solutions in the pyrite (for example, Hausen 1981 and page 102). Gold solid solutions in pyrite occur, but they are not common and they are usually in amounts which are smaller than 1 ppm (as per ion probe determinations discussed on page 100). Conclusions by the author that gold solid solutions in pyrite are far less common than they are reported, are largely based on her experience with gold-pyrite ores. Accurate microanalytical studies of thousands of polished mounts have always located sufficient particles of visible gold to account for the values of the chemical assay. Careful examination of the only ore where no gold was observed, although rich in pyrite,

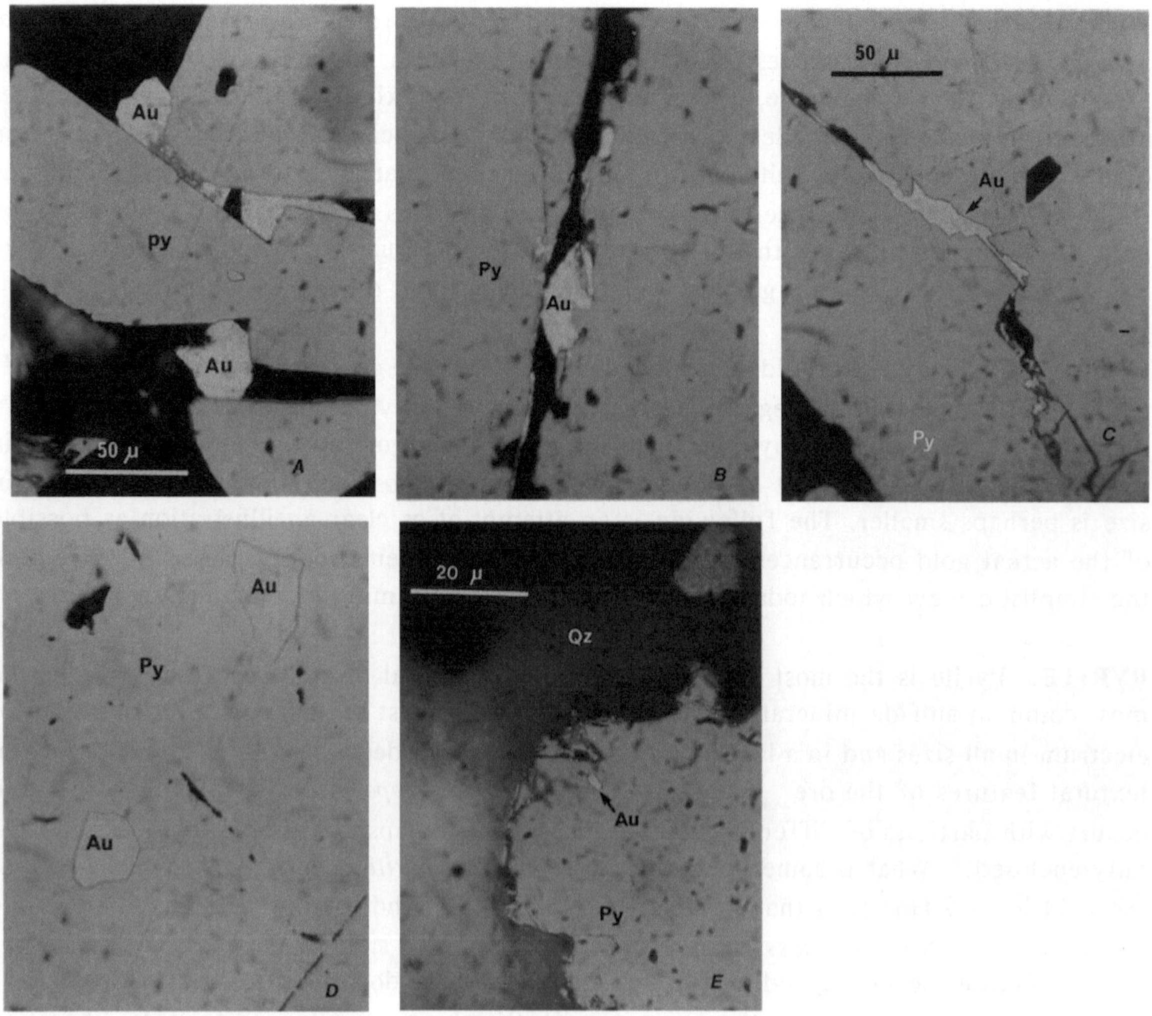

Fig. 21. *Gold-pyrite associations. A: native gold distributed on the outside of and in large gaps between pyrite grains. B: native gold in a medium size fracture of pyrite. C: native gold in a fine fracture of pyrite. D: native gold totally enclosed in pyrite and forming relatively coarse particles. E: native gold totally enclosed in pyrite and forming a very fine particle. Different qualities in recovery can be expected, depending on the type of gold/pyrite association. The gold in E will give the poorest recoveries by the standard techniques. The magnification is 500x for A, B, C and D and 1250 for E.*

revealed the presence of a clay/invisible gold association which was in no way related to the pyrite (see chart 16 on page 148 at the end of the section on metallurgy). In addition, experimental studies by Kallmann (1986) indicate that gold sulfide Au_2S forms solid solutions or is isomorphous with copper sulfide Cu_2S, but not with nickel sulfide.

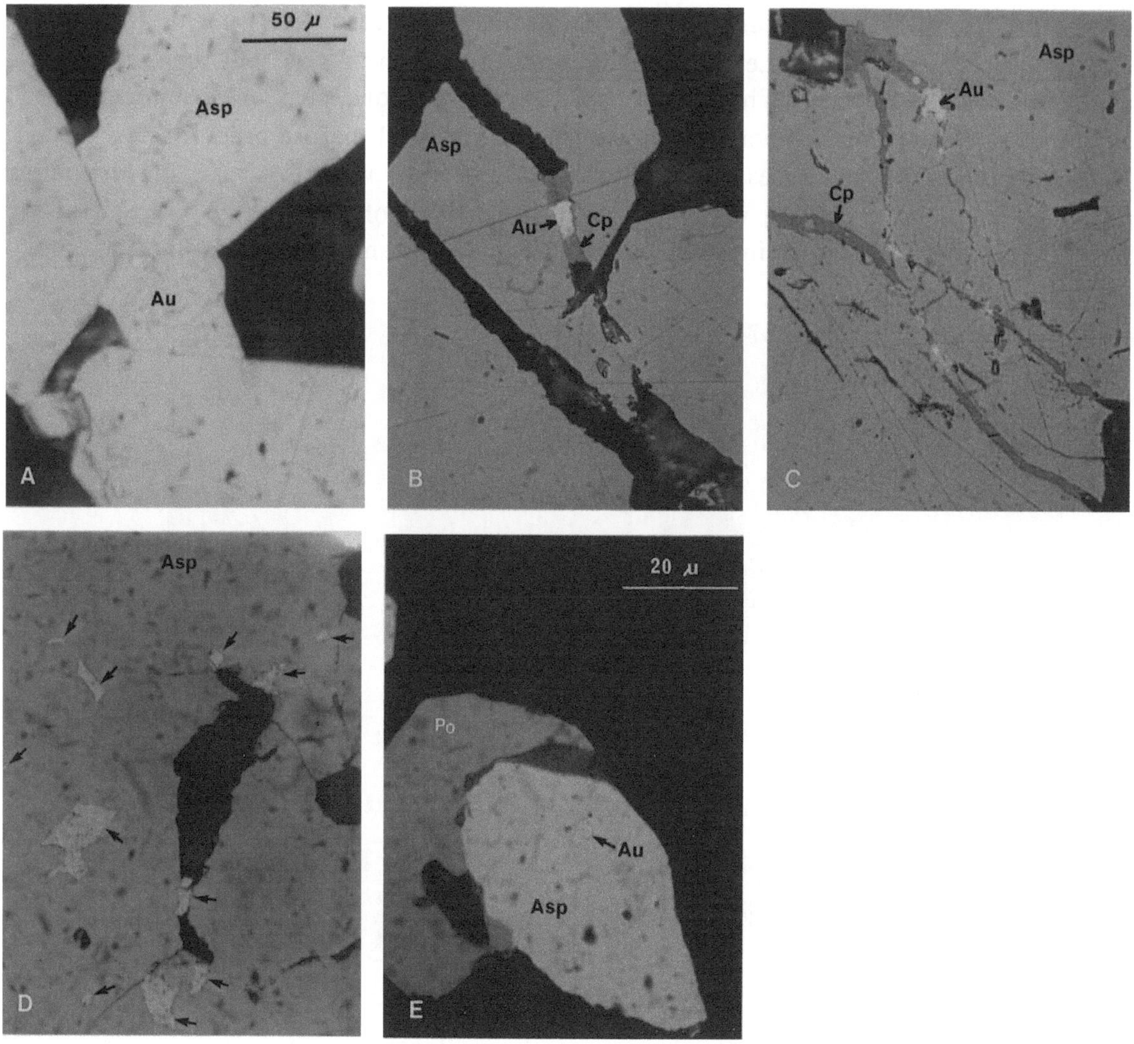

Fig. 22. *Gold-arsenopyrite associations. These associations are very similar to those of pyrite (compare with figure 21).*

And studies by G.Moh (1991, personal communication by letter) reveal that the copper bearing pyrite **villamaninite** contains variable low amounts of gold which increase with increased copper contents. These findings are consistent with the copper/gold substitutions observed in minerals and with the similar electronic configuration and crystal structure of the native gold and of the native copper (face centered cubic lattice). The same findings suggest that although Au may substitute for Cu or occur in other ways in solid solution in copper minerals, it does not substitute for Ni **or Fe (in pyrite)** in any appreciable amounts.

When studying sulfide ores which do not yield satisfactory recoveries, positive identification of the gold occurrence, i.e., in solid solution or in micron-size, but separate particles of native gold, is essential. Where dealing with solid solutions, only one chemical agent is sufficient to dissolve or in other way break the host mineral, before precipitating the gold: the same process which causes the dissolution of the host, will also cause the dissolution of the gold. Where dealing with fine discrete particles of native gold, two agents are necessary, one to dissolve or in other way break the host, and one to dissolve the gold.

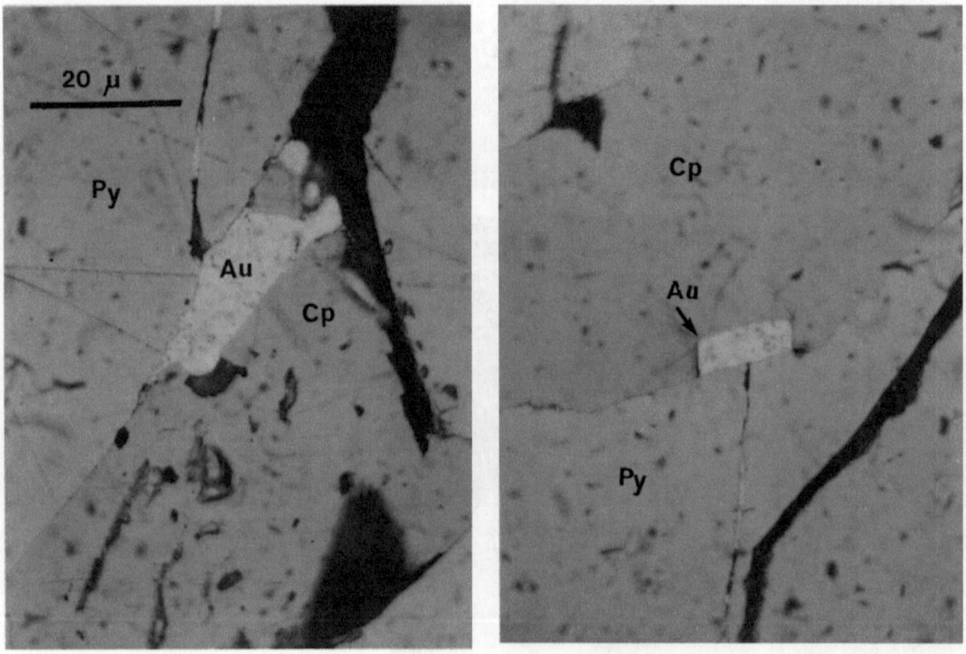

Fig. 23. *Native gold at the border between pyrite and chalcopyrite.*

ARSENOPYRITE. Arsenopyrite is a relatively common host to gold, probably second after pyrite.

Gold-arsenopyrite associations, shown in figure 22, are very similar to gold-pyrite associations, as they include distributions in fractures and gaps and totally enclosed particles of different sizes.

When studying pyrite-arsenopyrite ores, there is a tendency among some geologists/mineralogists, to assume that the gold is associated with the arsenopyrite. If not readily located by an optical microscope, there is then a tendency to assume that the gold is in solid solution in the same mineral. Neither assumption is correct. There are pyrite-arsenopyrite ores in which the gold is with the pyrite and there are pyrite-arsenopyrite ores in which the gold is with the arsenopyrite. The most reliable means of obtaining an accurate description of the gold occurrence in an arsenopyrite ore is through direct observations.

Gold solid solutions in arsenopyrite, where gold replaces arsenic in the crystal lattice of the mineral, are probably more common and are represented by higher concentrations than those detected in the pyrite (page 100).

REALGAR AND ORPIMENT. Occurrences of gold with realgar and orpiment are typical of the Nevada type disseminated gold deposits. Another element common in those deposits is mercury, found in the form of cinnabar, and associated with other sulfide/sulfosalts of antimony and arsenic, sometimes silver. Although gold in this type of environment is often reported to be *invisible* (for example, Hausen 1981), Bart Cannon (1991, personal communication by letter) states that he has seen discrete particles of native gold in several Nevada type deposits, in the form of inclusions in the pyrite and as suspended grains in clay altered feldspar.

CHALCOPYRITE. Gold-chalcopyrite associations are not as common as those with the pyrite and arsenopyrite, the most frequent occurrences being limited to those within the porphyry copper deposits. The occurrence and type of association are similar to those of the other two sulfides and largely dependant on the general textural features of the ore. In a pyrite-chalcopyrite ore where the pyrite is predominant, or where the pyrite to chalcopyrite ratio is large, gold is often either with the pyrite or at the border between pyrite and chalcopyrite (figure 23). In many such instances, the chalcopyrite grains are in turn completely enclosed in the pyrite. If the chalcopyrite grains associated with the gold are too fine to become liberated or exposed after the grinding process, this type of gold-pyrite-chalcopyrite association becomes a gold-pyrite association for metallurgical purposes.

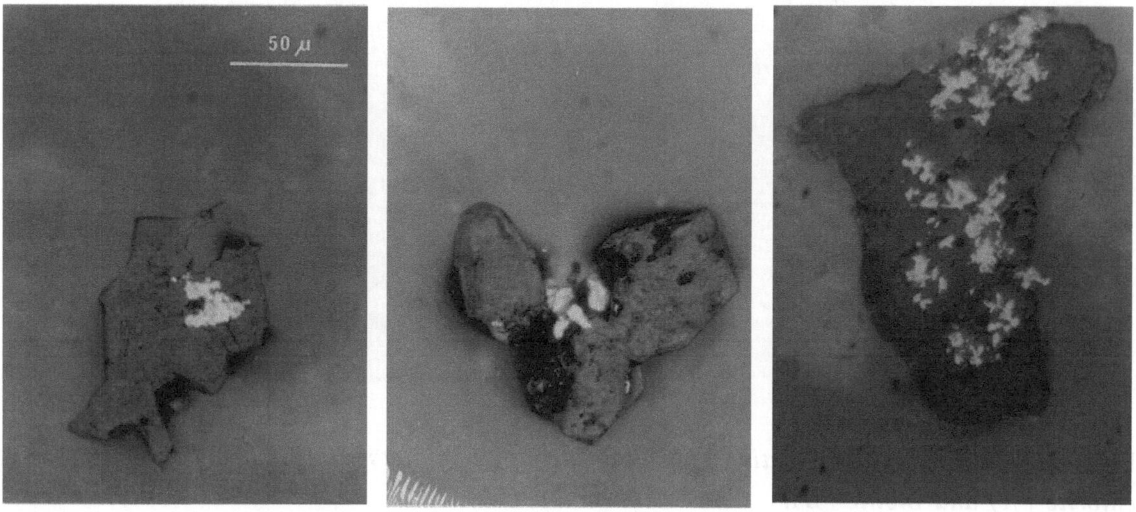

Fig. 24. *Native gold associated with secondary oxides of iron.*

SECONDARY COPPER MINERALS, bornite, chalcocite, covellite, digenite, djurleite, native copper, azurite and malachite. Gold may occur in altered, secondary, or weathered ores associated with any or all of these minerals, when they represent partial or total replacements of the chalcopyrite. The type of association is similar to that with the chalcopyrite, probably with a tendency of the gold to occur more often in fractures and gaps, due to the less coherent state common among many secondary products.

OTHER SULFIDES AND SULFOSALTS. Descriptions of the manner of occurrence of the gold in other sulfides and sulfosalts, such as pyrrhotite, marcasite, galena, sphalerite, stibnite, antimony-sulfur-bismuth-lead sulfosalts, nickel sulfides, arsenides and sulfoarsenides, are given by Boyle (1979). The manner of occurrence of the gold is similar to the one described and illustrated for the pyrite and arsenopyrite and is largely dependent on the relative abundance of the different minerals and on the general textural features of the ore.

OXIDES OF IRON. Magnetite, hematite and secondary iron oxides, such as goethite and limonite, are relatively common hosts to gold in oxidized environments. The type of association is again varied and may include grains of native gold coated by fine rimmings of secondary iron oxides. Figure 24 shows particles of gold associated with secondary oxides of iron.

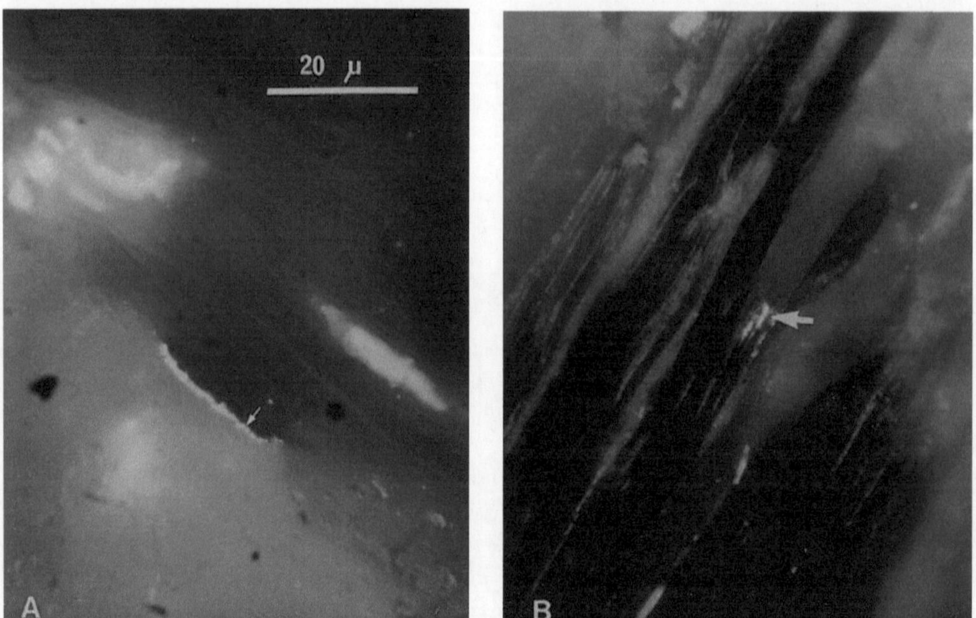

Fig. 25. *Flat particles of native gold (appearing as veinlets) distributed between plates of chlorite (A) and biotite (B).*

The preferential association of the gold with the iron oxides in oxidized environments has already been mentioned (page 105).

URANIUM MINERALS. Occurrences of gold in uranium ores are common in South Africa (De Waal and Herzberg 1969, Jochens and Laxen 1975). These consist of fine grains of native gold totally enclosed or in other ways associated with the larger uranium minerals. According to Boyle (lecture given in Toronto in 1979), it is possible that gold-uranium occurrences are not reported as frequently in North America as they are in South Africa, because uranium ores are not routinely assayed for gold.

QUARTZ, SILICATES, CARBONATES AND SULFATES. Quartz, like pyrite, is a very common host to gold. Differences are in the gold grain sizes which are often coarser when in a quartz matrix, and in the frequent distributions along grain boundaries rather than within. Gold associated with other gangue minerals, again usually distributed along grain boundaries, can be expected in ores where sulfides and oxides are scarce or absent, but, as already observed (page 105), gold **can** be associated with the gangue even when an abundant sulfide/oxide mineralization is present. When in the phyllosilicates, such as muscovite, biotite and chlorite, gold occurs in the form of veinlets and flat particles distributed between plates (figure 25). The coarse gold in the barite from the Tambo deposit in Chile, where so many new minerals were found, occurs distributed among the barite grains. Sulfide mineralization in this deposit is scarce. The clays are discussed on pages 134-136.

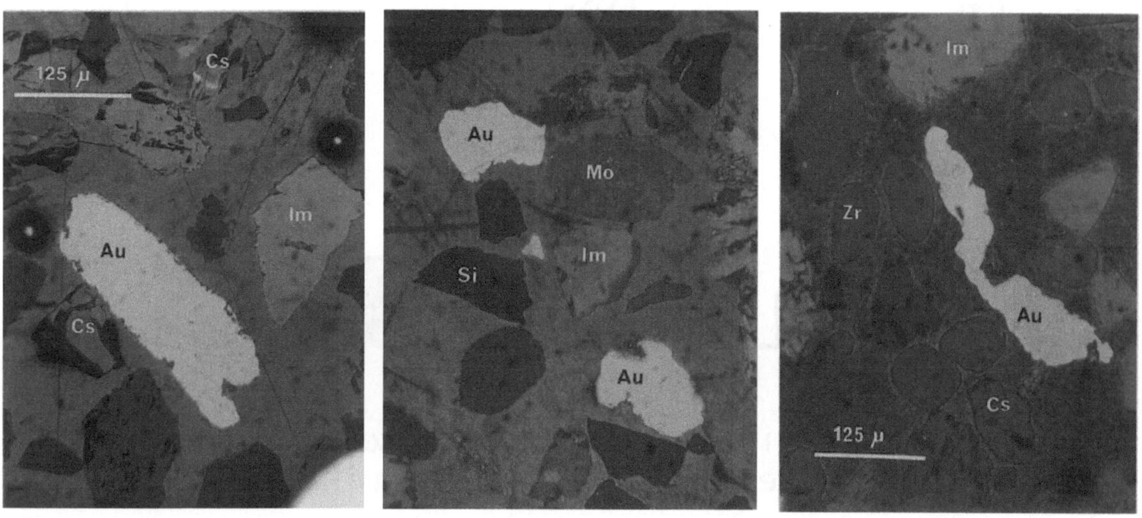

Fig. 26. *Placer or alluvial native gold showing coarser grain sizes than those of the gold in the other photomicrographs (note scales). Represented in the same photomicrographs are ilmenite (Im), cassiterite (Cs), monazite (Mo), zircon (Zr) and quartz (Si).*

CARBONACEOUS MATERIAL (GRAPHITE/CARBON). Gold is found associated with graphite and other carbonaceous matter, forming fine disseminated particles of native gold in reduced ores rich in carbon.

SANDS AND GRAVELS (PLACER GOLD). In this association the gold occurs in loose sediments of variable mineralogical composition, which may include rutile, ilmenite, monazite, cassiterite, iron oxides, quartz, staurolite, sometimes platinum group minerals. The gold particles are detrital and may occur in the form of dust, larger grains, irregularly shaped masses, and occasional nuggets dispersed through the sand or caught in rock crevices. The size is often coarse, ranging between 100-200 microns (figure 26) and several mm, the larger grains being more unusual. Placer or alluvial gold is the gold mined during the great gold strikes in Colorado, California and Alaska and recovered by panning. In the case of the California placer deposits, discovered in 1848, these "were traced upstream to the source whence they had been eroded, which led to the discovery of the Mother Lode, or main undestructed body, a rich quartz vein of unusual length in the Sierra Nevada Mountains. Gold was later discovered in Alaska in 1880 on the beaches and in streams and in

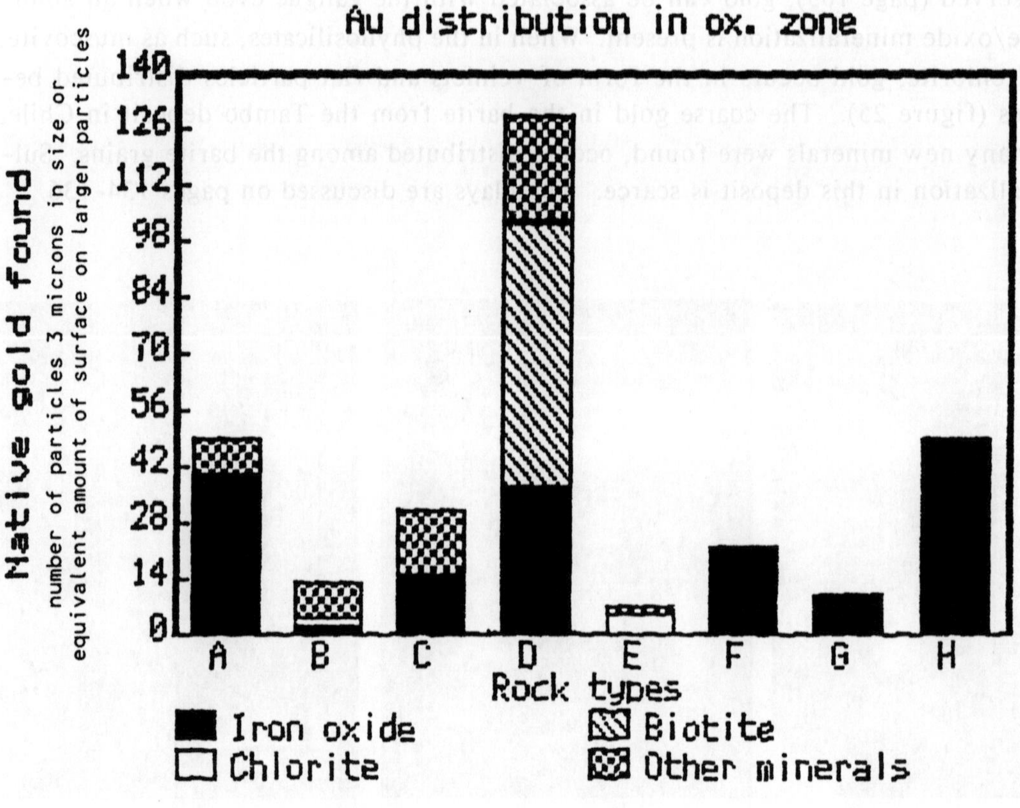

Fig. 27. *Gold distribution and associations in different rock types within the same ore. A: granitic rocks; B: biotite granites; C: muscovite granites; D: biotite gneisses; E: chlorite schists; F: other metamorphosed rocks; G: chlorite-muscovite schists; H: augen gneisses. Iron oxide: all primary and secondary iron oxides; other minerals: quartz, orthoclase, albite, muscovite, calcite, rutile*

vein deposits in the mountains inland in both Alaska and British Columbia." (Clark 1988, p. 340).

Mineral associations

As described and as shown in the photomicrographs, the host minerals and their type of association with the gold are very varied. In ores consisting of many rock types, very often the gold, in addition to showing different grain sizes and associations, may occur within different host minerals in each rock type.

Case study 10. Gold-host mineral associations. A detailed study was conducted on approximately 100 samples from different portions and depths of a gold deposit in the Western United States. The deposit consisted of sedimentary and metamorphic, oxidized and unoxidized rocks. The study revealed some very interesting mineral associations which were directly related to the rock type and degree of oxidation of the ore. In the unoxidized portion (or zone), the native gold was associated almost exclusively with the pyrite. In the oxidized portion, the gold showed a tendency to be associated with the iron oxide, except for the biotite gneisses where more than one third of the metal was associated with the biotite. As opposed to the unoxidized samples, characterized by an almost exclusive pyrite association, the oxidized samples had some gold associated with almost all of the major minerals (figure 27). Other differences between the native gold of the oxidized and unoxidized portions of the deposit were found to be in:

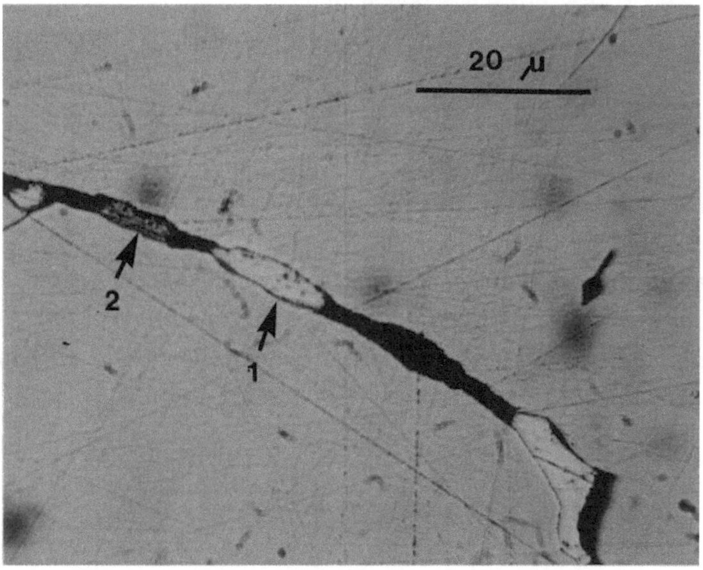

Fig. 28. *Native gold (1) and smaller amounts of the gold-silver telluride sylvanite (2) in a gap between two grains of pyrite.*

- The silver content of the particles. This was variable and up to 45 per cent in the unoxidized zone and absent from the oxidized zone.

- The size of the particles. Coarse, up to 60 microns in diameter in the unoxidized zone and fine 1 to 10 microns in diameter in the oxidized zone.

Gold/host mineral associations are very similar to the *type* associations shown for the pyrite and arsenopyrite in figures 21 and 22 and include totally enclosed particles of different sizes, and particles distributed in fractures and at the border between grains of the same or of two different minerals. Intergrowths are not as common, due to the usually high ratio of the host mineral to native gold.

Tellurides, if present, are often associated with the native gold and other minor minerals in composite grains and intergrowths, or occur in fine discrete particles not far from the native gold (figure 28).

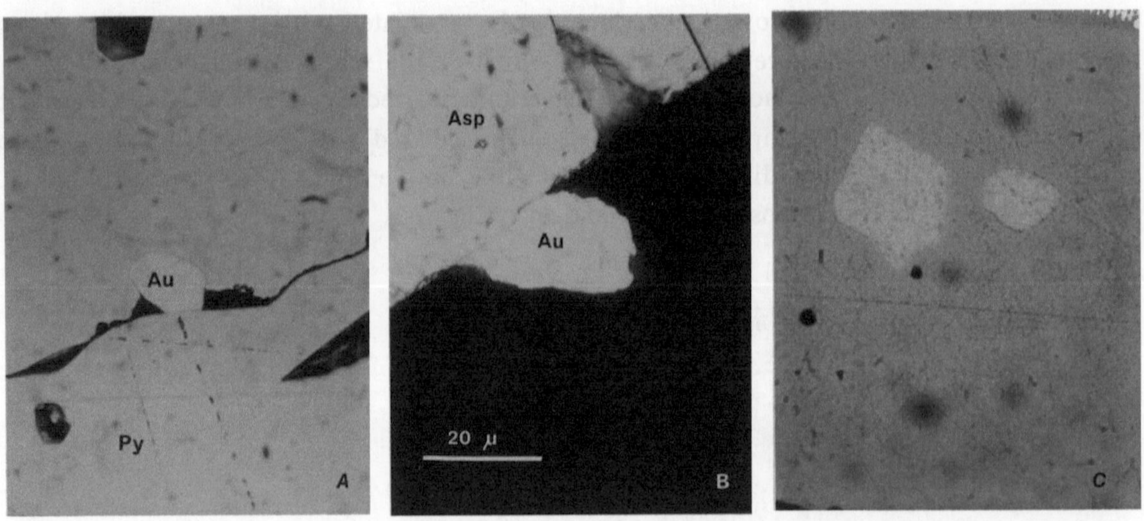

Fig. 29. *Different types of native gold/host mineral associations which affect differently the process of extraction. A: gold at the border between grains of the same mineral (pyrite); B: gold at the border between grains of two different minerals (arsenopyrite and silicate); C: gold totally enclosed in its host mineral (pyrite).*

For metallurgical purposes, gold/host mineral associations can be summarized as follows:

- Gold distributed in fractures or at the border between grains of the **same** mineral (figure 29 A).

- Gold distributed at the border between grains of **two different** minerals, for example two sulfides, a sulfide and a silicate (Fig. 29 B), or a sulfide and an oxide.

- Gold totally enclosed in the host mineral (figure 29 C).

- Solid solutions, in some host minerals only, such as copper minerals, possibly arsenopyrite, realgar/orpiment, other phases capable of accommodating the gold atom.

The metallurgical significance of these associations is discussed in different occasions, for example, pages 26, 67, 108, 120-121, 125-126.

An explanation for the **possible** presence of solid solutions and of discrete particles of native gold within the arsenopyrite, is provided by Boyle (1979, p. 32). "One can postulate that early formed high temperature arsenopyrite probably takes up gold and silver largely in solid solution, or as atomic layer on the growing faces of the sulfide minerals.... At lower temperatures, the presence of foreign constituents such as gold and silver, distort the pyrite and arsenopyrite lattices, with a consequent rise in the free energy of the crystal. In order to reduce the free energy of the system, to a minimum under the prevailing conditions, gold and silver migrate to nearby low chemical potential sites such as fractures and grain boundaries where they crystallize as native gold (and electrum). This appears to explain the frequent occurrence of much invisible gold and silver in relatively unfractured and unrecrystallized arsenopyrite; where reworking, recrystallization and fracturing are evident the gold is largely present in the native form.... The exsolution phenomenon described above is probably largely the result of diffusion processes, a conclusion supported by the experimental work of Clark (1960). "

Minerals other than the host mineral

Minerals which are not directly associated with the gold, but which may interfere with the recovery process are:

- Cyanicides. The term refers to any mineral unstable in cyanide, for example, chalcocite, covellite, pyrrhotite, secondary iron oxides.

- Clays/sericite.

- Carbonaceous material (graphite/carbon).

- Mercury minerals.

These are all discussed in the section on metallurgy (pages 134-139).

Mineral stability

Except for electrum rich in silver (30 per cent or more), gold minerals are stable.

METHOD OF STUDY

Despite the widespread belief that gold in amounts below 1 oz/ton, or 34.2 ppm cannot be seen, gold can be seen and can and should be located in a sufficient number of samples and of occurrences per sample (10 to 15), for a good indication of its correct distribution throughout the ore.

If using the correct techniques, gold distributions can be determined accurately in ores containing down to 0.1 oz/ton or 3.42 ppm Au. Lower values require the preparation and observation of a very large number of polished mounts and may be too time consuming. In such a situation, preparation of concentrates, and fire assay analyses of the different fractions prior to the microanalytical studies, may represent an easier, faster alternative.

Microanalytical studies are essential prior and during gold recoveries, as well as in the evaluation of the deposit, because of:

- The irregular gold distributions which need to be correctly defined for the efficient exploitation of the ore.
- The variable, but often fine grain sizes which make all decisions concerning grinding sizes critical.
- The common low contents in the host rock, requiring higher accuracy of approach.
- The high gold market value which calls for as complete an extraction as possible.

The main difficulty in the study of gold ores is not in the identification of the gold minerals, most of which are very easily recognized by their high reflectivity and bright orange, yellow and white colors. The difficulty is in the location of the gold minerals. Economic amounts of gold, in the range of 0.1 ounce per ton, or 3.4 ppm, are not readily observable on the surface of polished mounts.

The most important reasons for the scarcity of good microanalytical reports on gold occurrences, can be summarized as follows:

- Most investigators only use one or two polished mounts 1" in diameter (rather than six, as recommended in chapter 4, pages 54-55) for the description of the ore. The chances of finding gold in ppm concentrations on such a limited amount of surface, are very small.
- If more polished mounts are prepared and are available, identification of a sufficiently large number of gold particles for a good indication of the manner of occurrence of the gold in the sample requires the accurate examination of the entire surface of all of the mounts. Very few investigators believe that such a detailed study is necessary.

- If studying hand or core samples (rather than crushed material), due to the gold irregular distributions on a macro as well as on a microscopic level, gold may be altogether absent from the surface of the polished mounts examined, regardless of the fire assay values of other samples collected from the same general area. Some samples may be barren, even if derived from the portions of the ore rich in mineralization, These samples are sometimes examined without performing a fire assay analysis first, with the result that no gold is found, despite the accuracy of the microanalytical study.
- Concentrations in the ppm range may be coupled with very fine grain sizes. These sizes are not readily observed when using the more common ore microscopes equipped with magnifications of 100x or 200x. If using higher magnifications, unless these are attached to good quality microscopes, the resolution at those magnifications may not be as high as needed to observe the optical properties of a particle which is 1 or 2 microns in size.
- An EDS-SEM or electron microprobe may not be available.
- Experienced individuals whose knowledge extends to ore microscopy, as well as SEM/microprobe techniques are scarce. The frequent use of mineralogists instead of microanalysts, on the part of the mining companies, because of their more reasonable fees, has already been mentioned (page 14).

Keeping all these considerations in mind, an accurate study of an ore giving the correct distribution of the gold; in other words, a study aimed at locating a sufficiently large number of particles to be statistically representative of the gold occurrence in the same ore, and including gold-bearing minerals, average grain sizes, host mineral(s), and type(s) of association with the host mineral(s), will require the following:

- The availability of an experienced microanalyst.
- The use of a high magnification-high resolution optical microscope of good quality.
- The use of an EDS-SEM or a microprobe.
- Careful examinations of polished mounts cut from samples representative of (i) the average gold values for the deposit and of (ii) the different rock types, as selected based on the petrographic descriptions and fire assay (details are on pages 54-56 and chart 4). The rejects from the fire assay will probably provide the most accurate answers, because they are representative of the exact chemical value for the sample. The importance of the study of samples representative of average gold values, rather than of high grade ones, has already been discussed (case study 1 on page 18 and pages 54-55).

Only if all of the above conditions are satisfied, will the systematic study, as described in chapter 4, pages 54-62, give positive results.

Where studying gold, although the microprobe will provide satisfactory data, use of the EDS-SEM is preferable because:

- It allows for easy location of the gold particles, when these are very fine, because their high mean atomic number makes them clearly visible on the display screen. Location of the same particles is considerably more difficult when using the *average* microprobe.

- Accurate quantitative analyses are not usually needed.

If the presence of minor amounts of mercury, arsenic or other elements is suspected in the native gold and needs to be determined, then the use of the microprobe becomes necessary.

Due to their distinctive optical properties, native gold and electrum, if occurring in particles which are above 10 microns in size, are usually positively identified by the ore microscope alone. The use of EDS-SEM and/or microprobe for the study of most gold ores is necessary to: (i) confirm the identification of the finer particles which are not usually as easily identified, (ii) determine their silver content, (iii) determine the other minor constituents, and (iiii) positively identify the host and the other minerals.

Minerals which can be confused with native gold, electrum and sometimes other gold-bearing minerals (particularly when occurring in finer sizes) are native copper, native bismuth, bismuthinite, and any other metal or compound showing high reflectivity. Fine flakes of aluminum metal, a contaminant from the polishing process, may sometimes be confused with native gold, when the investigator is inexperienced and not entirely familiar with the optical properties of the gold bearing minerals.

Techniques for the location and quantitative determination of gold particles by computer are occasionally used (H.A.Hiemstra 1988, personal communication by letter). Besides speed, advantages of a computerized technique are in:

- The higher accuracy. If properly calibrated, the image analyzer will not miss any gold particle, while the human eye might.

- The determination of the exact amount of surface of gold exposed. This is a very difficult estimate, if attempted by the human eye using a conventional ore microscope.

Taking into consideration the high degree of accuracy and the volume of data which can be produced without as direct a human involvement, a technique based on a computerized system which uses image analysis and microanalytical instruments, represents the logical direction towards which the ore microscope-SEM/microprobe method should develop.

Given the computerized state of modern technology, the very limited availability of similar instruments and techniques for the study of fine particles in ores represents an anachronism, the modernization of which is long overdue.

METALLURGY

Although new process are being proposed on a regular basis (see page 71 and Yarar 1993), there have in fact been no dramatic changes in the metallurgical techniques for gold extraction since the introduction of the **cyanide process** (**cyanide leaching** or **cyanidation**) by McArthur and Forrester in 1887 (Rose 1986, Hamilton 1920, Taggart 1953, Hedley and Tabachnick 1968). The process involves the dissolution of gold (and of any silver present in soluble form) from the ground ore in a dilute cyanide solution and in the presence of lime and oxygen (Habashi 1966, Haque 1992), according to the reactions:

$$2Au + 4KCN + O_2 + 2H_2O = 2KAu(CN)_2 + 2KOH + H_2O_2$$

$$2Au + 4KCN + H_2O_2 = 2KAu(CN)_2 + 2KOH$$

Gold (along with silver) is then precipitated with zinc dust according to the reaction:

$$2KAu(CN)_2 + Zn = K_2Zn(CN)_4 + 2Au$$

The process known as **carbon in pulp**, or charcoal in pulp or **CIP** controls the gold precipitation from the cyanide solution by use of activated charcoal (carbon). "The technique involves contacting the leached pulp with granular carbon (about -8 to +20 mesh) in a series of gently agitating tanks with a sufficient retention time. The carbon is recycled through the circuit to build up the loading to 8-10 per cent by weight. The loaded charcoal is then separated from the pulp on a suitable vibrating screen, coarse enough to retain the carbon, but fine enough to allow the pulp to pass through. The carbon is next sent to the stripping column for desorption and regeneration" (Bhappu 1985, p. 13-49). The technique is used on low grade gold and silver ores in several processing operations in the Western United States, its main purpose being for slime treatment.

After precipitation, the product is treated with dilute sulfuric acid to dissolve residual zinc and most of the copper (if any is present). The residue is washed, dried and melted with fluxes. The remaining gold and silver alloy, called *dore'*, is cast into molds for assay (Wise 1988). Refining is accomplished by **electrolysis** during which silver and any platinum group elements are also separated and recovered. Another method of separating gold from silver in the dore' is by **parting**, where hot concentrated sulfuric or nitric acids are used to differentially dissolve the silver, while the gold is recovered from the residue.

Gold used in jewelry is alloyed with variable amounts of silver and copper which allow for a wide variation in the color of the finished artifact, from white to yellow to pink. Nickel and zinc are used in smaller amounts for increased hardness and durability and for color variations. Relatively small amounts of palladium alloyed with gold yield the best white gold. The gold content of the alloy is expressed in fractions of 24, each called caledaledat (carat in the UK). A 24 karat gold object consists of pure gold, a 12 karat object consists of 50 per cent gold, etc.

"**Thiourea leaching** was developed as a potential substitute for cyanidation (of gold as well as silver) by Soviet workers in 1960 (Plaskin and Kozhukhova 1941, and Plaskin and Kozhukhova 1960). Thiourea's lower toxicity and greater rate of gold and silver dissolution compared to cyanide give it a real advantage in reaching the commercial application before other non-commercial lixiviants. The lower sensitivity to base metals (As,Pb,Cu,Sb,Zn) or impurities render possible the use of this process on many refractory gold ores. Thiourea leaching has been tried on different materials and ores with success. Pre-roasted materials were leached with gold extractions up to 99 per cent with leaching time as short as 30 minutes. Recovery of gold from pyrite and chalcopyrite concentrate has given 96 per cent extraction. Extraction of 90 per cent is reported on carbonaceous material. Commercial application of thiourea in the extraction of gold is performed in Australia on antimony concentrate, on carbonaceous concentrate in China and in roasted zinc ores in France. Most of the work done to recover gold with thiourea is on the leaching part of the process. There are few results available on the recovery of gold and silver from the pregnant solution. Most of the references are Soviet papers with few details. Other countries such as South Africa, China, Australia and the United States have reported some work, but there are only limited results available from these papers" (Deschenes 1986, p. 76). Lan Xinzhe et al (1992) provide some additional information on the same subject.

Amalgamation is an ancient process which involves the alloying of the gold particles with metallic mercury to form *amalgam* and then the separation of the gold from the mercury by heating in retorts until the mercury is distilled off. The method is used for the treatment of coarser gold (30 microns in diameter or greater). **Panning**, discussed in chapter 5, page 69, is used for the separation of alluvial or placer gold and is probably the oldest technique for the recovery of gold.

Other techniques for the treatment of coarse gold include **corduroy tables** (page 69, and Richards and Locke 1925), **jigs** (page 69, and Gaudin 1939) and **unit cells** (Pryor 1965) all producing a mechanical separation of the gold based on its greater specific gravity. Like amalgamation, mechanical separation is applied to the higher grade ores or to the coarser gold within the same ore.

Many mines in North America use amalgamation for the separation of the coarse gold, followed by cyanidation of the finer particles. Gold recoveries in South Africa are obtained almost exclusively by cyanidation, due to the fine size of most of the gold.

Rock types hosting gold

The cyanide process may be applied to the ore by **heap leaching** techniques (pages 75-76), provided that the rock and host minerals are porous enough to allow penetration of the solutions and contact with the gold particles, without the need for similar grinding sizes to those of the native gold. The ore is in this case only coarsely ground and the leaching is accomplished by downward percolation of cyanide solutions fortified with lime. This technique lends itself well and is applied to many gold deposits in Nevada and other Western United States. Case studies 6 and 7 on pages 32-35, show that heap leaching is successful on fine native gold down to a few microns in size, provided that the gold is not enclosed in a mineral which is impermeable to cyanide, such as compact pyrite or arsenopyrite.

A more accurate technique such as **agitation leaching** (pages 76-77), is applied to less amenable ores which require finer grinding than heap leaching, for example, those where the gold is coarse enough for liberation or exposure by physical means, but occurs in a less permeable host, such as a rock containing fresh pyrite or arsenopyrite. These types of rocks could be exemplified by the unoxidized samples of case studies 6 and 7, or better still by ores where the gold occurs in gaps rather than within grains of an impermeable host mineral (for example figures 21A and B and 22A and B on pages 106 and 107).

Pressure leaching (page 77), and **pressure cyanidation** (at high temperature, page 77) are used to promote cyanide penetration in compact ores where the gold occurs in fine fractures or gaps (for example, figures 21C and 22C). For compact ores where the gold particles are totally enclosed and too fine for exposure by the finest grinding (for example, figures 21E and 22E), methods are employed such as **oxidative acid leaching** (page 77), **roasting** (pages 72-74, 127 and 129), **pressure oxidation** (page 128), **bacterial leaching** (pages 77-78 and 128), with the purpose of breaking or converting impermeable host minerals into soluble compounds, for exposure of the micron-size gold particles

Gold-bearing minerals

NATIVE GOLD AND LOW-SILVER ELECTRUM. Liberated native gold and electrum, containing less than 30 per cent of silver dissolve readily in the cyanide solution. A differential leaching of the silver may occur naturally and is often observed on placer gold (figure 30, Krupp and Weiser 1992) and in gold occurring in oxidized ores, like the one of the next case study.

Case study 11. Porous gold. Some gold occurrences consisting of silver-free and silver-bearing native gold associated together in an asymmetrical fashion, were found during the course of two different studies of an oxidized ore from the Western United States. The first study was conducted on gold isolated from its gangue by hydrofluoric acid. The second study was conducted on the original samples. After examination of the preliminary data, the occurrences were interpreted as *zoned gold*, the result of consecutive depositions. Following the examination of the data from the second study, asymmetrical distributions of the silver-free portions, distributions on the particles' surface and around holes, indicated that such portions were in fact the result of the natural leaching of the silver by solutions in contact with the surface of the native gold, or penetrated through the holes.

Interpretation of the same occurrence by *the overbearing consultant in gold*, based on his observations of the results of the first study only, was as follows: silver-free native gold very seldom occurs (a very personal opinion not supported by direct microanalytical observations), therefore when found in polished mounts, it must be an artificial product. In this particular case, such artificial product must have been the result of the leaching of the silver by the hydrofluoric acid used to separate the native gold from the silicate gangue.

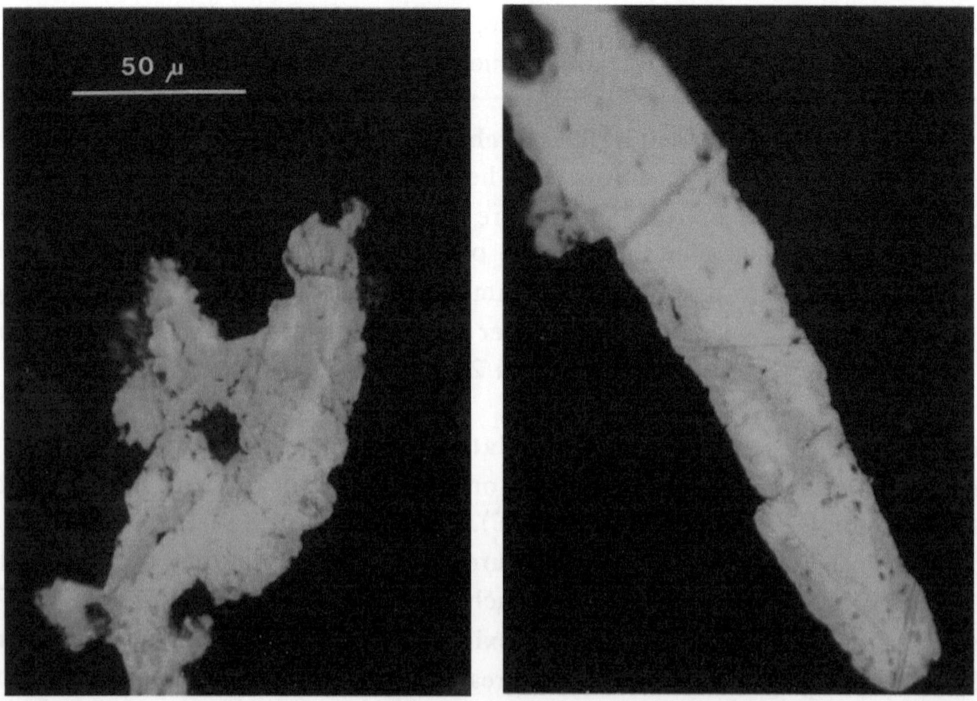

Fig. 30. *Large particles of "zoned" native gold where the silver-free portions (orange) predominate on the outside surface and around a large hole. Occurrences such as these suggest a partial leaching of the silver by natural processes, rather than consecutive depositions.*

Before reaching such a definitive conclusion, the man did not consider the logical approach consisting of the observation of those same gold occurrences in their natural form. The study of a number of unetched samples would have given him the correct answers.

Perhaps the reader will recognize the dangers of the very confident *I know it all* attitude, common among **some** gold metallurgists, where conclusions are reached based on very superficial, if existing at all, observations of small amounts of data. In this particular case, the information was largely of academic interest. Yet, it is that same attitude that is responsible for the planning of many recovery strategies, and unfortunately, for extractions which are well below the potential of the deposit.

HIGH-SILVER ELECTRUM. Electrum containing 30 per cent of silver or more, is not as amenable to cyanidation. The mineral tarnishes (figure 31A), then develops a coating of secondary silver sulfide (figure 31B) which can be up to several microns thick. Such coating attracts and mixes with the finer particles in the product (figure 31C), thus turning into an impermeable barrier to the cyanide. These fine particles are selected on the basis of their size, rather than their chemical composition, and may consist of both sulfide and silicate material. Where the silicate component is predominant, this may give the electrum particle a non-metallic character and keep it from floating during physical separation.

This phenomenon was first described by Head (1936) who discovered the coatings on particles of what he identified as native gold by using the optical microscope alone. Because no microanalytical instruments were at that time available, for an accurate determination of the chemistry and nature of the coatings, he referred to the particles simply as *oxidized gold*.

Very little is still known about these coatings, but, because they are limited to high silver, electrum, it may be safe to assume that a relation exists between silver content and tarnishing. If this is so, where dealing with electrum and native gold in the same ore, a partial loss of gold may be experienced, which is directly related to the native gold to electrum ratio.

Although the coatings of secondary silver sulfide may occasionally reach thicknesses of several microns, they are in most metallurgical products much finer, and therefore difficult to observe by a conventional optical microscope. They are also difficult to detect when using the SEM because: (i) the contrast created by the backscattered electrons and resulting from the difference in mean atomic number between electrum and coating is not sufficient to reveal an occurrence of such fine size, and (ii) the primary x-rays emitted by the microarea around the coatings and analyzed by the detector, include a considerable amount of radiation from the adjoining unaltered electrum, and are therefore difficult to interpret as produced by a separate phase. If the impermeable coatings are not readily detectable, silver contents of the native gold above 30 per cent, provide probably the best explanations for poor gold recoveries from this type of ore.

Systematic studies are necessary for the correct definition and understanding of the silver sulfide coatings, although metallurgists **are** aware of the problems related to gold extractions from electrum. B. Wyslouzil (1986, personal communication), controls such coatings by using a high pH solution during the process of cyanidation.

In addition to the impermeable coatings which develop after the grinding process, fine rimmings of minerals such as iron oxides, chalcocite and covellite are also reported to occur over particles of native gold and electrum. *Dirty gold* is a common expression used, when referring to these contaminant minerals and often poor recoveries have been linked to their presence. Again, there are no experimental data corroborating such an assumption. In fact ready solubility of all of these minerals in cyanide suggests that recovery problems are not caused by the coatings themselves. Unless they are associated with other insoluble material, these minerals should cause no problems, other than those derived from the consumption of some of the recovery agent. Figures 31D and E show a particle of gold coated by secondary iron oxide, as photographed in the ore microscope and on the SEM display screen. The dramatic increase in clarity is due to the difference in mean atomic number between the native gold and the iron oxide. Figure 31F is the x-ray spectrum of the iron oxide revealing the presence of minor amounts of silicon, phosphorus, sulfur, chlorine and calcium.

TELLURIDES. The information on the solubility of gold telluride minerals in cyanide is rather confusing. For many years **all** tellurides of gold were considered insoluble and many poor recoveries were linked to the presence of such minerals, usually even before accurate tellurium determinations were obtained for the ore. In fact in some instances the presence

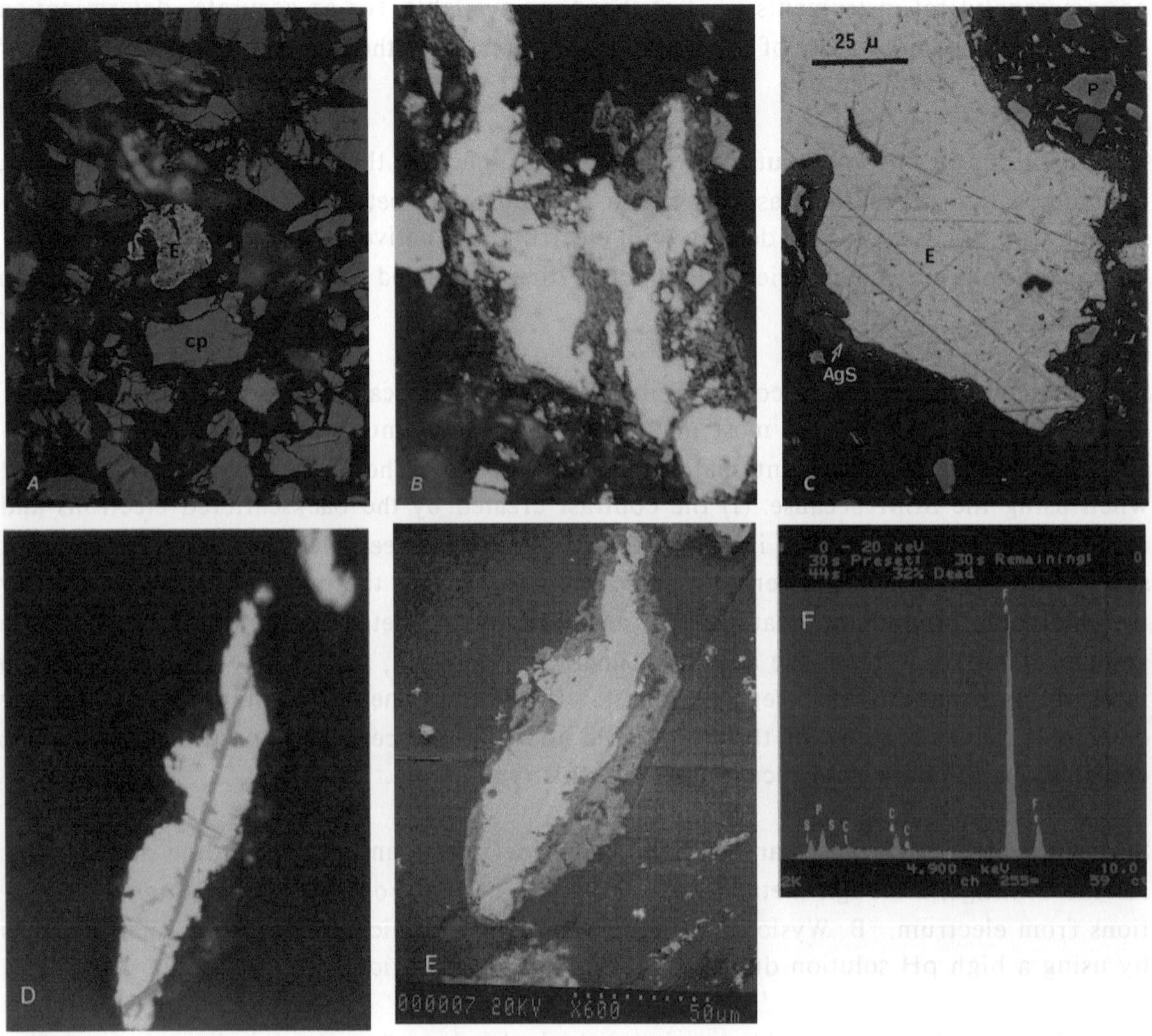

Fig 31. *Tarnishing and coatings over the surface of electrum and native gold, developed after the grinding process (A, B and C) and natural (D). A: electrum showing surface tarnishing. B: electrum coated by secondary silver sulfide. C: electrum coated by secondary silver sulfide mixed with fine particles of pyrite. D: native gold coated by iron oxide containing minor amounts of silicon, phosphorus, sulphur, chlorine and calcium. E: the same particle of native gold magnified and photographed on the SEM screen - Note the dramatic improvement in clarity of the coating. F: SEM x-ray spectrum of the iron oxide coating showing the various minor elements. (SEM photograph and spectrum provided by Fred Neub. University of Toronto).*

of tellurium was assumed and taken for granted every time poor gold recoveries were experienced. In consideration of the fact that ores which yield poor recoveries are relatively common, such an assumption dramatically increased the number of *presumed* tellurium bearing gold deposits. In more recent years, microanalytical studies and metallurgical test work have determined that, in addition to being far from common, gold telluride minerals are not generally insoluble. For example, calaverite is reported to be readily soluble by Tremblay (1983) and Habashi (1967, see table 4 on page 133). Tellurides are reported to be soluble *only if finely ground* by Sarbutt (1983).

The fields is again open to systematic investigations. We can however speculate for the time being that the answer to solubility of tellurides of gold with and without silver is probably that some of them are soluble, some are poorly soluble (Johnston 1933) and some are not soluble at all. Considering that eleven compounds of gold with tellurium are recognized as minerals, and that several more are probably unreported, it is safe to assume that the solubility in cyanide varies for each of these minerals.

Insoluble tellurides may need oxidation prior to the cyanidation step. A paper on the treatment of **some** gold tellurides has been published by Cornwell and Hisshion (1976).

OTHER GOLD-BEARING MINERALS. Little information is available on the response of the other gold-bearing minerals to cyanide. These however are so rare and, when present, usually only account for such a small portion of the total gold, that, for metallurgical purposes, knowledge of their correct solubility is seldom necessary. The new minerals from the Tambo deposit in Chile leached easily, thus posing no recovery problems.

Grain size and mineral associations

When leaching native gold and electrum, containing less than 30 per cent of silver and occurring in impermeable hosts, the size of the particles should be coarse enough to be liberated or exposed by the grinding process, but not too coarse to slow down the actual cyanidation step. This size ranges between 10 and 30 microns. Of course we have seen that finer sizes are leached, when occurring in permeable hosts (case studies 6 and 7 on pages 32-35 and pages 120-121).

Coarser sizes may lend themselves to cyanidation, if the gold particle is porous and/or has uneven contours, because such feature provides for a larger amount of surface exposure to the dissolving agent.

In addition to the grain size, the type of association with the host mineral plays an important role. Native gold and low-silver electrum, occurring in finer particles than the finest to which the ore can be ground and in *relatively* compact ores, will still be recovered by standard cyanidation, if associated with their host minerals in a manner which will favor exposure, following the grinding process. Such associations include distributions in fractures and distributions at the border between grains of the **same** mineral. For example, figures 22B and C (page 107) and figure 29A (page 114). In similar situations, when using some of the more accurate techniques for extraction (for example, agitation leaching or pressure leaching), the cyanide solution comes into contact and dissolves the gold particles, even if still associated with the host, because fractures and gaps promote its penetration and because, being fractures directions of weak resistance, particle exposure is promoted during the crushing/grinding processes. On the other hand, an ore in which the gold is distributed between grains of **two different** minerals, such as pyrite-chalcopyrite (figure 23 on page 108) or pyrite-silicate (figure 29B on page 114), may not be as easy to treat, particularly if concentrating the economic mineral, for example the chalcopyrite. After the grinding process, part of the gold remains attached to the pyrite and part remains attached to the chalcopyrite (figure 32), thus resulting in gold losses to the pyrite tailings, following concentration of the chalcopyrite. Improved recoveries will be obtained once the problem has been defined by the microanalytical studies, by differentially floating and leaching the pyrite product, as well as the chalcopyrite concentrate.

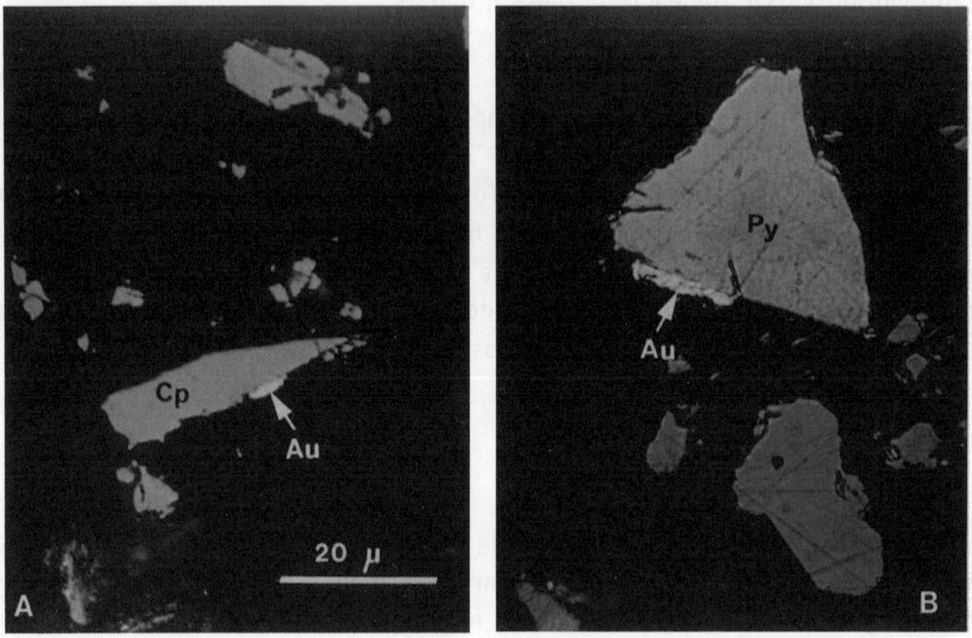

Fig. 32. *Veinlets of native gold attached to grains of chalcopyrite (A) and pyrite (B). The veinlets were identified in the copper concentrate and tailings derived from an ore in which the gold was distributed at the border between pyrite and chalcopyrite. In similar assemblages, if concentrating the copper, a portion of the gold is likely to be lost with the pyrite tailings.*

An interesting situation is that of a high grade ore in which the gold is within the sulfide, along the silicate-sulfide border and in the silicate, but in close proximity of the sulfide. An efficient recovery strategy could be developed where:

- The ore is coarsely ground and a rougher sulfide concentrate is floated containing the sulfide-silicate middlings with the gold still unliberated.

- The silicate-sulfide middlings are finely ground and the gold is liberated or exposed, ready for the cyanide process.

Very fine grain sizes (less than 10 microns), where associated with an impermeable host in a manner which does not allow easy liberation or exposure (for example figures 21E and 22E on pages 106 and 107) result in poor gold recoveries. Poorest recoveries happen when **most** of the gold remains totally enclosed in an impermeable host, even after the finest grinding. Identification of the host mineral(s), determination of the correct average size of the particles of native gold and knowledge of their association and distribution within the host mineral(s), become essential in such situations, for an indication of the quality of recovery to be expected and/or for the development of a plan for exposure of the gold prior to cyanidation.

Host minerals

Listed below are most of the common host minerals to gold, along with their reaction, if any, to the cyanide and some of the treatments applied to optimize the recoveries.

PYRITE. This is a stable mineral during cyanidation. Cyanide consumption, sometimes experienced during the leaching of pyrite ores, is in fact very likely due to minor amounts of secondary iron oxides replacing the pyrite itself. Although an advantage when dealing with coarser gold or with gold distributed in fractures, stability becomes a problem when pyrite is host to fine gold: the mineral is difficult to break by chemical methods.

Sometimes *pre-aeration* prior to the cyanidation step is sufficient to control impermeability of an otherwise permeable pyrite (showing fracturing and/or porosity), that means when impermeability is due to hydroxyl ions adsorbed on to its surface.

Positive results in the liberation of fine gold from impermeable pyrite (free of fracturing and porosity) are obtained by roasting the product in furnaces under a variety of temperatures (in the range of 650°C or higher) and environment conditions. The purpose of the roasting process is to transform the mineral into a more permeable compound from which the gold can then be separated by the conventional techniques. Pyrite is roasted in a furnace under oxidizing conditions to produce iron oxides and under neutral conditions, in air, to produce troilite (Freeman, Hartley and Wick, 1981). The reasons why the roasting process renders the ore amenable were discussed on page 35. Environmental restrictions controlling the amount of sulfur dioxide allowed into the air and produced during the roasting process, contribute to the difficulty in the treatment of these gold ores.

Case study 12. Fine gold in pyrite. Over 50 per cent of the gold in a pyrite concentrate was being lost during cyanidation, and the presence of gold minerals other than native gold or electrum, such as insoluble tellurides, was at first suspected as the reason for poor recovery. An accurate microanalytical study showed that the average size of the gold (1 to 2 microns) was much finer than the average size to which the pyrite had been ground (20 to 25 microns, figure 33) and that the gold which did not dissolve was not itself insoluble, but was still enclosed in the host mineral, thus never coming into contact with the cyanide solution. Subsequent to the study, roasting of the pyrite, rather than telluride oxidation, was used before the cyanidation step, resulting in improved gold recoveries.

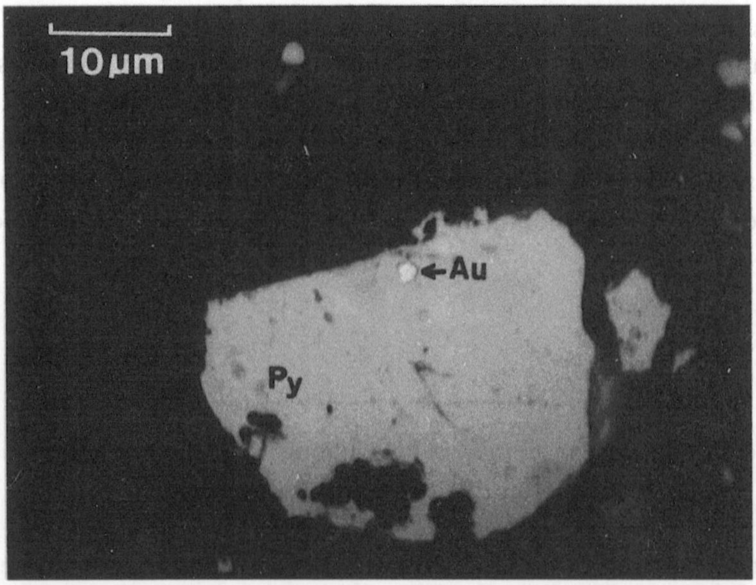

Fig. 33. *Fine native gold in a coarser grain of pyrite from a concentrate giving poor gold recoveries. Gold occurrences this fine require roasting of the pyrite prior to cyanidation.*

A method of oxidizing pyrite without roasting, thus eliminating the environmental problems, involves **hot aqueous digestion** applied to the finely ground ore (Hausen 1981). Another method of oxidation in aqueous medium is the relatively new **pressure oxidation**, carried out in autoclave using high pressure and temperatures. This oxidation is conducted in water and in the presence of air and leads to the formation of ferric hydroxide and sulfuric acid. **Natural oxidation** of the pyrite is produced by the weathering process, resulting in gold recoveries from tailings after six months to one year of their disposal. The oxidation time depends in this instance on the atmospheric conditions as well as on the textural features of the ore. Humid weather, fractures and porosity expedite the process.

Another method of oxidizing pyrite is by **bacterial leaching** (pages 77-78) During the late 80's, bacterial leaching was used routinely in North America for the extraction of uranium and copper and was still in an early experimental stage for the extraction of gold. The process was being used in South Africa for the first time by Gencor to recover gold from pyritic ores, according to the South African Trade News Magazine (February 1988, p. 4): "Gencor has become the first company in the world to establish a commercial plant and hopes to engage in joint manufacturing ventures with other companies who are yet to exploit their refractory ore deposits. Investigations into the bioleaching technology using the bacteria Thiobacillus Ferroxidans began 10 years ago at Gencor's laboratories near Johannesburg. The new process is now producing designated tonnages and is set to replace conventional roasters at Gencor's Barberton Mine within two years."

ARSENOPYRITE. Arsenopyrite is a very stable mineral even more so than pyrite and therefore similar to pyrite in gold metallurgy. Coarser gold and gold that becomes liberated or exposed by grinding, when associated with arsenopyrite, is extracted by cyanidation without major problems. But finer gold or gold in a compact arsenopyrite is not easily liberated. Oxidation of arsenopyrite is accomplished by roasting, to produce arsenic oxides. Important in the roasting of arsenopyrite are the air/oxygen supply, which should be limited, and the temperature which must be kept lower than the one used for pyrite (450°C). Quick temperature increases produce undecomposable impermeable arsenates over the gold particles. These are controlled *to some extent only* by pre-aeration. Environmental restrictions imposing limitations on the arsenic release into the air (as well as the sulfur dioxide) contribute to the difficulty in treating the mineral when host to fine gold.

Weathering may again prove effective on tailings, roasted residues and leach residues, like in the case of the pyrite.

REALGAR AND ORPIMENT. Realgar and perhaps more so orpiment react in the cyanide by forming complex reducing compounds that retard or prevent the gold dissolution. Active inhibitors are probably the thioarsenite and sulfide ions that adsorb at the surface of the gold. A way of controlling these minerals is suggested by Sarbutt (1983) who adds lead ions to the cyanide-gold solution. The lead ions remove the sulfide as lead sulfides and accelerate the rate of oxidation of the thioarsenite to thioarsenate, thus freeing the surface of the gold. Operations at low pH and pre-aeration are also beneficial.

CHALCOPYRITE AND SECONDARY COPPER MINERALS. One technique for the separation of gold from chalcopyrite and other copper ores exploited primarily for copper, involves the preparation of a copper product by flotation, smelting and then recovery by electrolytic refining processes. In ores which are exploited for gold only, the presence of chalcopyrite and more so the presence of secondary soluble copper minerals such as chalcocite, covellite, digenite, djurleite and azurite-malachite, complicates the process of cyanidation because these minerals form copper ions in solution along with the gold. The process not only causes excessive cyanide consumption, it also results in high copper contents of the pregnant solution. Soluble copper will also precipitate on zinc dust, forming a coating which retards the effective deposition of the gold. Where the amount of copper soluble in cyanide is great, the cyanidation process may precipitate considerable quantities of copper. Frequent clean-ups and the use of expensive acid washes of the precipitate to remove the excess copper become then necessary.

Dorr and Bosqui(1950) recommend to maintain a ratio of total cyanide to copper of 4 to 1 to avoid serious loss of the dissolving power. In situations where the secondary copper minerals are abundant and not directly associated with the gold, a sensible approach might involve their separation, prior to the leaching step.

PYRRHOTITE. Pyrrhotite reacts in the cyanide solution, resulting in consumption of large amounts of cyanide and oxygen. An effective treatment of pyrrhotite includes aeration in alkaline solution, running off after thickening or filtering of the solution and subsequent cyanidation. According to Tremblay (1983), pyrrhotite, besides consuming oxygen and cyanide, causes the formation of hydroxyl ions which may adsorb on to the gold surface and reduce its rate of dissolution. The corrections to be applied to pyrrhotite ores are extremely varied, largely depending on the ore itself and may include:

- High alkalinity, strong aeration, then cyanidation.
- Low alkalinity, strong aeration, then cyanidation.
- High alkalinity, high cyanide strength and strong pre-aeration.
- Pre-aeration in cyanide only.
- Low alkalinity and low cyanide strength, with strong aeration, then discarding of the solution.
- Additions of about 0.2 lbs/ton of litharge (lead monoxide) or of lead nitrate and low alkalinity during the leaching step.
- Small additions of zinc or mercury ions.
- Roasting prior to the leaching step.

MARCASITE. Marcasite is reported to be more soluble than pyrite and less soluble than pyrrhotite. Microanalytical studies of marcasite head samples and cyanided tails show no difference in the appearance of the mineral before and after cyanidation. Factors such as cyanide strength or pH may be involved in the active dissolution, if the mineral is in fact soluble. Secondary iron oxides, common in marcasite rich ores, are also responsible for the cyanide consumption, probably more so than the marcasite itself.

GALENA AND OTHER LEAD MINERALS. According to Tremblay (1983), galena and anglesite are the only important lead minerals which interfere with cyanidation. Because galena oxidizes easily to the sulfate form, the problems derived from the presence of the two minerals in the cyanide solution, are essentially the same. Stability of the lead minerals during cyanidation is controlled through the pH of the solution. If alkalinity is kept low, the formation of plumbite which reacts with the cyanide to form insoluble basic lead compounds, is reduced. Ores containing oxides of lead can be advantageous in some situations, especially if the pulp consists of soluble sulfide ions, such as those found in ores rich in realgar and orpiment.

A particularly troublesome type of ore is the one containing galena with compact impermeable pyrite and/or arsenopyrite which need to be roasted. The roasting process may in these situations produce coatings of oxidized lead over the gold particles, resulting in poorer recoveries than those experienced from the unroasted product. It would be wise to eliminate the galena before roasting these types of ores.

SPHALERITE. "Sphalerite and other zinc minerals are fairly soluble in the cyanide solution. The zinc ions are cyanicides and dissolved oxygen consumers. Nevertheless with a total cyanide concentration to zinc ion concentration ratio of at least three and a high alkalinity, zinc can actually increase the gold extraction rate. One obnoxious effect of zinc ions is on the cyanide ion determination with silver nitrate which reads high. In their presence or that of copper ions, an addition of about 0.5 grams of litharge precipitates the ion as a sulfide. After filtration it is possible to obtain a reliable free cyanide assay". (Tremblay 1983, p. 30).

If gold is recovered as a byproduct of lead and zinc, then flotation, smelting and separation during refining are applied in a similar manner as when treating copper concentrates. Additional information on the separation of gold from lead-zinc ores is on page 260.

STIBNITE AND ANTIMONY SULFOSALTS. According to Sarbutt (1983) stibnite and other antimony minerals have deleterious effects on cyanidation similar to those caused by the orpiment. The chemistry involved is thought to be very similar. The stibnite is oxidized to thioantimonite and sulfide. As with the arsenic minerals, inhibition of the cyanidation process is brought about by the action of the thioantimonite and sulfide ions adsorbed at the surface of the gold and is controlled by similar means.

When roasted, antimony minerals may give similar problems to those described for the arsenopyrite: high temperatures or quick temperature increases may produce impermeable coatings of antimonate compounds over the surface of the gold. They are treated in a similar manner as the arsenopyrite ores: low temperature conditions with limited amounts of air.

OTHER SULFIDES AND SULFOSALTS. Reactions of most of the other sulfides/sulfosalts to cyanide are seldom reported in the literature. In many deposits, these minerals are not hosts to gold, or, when they are, they only host a fraction of the total and/or do not occur in sufficiently high quantities to cause significant problems.

OXIDES OF IRON (AND OTHER OXIDIZED MINERALS). Hematite and magnetite are sparingly soluble or insoluble. Goethite and limonite and many minerals derived from the secondary oxidation of sulfides are more soluble and may cause some cyanide consumption. They however have the advantage of promoting cyanide penetration, as demonstrated in case studies 6 and 7, on pages 32-35, and are therefore particularly well suited for heap leaching where only a coarse grinding is used.

URANIUM MINERALS. Separation of fine gold from some of the uranium minerals (Corrans and Levin 1979) such as uraninite and thucholite is practiced at Mintek in South Africa by a *reverse leach* process (Boydell et al 1979). The process involves an acid treatment for the extraction of the uranium prior to the cyanidation step.

QUARTZ, SILICATES, CARBONATES AND SULFATES. Most silicate, carbonate and sulfate minerals, when hosts to gold, are amenable to cyanidation, because stable during the process. Gold coarser sizes and distributions along grain boundaries, or between plates (like in the case of the phyllosilicates, figure 25 on page 110), rather than within, promote cyanide penetration and particle contact, thus presenting no obstacles to the process of dissolution.

CARBONACEOUS MATERIAL. Carbonaceous material is deleterious, whether directly associated with the gold or not. It is discussed in the next section on minerals other than the host mineral.

SANDS AND GRAVELS (PLACER GOLD). This is the easiest gold to recover, because, due to the liberated state of the majority of the particles, no comminution is needed, and because recoveries largely involve simple techniques such as panning and other methods of gravity concentration, with or without amalgamation.

Tables 3 and 4, modified after Henley (1975) and after Habashi (1967) show the solubility of some of the more common minerals in cyanide.

TABLE 3 - SOLUBILITY OF COPPER, ZINC, ARSENIC AND ANTIMONY MINERALS IN CYANIDE

(Modified after Henley, 1975)

MAIN ELEMENT	MINERAL	% EXTRACTION 23°C	45°C
Copper	Chalcopyrite	5.6	8.2
	Chrysocolla	11.8	15.7
	Tetrahedrite	21.9	43.7
	Enargite	65.8	75.1
	Bornite	70.0	100.0
	Cuprite	85.5	100.0
	Metallic Cu	90.0	100.0
	Chalcocite	90.2	100.0
	Malachite	90.2	100.0
	Azurite	94.5	100.0
Zinc	Willemite	13.1	
	Hemimorphite	13.4	
	Sphalerite	18.4	
	Franklinite	20.2	
	Hydrozincite	35.1	
	Zincite	35.2	
	Smithsonite	40.2	
Arsenic	Arsenopyrite	0.9	
	Realgar	9.4	
	Orpiment	73.0	
Antimony	Stibnite	21.1	

TABLE 4 - SOLUBILITY OF SOME OF THE MORE COMMON MINERALS IN CYANIDE

(Modified after Habashi, 1967)

MAIN ELEMENT	MINERAL	% DISSOLVED IN 24 HOURS
Gold	Calaverite	Readily soluble
Silver	Argentite	Readily soluble
	Cerargyrite	Readily soluble
	Proustite	Sparingly soluble
	Pyrargyrite	Sparingly soluble
Copper	Azurite	94.5
	Malachite	90.2
	Chalcocite	90.2
	Cuprite	85.5
	Bornite	70.0
	Enargite	65.8
	Tetrahedrite	21.9
	Chysocolla	11.8
	Chalcopyrite	5.6
Zinc	Smithsonite	40.2
	Zincite	35.2
	Hydrozincite	35.1
	Frlanklinite	20.2
	Sphalerite	18.4
	Gelamine	13.4
	Willemite	13.1
Iron	Pyrrhotite	Readily soluble
	Pyrite	Sparingly soluble
	Hematite	Sparingly soluble
	Magnetite	Practically insoluble
	Siderite	Practically insoluble
Arsenic	Orpiment	70.3
	Realgar	9.4
	Arsenopyrite	0.9
Antimony	Stibnite	21.1
Lead	Galena	Soluble at high pH
	Anglesite	Soluble at high pH

Minerals other than the host mineral

- **CYANICIDES**. These minerals dissolve during the leaching process, thus consuming cyanide and oxygen. Or they form complex ions which adsorb onto the surface of the gold and render it impermeable. Some of the cyanicides are pyrrhotite, orpiment, copper minerals, sphalerite, stibnite and antimony sulfosalts, and some iron oxides, all already discussed. If not directly associated with the gold, elimination prior to cyanidation represents an efficient way of coping. Ores high in some cyanicides are not suitable to heap leaching.
- **CLAY AND SERICITE** (clays). Clays tend to hinder the cyanidation process because: (i) they form impermeable coatings over the surface of the gold which develop after grinding; (ii) the gold is sometimes fine enough to be in a colloidal

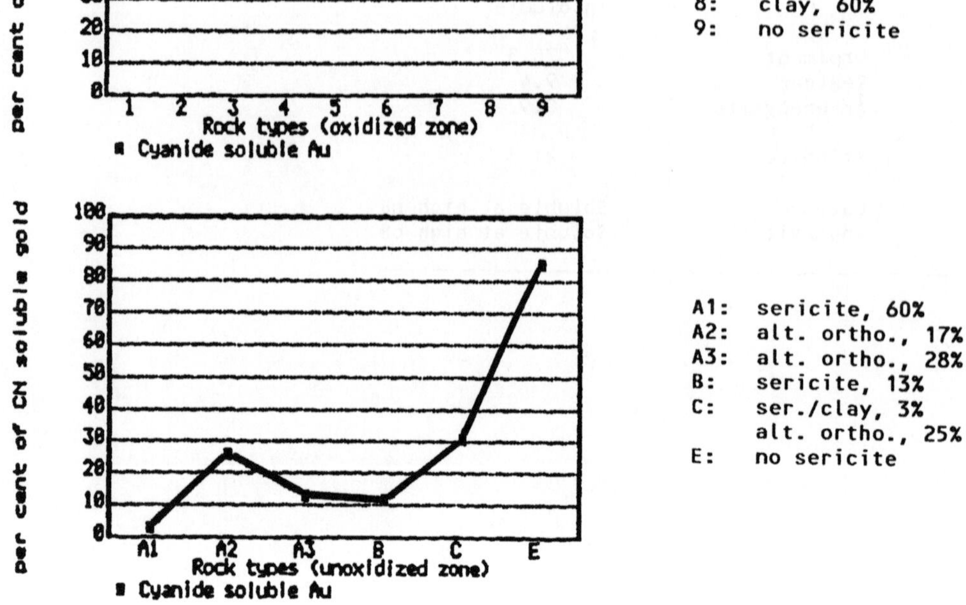

Fig. 34. *Diagrams showing an inverse relation between clay/sericite content of the ore and gold solubility.*

state and therefore impossible to separate. Whether directly associated with the gold, or just present in the ore, clay minerals reduce the gold dissolution rates. Like in the case of the cyanicides, elimination prior to cyanidation of all the clays, when not directly associated with the gold, may represent the easiest way of improving recoveries. When gold is directly associated with clay or sericite, microanalytical studies are of little help, because the native gold particles are often of similar size to that of the clays and therefore invisible. If no gold is located by the microanalytical studies, one method of obtaining a positive answer as to gold associations could involve the concentration of the sulfides followed by fire assay of the heavy and light fractions. If the fire assay indicates that most of the gold has remained in the light fraction carrying the clays, then the conclusion can be drawn that the gold occurs in invisible particles of similar size to that of the clays. In such an instance, gold separation may not be possible by the conventional methods.

Case studies 13 and 14. Gold in clay ores with reduced recovery rates. Hinderance of gold recoveries by clays is confirmed by several microanalytical studies. Two such examples are in figure 34, showing a relation between percentage of gold insoluble in cyanide and percentage of clay/sericite, and in table 5, indicating that the samples giving the poorest recoveries all contain high clay. Table 5 also shows an inverse relation between gold visibility and presence of clay. Figure 34 proves that, although there is an inverse relation between clay

TABLE 5 - Microanalytical and chemical data on 7 gold-bearing samples from a locality in Nevada showing a direct relation between clay content, visible gold and gold solubility

SAMPLE	TRANSPARENT MINERALS	OZ/TON	NATIVE GOLD FOUND	CN SOL%
A	**CLAY, ABUNDANT;** quartz; adularia; chlorite; pyroxene; plagioclase	0.62	3 occurrences	11
B	**QUARTZ, ABUNDANT;** clay	0.464	7 occurrences	50
C	**CLAY, VERY ABUNDANT**	0.424	none	4
D	**QUARTZ, ABUNDANT;** clay	0.168	1 occurrence	42
E	Quartz; adularia; clay	0.068	none	21
F	**DOLOMITE, ABUNDANT;** quartz	0.052	none	25
G	**QUARTZ, ABUNDANT;** adularia	0.040	none	55

and gold solubility, the degree of oxidation of the host rock plays just as important a role: the average solubility is higher in the oxidized samples (top diagram).

The overbearing consultant in gold was involved in the project of table 5. After seeing the **preliminary** results, and before a final report was even completed, the man suggested a hot cyanide treatment for the ore which resulted in no improved recoveries. Again the readers will not only note the lack of understanding of the **partial** results of the microanalytical study, they will realize the unnecessary hurry with which the consultant reached his decision to use the hot cyanide. The complete description of the gold occurrence in the ore, which became available at the end of the microanalytical study, should have been carefully examined, before any steps in the development of the recovery plan were taken. And perhaps those readers who are metallurgists will also realize the importance of the completion of **all** of the microanalytical work, before any metallurgical test work is started, as well as the senselessness of pressuring the microanalyst for answers. Investment of a little extra time during the planning stages of recovery will not only lead to the understanding of the ore, it will save time later, when the project reaches its production stage.

Case study 15. Gold in clay ores with good recovery rates. Although the above case studies indicate an inverse relation between clay content of the ore and gold dissolution rates which is directly linked to a clay/gold interaction, clays may not always be responsible for poor recoveries. P.M.J. Gray (1988, personal communication by letter) states that he was involved in a gold extraction project related to a deposit in the South Pacific, whose upper zone consisted predominantly of clay and which "*surprisingly*" leached well by the standard cyanidation tests. "The clay was however an awful nuisance in every other way - mining, ore handling, liquid-solid separation, tailings disposal, environmental, etc. Because of these operating difficulties, gold recovery was not nearly as good as the laboratory tests would indicate." Gray also states that this case study represents a "good illustration of how the texture and composition of the **whole** ore and not just its values have to be interpreted by the metallurgist."

- **CARBONACEOUS MATERIAL (GRAPHITE AND CARBON).** The cyanidation of gold ores containing graphite and/or carbon is normally accompanied by erratic results due to the property of the carbon to cause early precipitation and absorption of the gold from the cyanide solution. The chemical processes causing the gold precipitations in the presence of carbon are not completely understood. Elimination of the graphite and carbon (if not directly associated with the gold) prior to the cyanidation step, is accomplished by flotation, or by oxidation by roasting or chlorination.

"The first recorded successful method of directly cyaniding ore containing high activated carbon material, was developed for the treatment of carbonaceous gold ores in Nevada through a joint effort by USBM Reno Station and Newmont Mining Corporation....In this process, approximately 30 pounds of gaseous chlorine per ton of ore is bubbled into the pulp at a temperature of 80 to 100°F in agitation tanks. Chlorine that escapes from the pulp is converted into sodium

TABLE 6 - COMMON METHODS FOR THE TREATMENT OF GOLD ORES

(Modified after McQuiston, 1985, p.18-16)

MINERAL OCCURRENCE	TREATMENT METHOD
Placer gold (native Au)	1. Gravity concentration 2. Amalgamation
Coarse gold in lode ores (native Au/electrum)	1. Gravity concentration 2. Amalgamation 3. Direct cyanidation-activated carbon in pulp
Coarse gold in sedi-mentary ores (native Au)	1. Direct cyanidation 2. Treatment of refractory carbon-direct cyanidation
Gold with pyrite and marcasite (native Au/ electrum)	1. Flotation-smelting of concentrates 2. Flotation-cyanidation of concentrate with or without roasting
Gold with arseno-pyrite (native Au/ electrum)	1. Direct cyanidation 2. Flotation-roasting of concentrate -cyanidation 3. Roasting of the ore-washing-cyanidation
Gold with copper ores (native Au/electrum)	1. Flotation-smelting of concentrate -recovery during electrolytic refining 2. Flotation-cyanidation of molybdenum concentrates
Gold with pyrrhotite (native Au/electrum)	1. Direct cyanidation with pre-aeration at low lime alkalinity 2. Direct cyanidation-flotation of cyanide tailings-regrinding and recyanidation of flotation concentrate or roasting and recyanidation
Gold with lead-zinc ores (native Au/ electrum)	1. Flotation-smelting of concentrates 2. Jigging-amalgamation-retorting
Gold in carbonaceous ores (native Au/electrum)	1. Roasting-cyanidation 2. Chlorination of ore-cyanidation 3. Flotation of graphitic material-cyanidation of tailings
Gold tellurides	1. Bulk flotation-roasting-cyanidation 2. Direct cyanidation-SO_2 treatment of filtered tailings-sulfide flotation-roasting of concentrates -cyanidation 3. Flotation-cyanidation of concentrate-roasting of residue-recyanidation 4. Direct cyanidation with added bromocyanide

hypochlorite in a scrubber. The *chlorination process* oxidizes carbon and organic carbon compounds to CO and CO_2, thereby allowing the resulting pulp to be fed directly to the existing cyanide circuit without deleterious effects. This carbonaceous ore would yield only 10 to 40 per cent gold extraction by direct cyanidation and could not be milled with non-carbonaceous ores due to the "robbing" of gold from solution by the carbon. After an oxidizing treatment, the cyanidation extraction is around 85 per cent." (McQuiston 1985, p. 18-15).

In addition to flotation or oxidation of the carbon, prior to the leaching step, Tremblay (1983) suggests kerosene addition to prevent any residual graphite from absorbing the gold.

- **MERCURY MINERALS.** These are cinnabar, various sulfosalts and various other minerals, including native gold, when containing minor amounts of the element in solid solution. Control of mercury during the processing of gold ores is necessary to avoid environmental contaminations resulting from its presence in the air and water: the element can be absorbed by the living environment with consequences that are sometimes lethal. Contaminant mercury is usual-

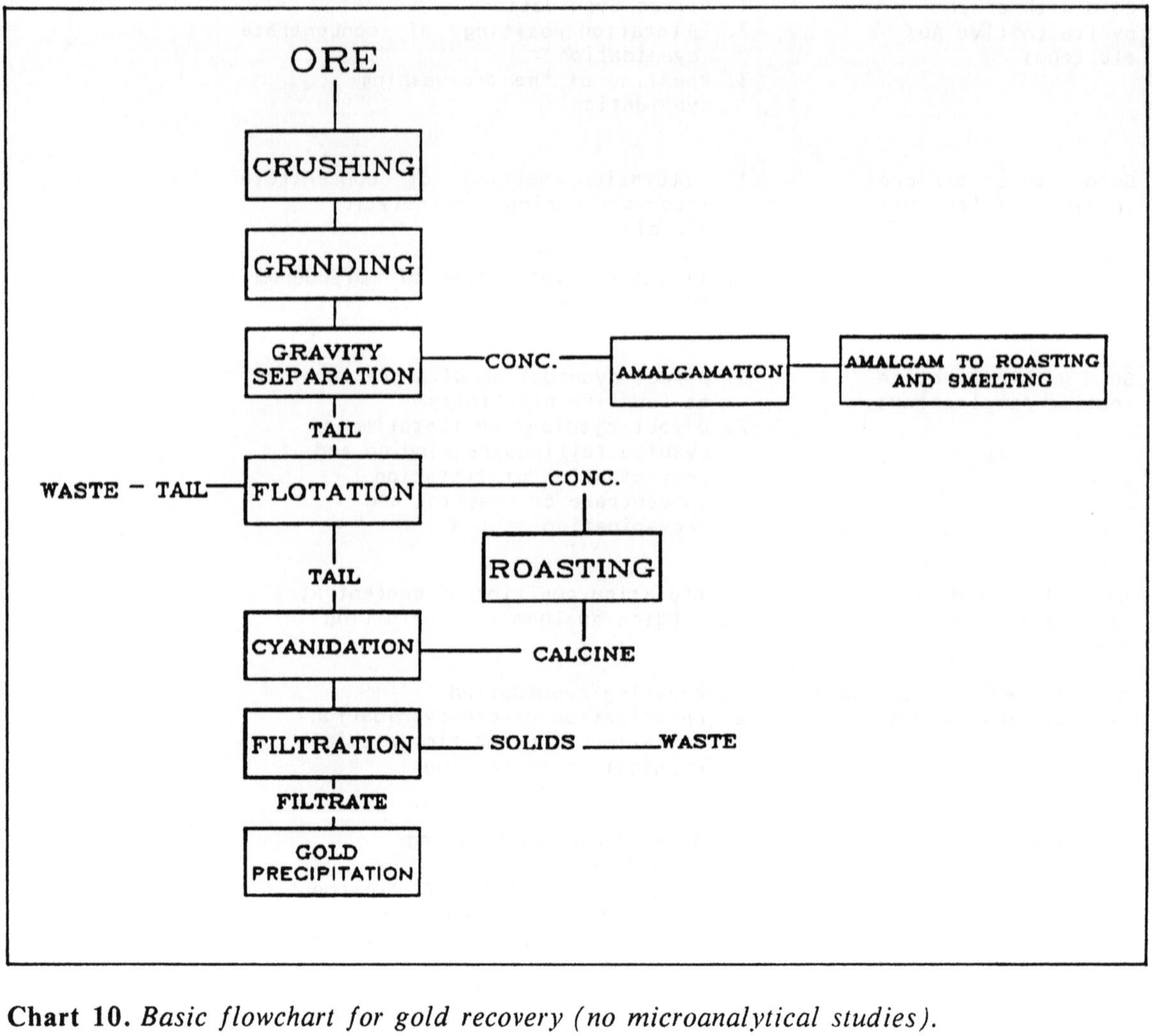

Chart 10. *Basic flowchart for gold recovery (no microanalytical studies).*

ly collected by retort systems and, if present in sufficiently high amounts, is used for commercial purposes.

Table 6, modified after McQuiston (1985), summarizes the treatments for the more common gold ores.

DISCUSSION AND CONCLUSIONS

Flowcharts for the study of gold are regularly published in scientific papers and reports. These usually make little mention of the microanalytical studies conducted for the understanding of the ore itself. Two of these flowcharts, which apply to very general plans, were published by Henley (1975). The more comprehensive of the two has been redrawn and is reproduced as chart 10. It was chosen for this book because published in an article on gold mineralogy.

Flowcharts will vary, depending on the needs and conditions of each individual ore, sometimes on the preferences of the person in charge of the project. Starting from Henley's chart, three very general charts were drawn for this book. They all include the fire assay and the microanalytical work, in addition to the various metallurgical processes. In the first flowchart (chart 11), the microanalytical studies are extensive and systematic, and are conducted at the very beginning of the project, immediately after the fire assay and before any metallurgical test work. They are conducted again on the light fraction from the gravity separation (or concentration) and on the tailings, if the fire assay of either product shows the presence of economic amounts of gold.

Microanalytical studies at a very early stage and before any recovery is attempted, assist in decisions concerning grinding sizes and uncover possible sources of problems such as:

- The presence of electrum. The presence of impermeable coatings over the surface of the electrum.
- The presence of tellurides.
- The presence of gold grain sizes which, coupled with an impermeable host, are too fine to be leached without roasting or other special preparation.
- The presence of cyanicides.
- The presence of clay/sericite.
- The presence of graphite/carbon.
- The presence of mercury.

Rock suitability to heap or other leaching technique is also determined at this early stage.

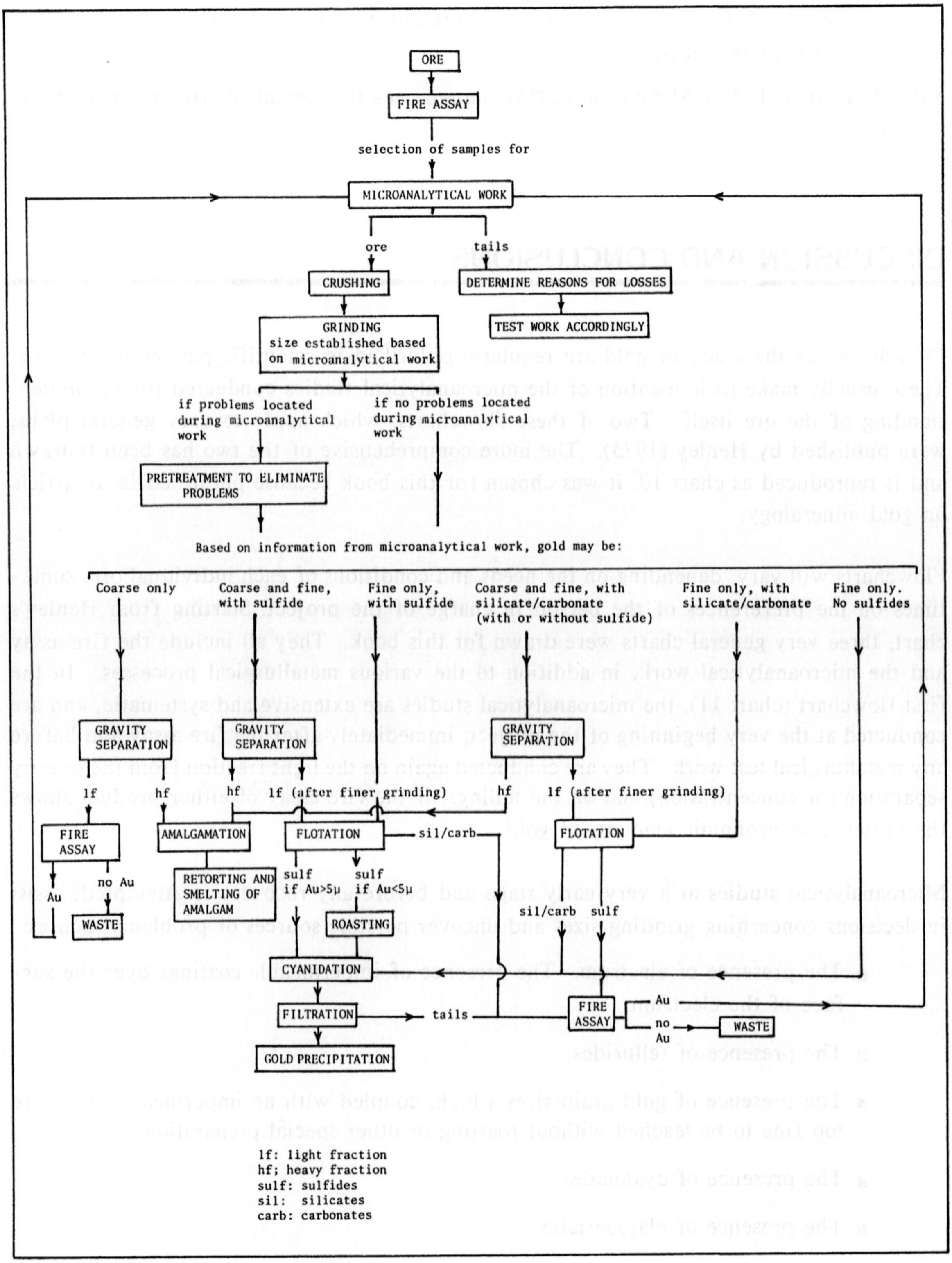

Chart 11. *Flowchart for gold recovery including extensive microanalytical studies.*

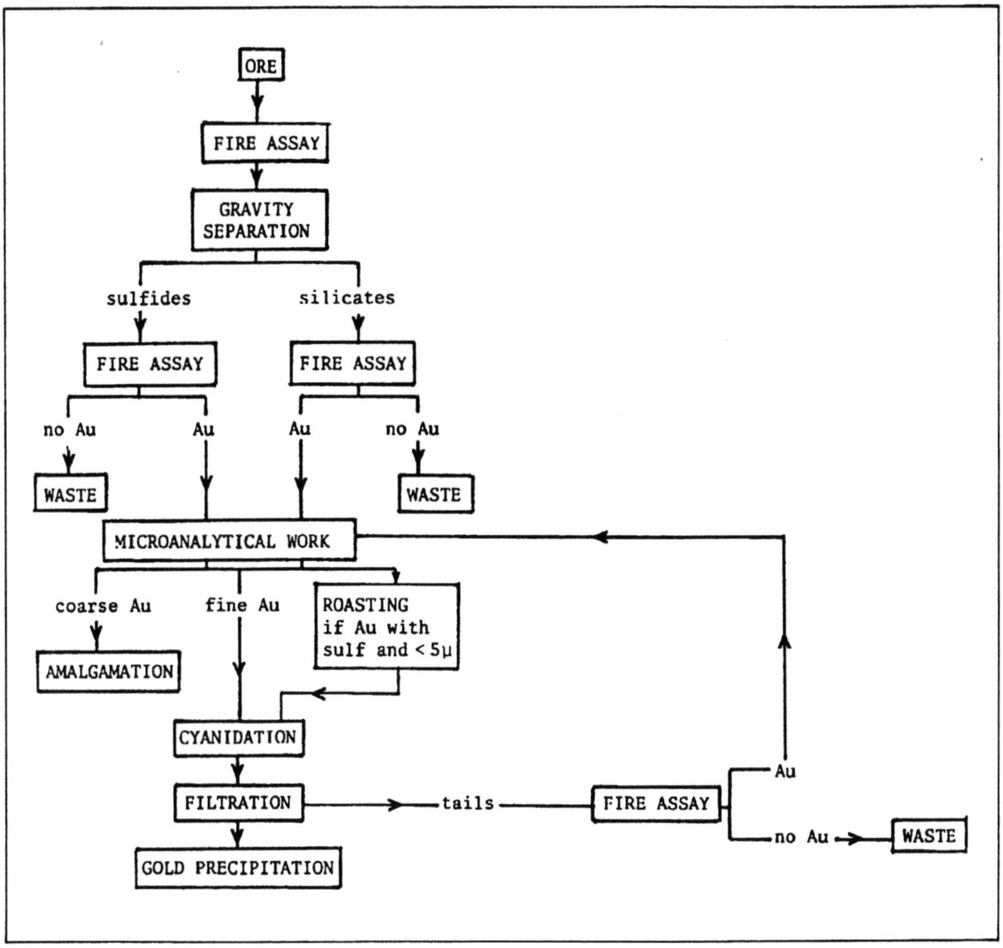

Chart 12. *Flowchart for gold recovery where the microanalytical studies are conducted on samples of gravity concentrates and tails, only if these are gold-bearing.*

Awareness of the presence of cyanicides, of clay/sericite, or of graphite/carbon leads to (i) their elimination prior to any recovery attempt, if they are not directly associated with the gold, or to (ii) the use of techniques for their control, if they are directly associated with the gold. Knowledge of the gold-bearing mineral(s), grain sizes, host mineral(s), and type of association, then provides for efficient planning of the recovery process. Decisions concerning correct grinding sizes, choice of the appropriate recovery techniques, such as amalgamation, cyanidation, roasting prior to cyanidation, preparation of concentrates, control of impermeable coatings, oxidation of insoluble tellurides, are all made beforehand based on the microanalytical data. Consequently, all blind attempts are eliminated, where techniques and laboratory conditions are applied with a minimum knowledge of the ore composition; where the reasons for poor recoveries are seldom understood; and where poor recoveries are accepted as the only alternative to no recovery at all.

Additional microanalytical studies conducted on the light fraction, if still gold-bearing after the gravity separation, give an indication of the reasons why not all the gold was separated with the heavy fraction.

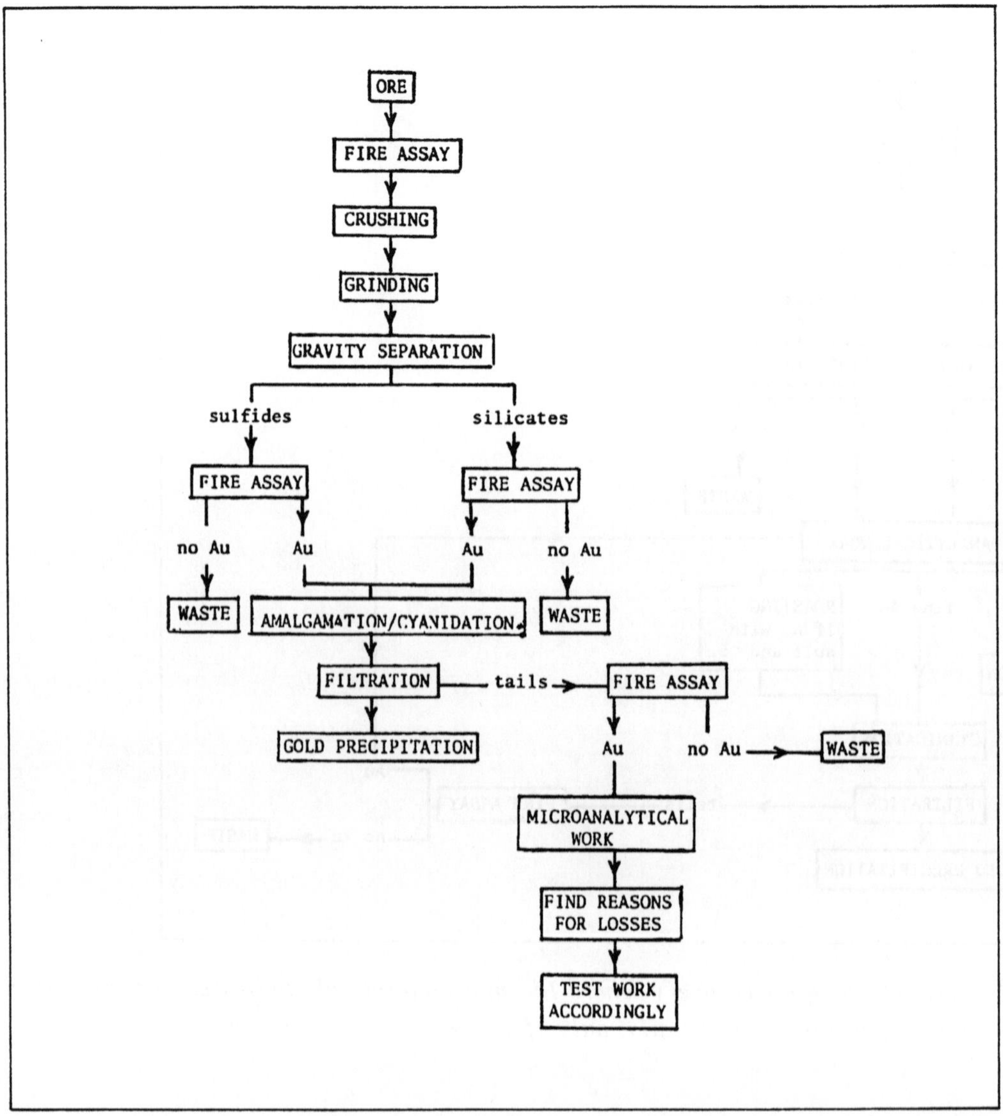

Chart 13. *Flowchart for gold recovery where the microanalytical studies are conducted only on the tails, if they are still gold-bearing.*

Microanalytical studies conducted on the tails, if still gold-bearing, determine the causes for losses which were not predicted during the initial study at the beginning of the project.

The second flowchart (chart 12) suggests microanalytical studies to be applied to the sulfide and/or silicate fractions, following the gravity separation, and only if these fractions contain gold. Such approach has the advantage of concentrating the gold in one product only, thus making its location easier, and is based upon the assumption that the gold is (i) coarse enough to be all liberated, or (ii) associated all with the sulfide, or (iii) associated all with the silicate. It is not however as thorough as the first plan because some gold may remain with one fraction in amounts which are still economic, yet too low to be located by the microanalytical studies. This approach does not give the same complete description of the ore. And, if the gold is coarse and liberated, it does not give any information on host minerals and mineral associations.

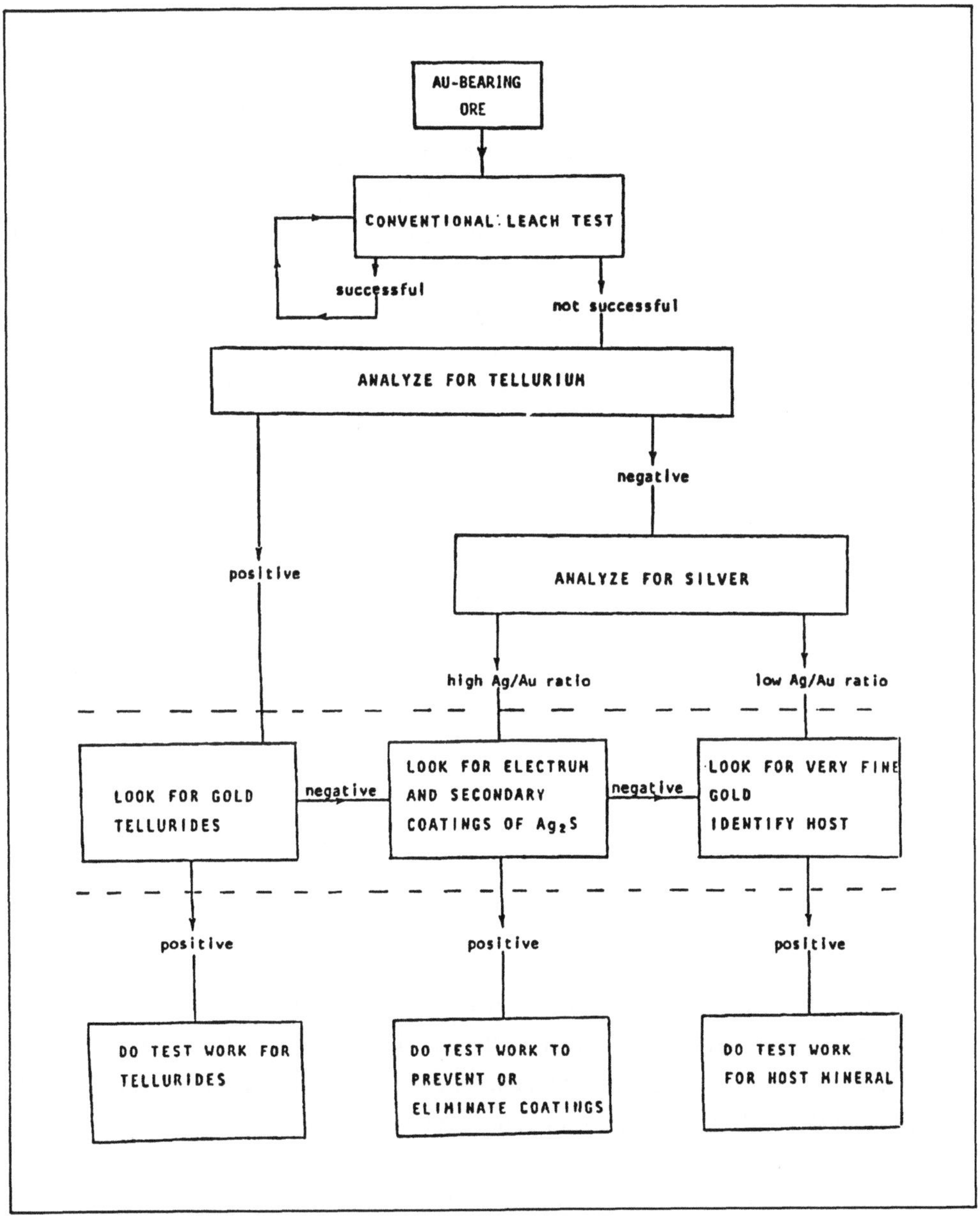

Chart 14. *A plan for the study of gold ores combining test work, chemical analyses and microanalytical studies. The dashed lines enclose the boxes indicating the stage of the project when the microanalytical studies are more effective.*

The third flowchart (chart 13) is probably the most popular one within the mining industry, and only in those situations where the microanalytical studies are in fact conducted. It involves the detailed observation of the tails, after separation has been attempted and recoveries have proven unsatisfactory. The plan is not as efficient as the previous two and may lead to considerable wastes in time and expenditures.

A plan of a different kind which combines test work, chemical analyses and microanalytical studies, and which only uses the microanalytical studies if the test work is unsuccessful, was published in a previous article (Gasparrini 1983), and is represented in chart 14. The plan could be used to determine the causes for low extractions or losses experienced during the early stages of the metallurgical test work. It suggests the following:

- A preliminary conventional leach test. If this is successful, there is no need for detailed microanalytical work, nor for additional chemical analyses.
- Determination of the presence or absence of tellurium in the ore by wet chemical analysis, **only if** the test work does not lead to satisfactory recoveries. A negative analysis for that element will narrow down the causes for poor recoveries to either impermeable coatings over the surface of the native gold/electrum or to fine grain sizes. A microanalytical study, if the tellurium analysis is positive will determine if all, part or none of the gold is present in one or more tellurides. A partial recovery may be due to part of the gold being combined with tellurium to form insoluble telluride minerals, the rest being in native gold and/or in soluble telluride minerals.
- Observation of the gold and silver contents of the ore, if the analysis for tellurium is positive, but the microanalytical studies fail to locate any gold tellurides, or if the analysis for tellurium is negative. Large silver-to-gold ratios coupled with strong correlations between silver and gold may indicate the presence of electrum. This again will be confirmed by the microanalytical studies. On the other hand, small silver-to-gold ratios, poor correlations between silver and gold and the absence of electrum from the polished mounts indicate that a poor recovery is almost certainly due to fine unliberated particles of native gold. The microanalytical studies will then determine the correct grain sizes, host minerals and associations with the gold.

Once it has been established that poor recoveries are due to fine unliberated grain sizes, and before conducting the microanalytical studies, the observation of the chemical analyses of the various metallurgical products, showing gold enrichments and losses, may give a good indication of the host minerals. Also, iron-sulfur and gold correlations may indicate a pyrite-gold association; arsenic and gold correlations may indicate an arsenopyrite-gold association. The microanalytical study will then confirm such findings.

Comparisons of the microcomposition of products giving acceptable and poor recoveries will also provide clues and explanations for gold losses.

And, if the ore or tails have too low grades for the microanalyst to be able to locate a sufficiently large number of gold particles to draw general conclusions as to correct distributions. Or if expertise, time or proper equipment are not available, gold occurrences *may* be guessed through the observation of the minor minerals other than gold. The study of the ore might then be conducted as follows:

After following the steps outlined in chart 14 and if the possibility of the presence of tellurides and electrum is eliminated by negative chemical analyses, instead of trying to locate, identify and determine the gold occurrence directly, a general indication of such occurrence can be obtained by looking at the other minor and trace minerals. For example, if dealing with an essentially pyritic or arsenopyritic ore where the chemical analyses have indicated a gold pyrite or a gold arsenopyrite association, minor amounts of chalcopyrite or galena distributed between grains or in fractures may indicate that gold is also distributed between grains or in fractures like in figures 35A and B. Chalcopyrite distributed at the border between silicate and sulfide may indicate that such is probably the manner of occurrence of the gold (figures 35C and D). And chalcopyrite and galena totally enclosed in the pyrite or arsenopyrite in the form of very fine particles, may indicate that the gold is totally enclosed in the pyrite or arsenopyrite in a similar fashion. A compact arsenopyrite

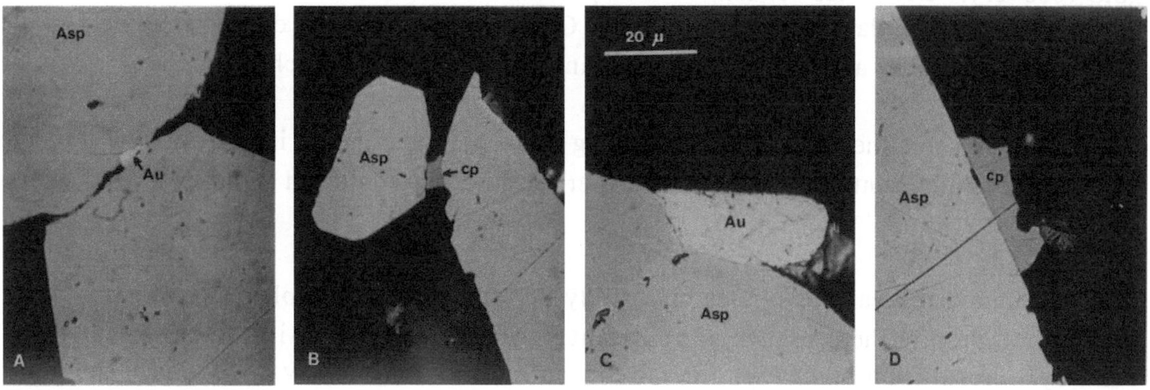

Fig. 35. *Minor minerals (for example chalcopyrite) may show a similar manner of occurrence to that of the gold. In the photomicrographs, native gold and chalcopyrite are distributed between grains (A and B) and along the outside (C and D) of grains of arsenopyrite. Native gold and chalcopyrite were photographed in polished mounts prepared from the same sample.*

free of fractures or inclusions, may indicate a gold solid solution in the arsenopyrite, or an absence of gold from that particular section of the deposit, or from the surface of that particular polished mount. Abundant clay may indicate that, at least a portion of the gold, is invisible and associated with the clay.

The above considerations may be applied to the manner of occurrence of the tellurides and electrum, if the chemical analyses indicate that these minerals may be present.

Results from preliminary test work may also give an indication of the manner of occurrence of the gold. For example, a gold-pyrite ore in which *framboidal* (porous) and *compact* types of pyrite of similar gold contents yielded different gold values to the cyanide treatment, was described by Hausen (1981), who did not provide any positive evidence regarding the actual gold occurrence. Despite the lack of microanalytical data, the poor and good recoveries suggest a totally enclosed state of some very fine gold in the compact impermeable pyrite, and an occurrence in fractures and gaps in the porous pyrite.

The indirect method *may* work. But on the other hand, it should probably only be used where dealing with very low grades (0.01-0.02 oz/ton or 0.34-0.68 ppm), when direct definitions of the gold occurrence are too time consuming. In fact the reader may find that, if he/she were to search for gold by looking directly at the polished mounts, after making assumptions based on the above considerations, the gold occurrence is very different from what he/she had predicted. One of the exciting aspects associated with the study of gold ores is in the unpredictable variety of occurrences of the metal.

Charts 10, 11, 12, 13 and 14 are general and can be used as guidelines when planning a gold recovery. Better flowcharts which apply to individual ores are drawn, as chemical and microanalytical data become available. Charts 1 and 2, at the end of chapter 2, pages 22-23, and charts 15 and 16, on the following pages, are some such examples.

Chart 15 represents the steps followed during the project described in case study 7 on pages 33-35, where different recoveries were experienced from unaltered (sample A) and altered (sample B) pyrite.

Chart 16 shows the various phases of a study of some altered samples where it was determined that, despite the presence of abundant pyrite, gold was invisible and associated with the clay. Preparation of heavy mineral concentrates and fire assay analyses for gold, indicating a gold enrichment in the light fraction, led to the conclusion concerning the gold-clay association, despite the invisible state of the native gold.

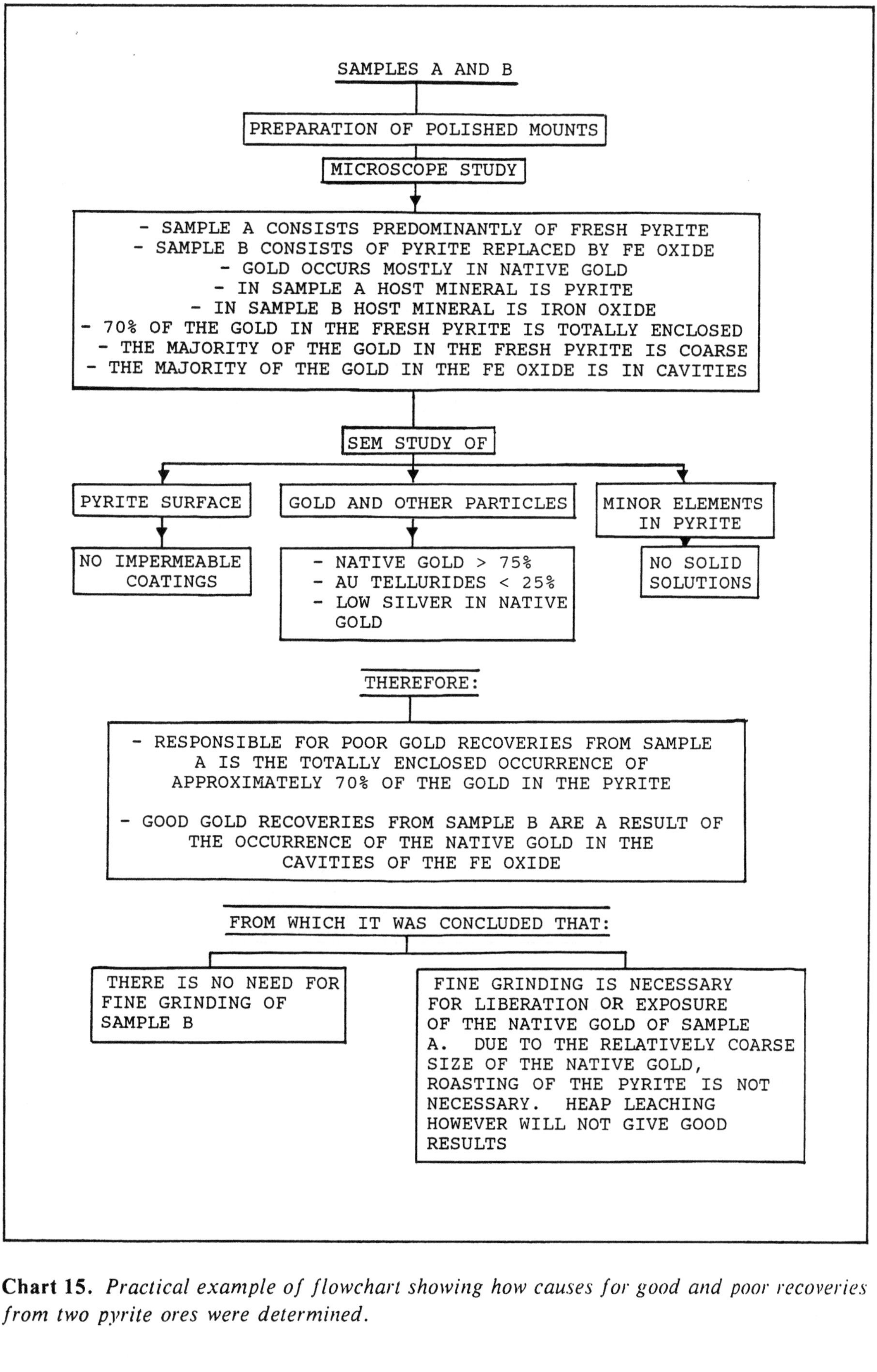

Chart 15. *Practical example of flowchart showing how causes for good and poor recoveries from two pyrite ores were determined.*

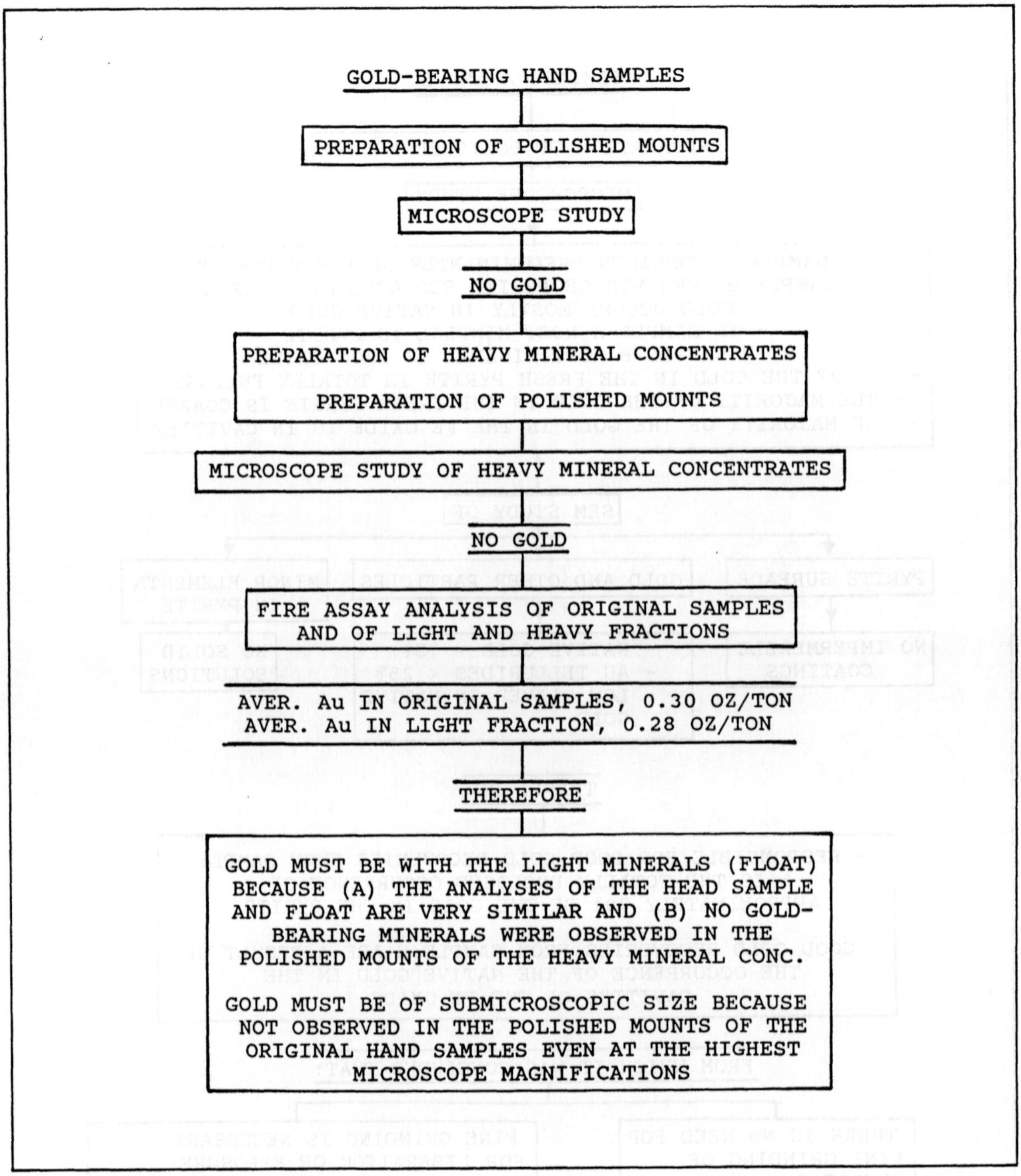

Chart 16. *Practical example of flowchart showing how a gold-clay association was confirmed, despite the invisible state of the gold.*

CHAPTER 7

SILVER

INTRODUCTION

The use of silver for jewelry and personal ornaments is probably as old as the use of gold: silver artifacts and decorations have been found in royal tombs dating as far back as 4000 BC. We cannot deny that a certain amount of mystique is associated with silver, yet the attraction is not as powerful or hypnotic as the one of gold. Alchemists have not wasted their lives trying to create silver. There have been silver rushes such as those related to the Comstock Lode strike in the Sierra Nevada during the second half of the 19th century, but somehow, hearing about these does not steer the same deep emotions produced by the gold rushes, nor cause the same amount of interest. The color of silver is beautiful, but cold. Silver is delicate, as it tarnishes easily, thereby losing its luster. Silver jewelry darkens the fingers or other parts of the body where it is worn, thus making it less appealing as a personal ornament. Silver combines with many other elements, and some of its compounds alter easily, thus defying the definition of noble metal.

On a mythological level, gold and silver symbolize the sun and the moon which is possibly a good description of their role in the universal order of things and of their difference in magnitude. On a more practical level, their market value, 6 to 16 times as high for gold, through the ages, and about 60 times as high at the approximate time when this book was being written, gives a good indication of the importance that populations have been giving to the two metals.

Microscopic observations of silver ores are far less exciting than those of gold ores. The particles are usually dull gray in color and may be confused with many uninteresting or common minerals including some of the secondary iron oxides. Probably variety in mineralogy is the only attractive aspect of the study of silver ores. Light tarnishing, where the particle glows and changes from gray to bright yellow colors under the microscope illumination, provides some entertaining results, but spectacular effects are not common.

Historically, a major use for silver has been monetary, and represented by reserves of silver bullion and by coins. By the 1960's however, the demand for silver for industrial purposes became great and exceeded the total annual world production. Because of its high electrical and thermal conductivity, silver is used in printed electrical circuits and as a coating for electronic conductors. It is also alloyed with copper and gold for use in electrical contacts. (Smith 1988). Modern uses of silver can be summarized to include: photography (40 per cent), electronics-electrical components (19-26 per cent), sterling tableware-jewelry and ornaments (15-19 per cent), catalytic converters (8 per cent), medical (6 per cent) and miscellaneous others (4 per cent). (Mohide 1985. page 222).

* * * * * *

Largest silver producing country today is Mexico, followed by the USSR, Peru, the United States, Canada and Australia. Over half of the silver in the Western world comes from North America. Peru, Mexico, the United States, Canada, and Australia produce about 70 per cent of the total silver in the Western world.

World primary silver reserves are highest in Canada and the USSR (19.0 per cent for each country), followed by the United States (18.6 per cent), Mexico (13 per cent), Australia (12.6 per cent) and Peru (7.5 per cent). According to a statement of the Mexican Chamber of Mines made in 1983, Mexico's ore reserves of silver could be exhausted within 18 years. (Mohide 1985).

Metal market observations show no signs of decreased interest in silver on the part of the free world, in the years to come. According to a newsletter published by the Silver Institute in Washington DC (October-November 1988), the amount of silver produced by the free world during 1987 was not sufficient to satisfy the demand, despite the increase in mine production by 2.4 per cent when compared to the production during 1986. Up until the present, silver has largely been a byproduct of the gold and base metal mining, which means that its production has risen and fallen with the demand for gold, lead, zinc and copper (T.P. Mohide 1988, personal communication). The increased interest suggests that exploitation of deposits primarily for silver may become common in the future.

MARKET VALUE

When compared to the other precious metals, silver is the lowest valued. Market value for silver was in the range of US$ 5.5 per ounce, in July 1989.

According to Mohide (1981 and 1985), the gold to silver price ratio was in the range of 2.5:1 in Egypt, about 3500 BC. It then varied between 9:1 in 2700 BC and 6:1 in 1800 BC under Babylonian king Hammurabi. By the 6th century BC it was 12:1 and remained between that value and 16:1 for many centuries. The same ratio became much higher during the 20th century: 70:1 in 1934, 45:1 and 30: 1 in 1973. It came down to about 26:1 during February 1974, then went back to 34:1 in 1978. The ratio decreased sharply during the period spanning from August 1979 through January 1980, when the silver market value rose to about US$ 50. After the market value dropped again, the ratio reached figures of 60:1. It remained in that range during the second half of the 80's, with a record high of 75:1. If we compare these values with the 10-15:1 ratio unchanged over several millennia, and with the 30-45:1 during a good portion of the 20th century, it is difficult to estimate whether it is the price of gold that is too high or the price of silver that is too low.

Some experts argue that the gold to silver price ratio is now irrelevant (T.P. Mohide 1988, personal communication). Yet, the smaller ratio for so many millennia suggests that a return to lower values may represent a realistic expectation.

GEOCHEMISTRY SYMBOL: Ag ATOMIC #: 47

As a member of group 1-B of the periodic system, silver forms extensive solid solutions with the other two elements in the same group, copper and gold. Cations which silver replaces to a lesser extent than copper and gold are iron, manganese, zinc, palladium, and probably nickel and cobalt. Silver also combines with sulfur, tellurium and selenium in covalent bonds and with chlorine, bromine and iodine in ionic bonds. Where the fluid composition, temperature and pressure during crystallization allow it, silver forms minerals with arsenic, antimony, lead, bismuth, tin, mercury, indium, thallium and germanium. Silver is detected in minor amounts in many sulfides, sulfosalts and native metals.

Silver has a moderately high melting point (960.8°C) and the highest electrical and thermal conductivity of any metal. When in the metallic form, silver is unaffected by water or oxygen, but is altered by ozone with the formation of the silver oxide, and by sulfur and other sulfur compounds with the formation of silver sulfide. The tarnishing commonly observed over the surface of many silver artifacts, consists largely of silver sulfide.

Silver abundance in the Earth's crust is 0.065 per 10^6 atoms of silicon (Brownlow 1979), 32.5 times as high as that of gold. Average economic values, for deposits exploited primarily for silver, are in the range of 5 oz/ton or 171 ppm. They may be higher, depending on the market value, and ease of recovery. Lower grade deposits are exploited. Byproduct silver occurs in amounts between traces and 1,000 oz/ton (or more than 1 per cent).

Irregular silver distributions, when present (figure 36), should probably be determined as accurately as those of gold. Exploitation of the sections of the ore showing the higher concentrations (5 oz/ton or more, depending on difficulty of recovery) may represent a more

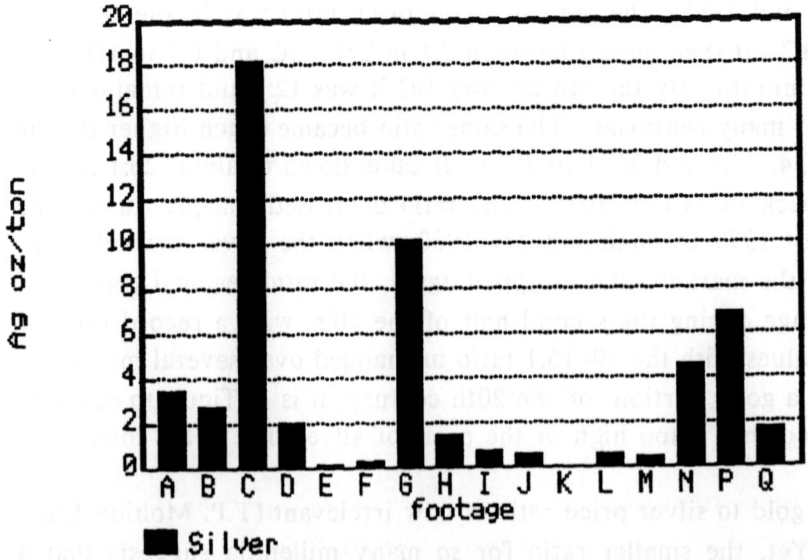

Fig. 36. *Irregular silver distributions detected at different depths/intervals of the same drill-hole.*

efficient alternative when compared to the flotation or leaching of the entire ore. Such an approach may in fact save time and unnecessary expenditures (see case study 19 on pages 190-191 at the end of the chapter).

Silver abundance and distribution in the ore are determined by fire assay techniques (pages 91-93). Lower quantities (below 0.2 oz/ton or 6.84 ppm) or amounts in the presence of high gold are detected by AAA, to which the metal appears to be particularly sensitive.

ROCK TYPES HOSTING SILVER

"Silver tends to occur in veins where the gangue is carbonate and/or barite. Especially characteristic is the frequent occurrence of manganiferous siderite, rhodochrosite and manganiferous calcite among the carbonates found in silver deposits." Boyle (1968, p. 99)

Boyle also states that greatest silver concentrations are in hypogene veins, lodes, stockworks and mantos. Appreciable secondary silver enrichments may occur in near-surface parts of some of these veins because of supergene oxidation processes. Silver is sometimes locally enriched in certain types of copper-bearing shales and in some copper and lead-bearing sandstones. In these occurrences, the metal is often associated with high concentrations of copper and lead, occasionally with uranium, because of their similar solubilities. Silver may also occur in copper-nickel deposits associated with mafic rocks and in skarn-type deposits. Manganese-silver ores tend to occur in the oxidized zone of silver deposits consisting of acid eruptive rocks such as rhyolites and dacites of late Tertiary age. In such deposits, the fresh deeper portions are usually manganese-free. Pegmatites contain generally minor silver which, if present, is associated with the few scattered sulfides in the bodies.

Silver derived from placer deposits occurs largely as the minor constituent of native gold. Microanalytical studies by the author of placer gold indicate variable silver contents ranging between the limits of detection of the analytical instrument and 25 per cent.

A classification of the silver deposits is provided at the end of the chapter in table 7, which also lists the type localities, the more common silver minerals for each locality, the host and other associated minerals, the typical elements and the metals mined. The table was constructed by using the data in Boyle (1968).

ORE MICROCOMPOSITION

Silver-bearing minerals

Because of the ease of substitution for many different elements, a large number of minerals exist in nature in which silver occurs in essential, minor and variable amounts. Over 100 are reported, in which silver is an essential constituent. In many more, silver is a minor/variable constituent. The correct number of existing silver-bearing minerals is not known.

While many gold deposits contain native gold (with variable amounts of silver) as the only or the predominant gold-bearing mineral, most silver occurrences consist of several different silver-bearing minerals. Five to six can usually be identified within the same ore, sometimes as many as ten.

When observed under the microscope, silver minerals show a great variety of optical properties. The colors range from white to light and dark gray, blue, light and dark brown, purple. The luster is metallic and non-metallic. The reflectivity and hardness vary from high to low. Consequently, correct identifications, usually possible by the ore microscope alone when studying gold ores, are seldom obtained without the aid of a microanalytical instrument, when describing silver occurrences.

Minerals containing silver as an essential constituent are listed in table 8. Minerals in which silver is reported as occurring in minor/variable amounts, but whose common varieties are

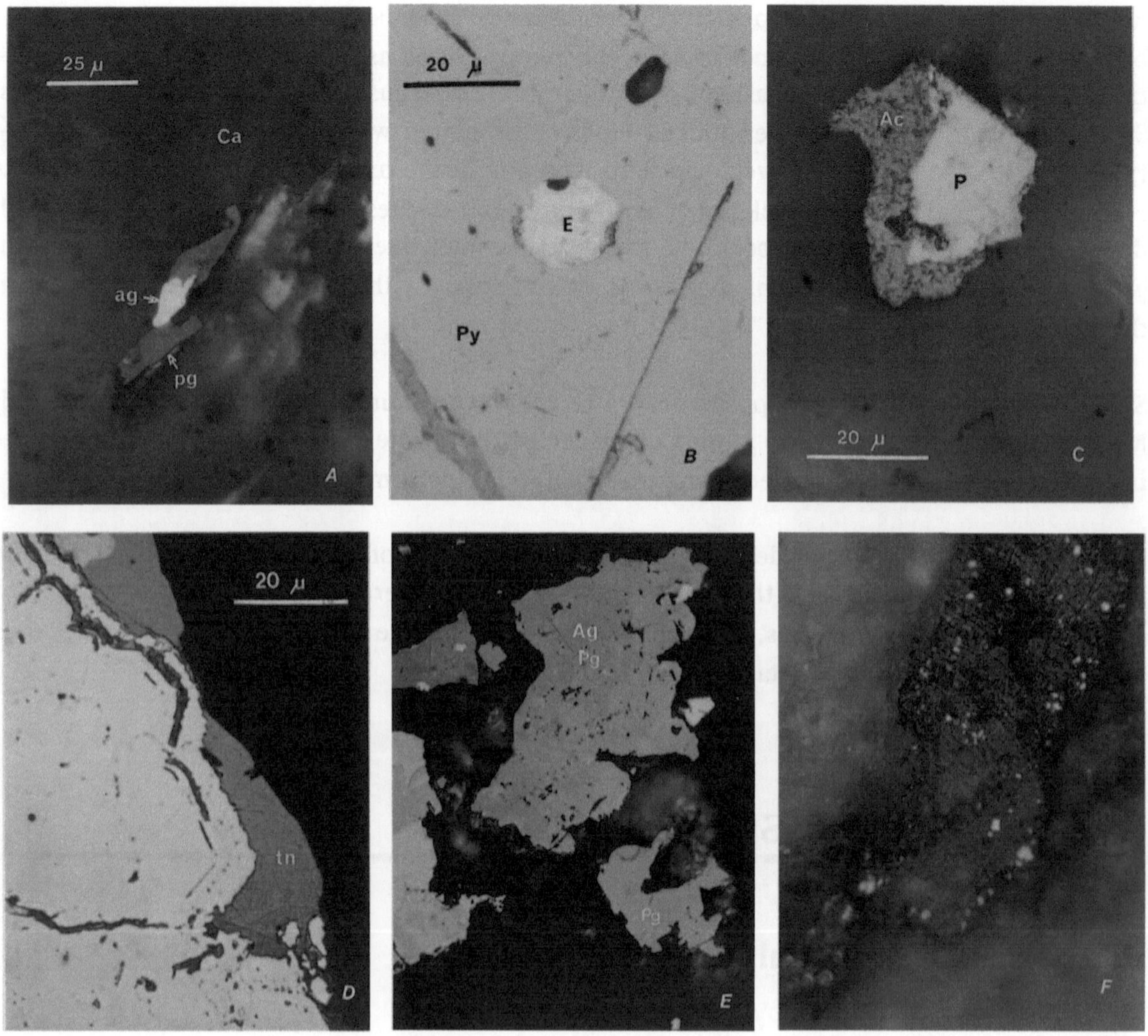

Fig. 37. *Some of the more common silver-bearing minerals. A: native silver (ag) and pyrargyrite (pg) in calcite (Ca). B: electrum (E) in pyrite. C: acanthite (Ac) attached to pyrite. D: tennantite (tn) coating some large crystals of pyrite. E: pyrargyrite in silicate. F: cerargyrite with fine inclusions of native gold.*

silver-free are listed in table 9. Tetrahedrite-tennantite were included in both tables because they are frequent silver carriers and yet they are also often found in their silver-free varieties. Some of the more common silver-bearing minerals are shown in figure 37.

NATIVE SILVER, the elemental form of the metal, although widely distributed in small amounts in many ores, is not as common as native gold. When present, it usually accounts for small portions only of the total silver, the remainder being concentrated in sulfides and sulfosalts. An exception is the native silver discovered in the Cobalt district of Northern

Ontario at the beginning of the 20th century, where the mineral was found in concentrations sometimes as high as 1000 ounces per ton. Contrary to native gold, which often contains more (or much more) than 5 per cent of silver, native silver does not usually contain gold in sufficiently high amounts to be readily detected by a routine EDS-SEM analysis (a few per cent). Minor mercury (less than 1 per cent) is often present in native silver of hypogene origin (Boyle 1968). Although the mineral may be primary, such as the one at Cobalt, more common occurrences of native silver are secondary and usually found in the upper portions of silver-bearing veins. Such occurrences are probably the result of the oxidation of silver sulfides and sulfosalts from the lower levels, by hot water vapors derived from hydrothermal solutions, or by oxygen from the air.

Placer occurrences of native silver, such as nuggets, and flakes are rare, but recorded in Alaska and the Yukon (Boyle 1968).

ELECTRUM AND KUSTELITE, the gold-silver alloys, are discussed in the chapter on gold, pages 100-101).

ACANTHITE, also known as *silver glance*, and its high temperature polymorph, **ARGENTITE** are reported to be the most common and the most important primary minerals of silver. (Dana 1963). Acanthite is known to form large masses, although the most frequent occurrences are represented by small particles associated with copper and other sulfides. Variable quantities of acanthite are identified in most silver ores. Acanthite is generally of hypogene origin, but it has been identified in the supergene portions of some deposits.

PYRARGYRITE and its arsenic equivalent **PROUSTITE** are also known as *ruby silvers* because of the red internal reflection which they display, when observed by high magnification optical microscopy. They are primary minerals found in the upper portions of silver veins where they have been deposited from late solutions. Supergene pyrargyrite, although less common, is reported.

TETRAHEDRITE, also called *freibergite* when rich in silver (more than 3 per cent), and its arsenic equivalent **TENNANTITE**, although not always silver-bearing, are two very common silver minerals. If occurring in a silver ore, the minerals may contain the element in variable amounts up to 34.8 per cent (reported by Boorman et al 1982, in tetrahedrite). And yet presence of tetrahedrite in a silver rich environment is not necessarily indicative of appreciable silver concentrations in its lattice. Silver-free tetrahedrite has been identified by the author in several silver deposits. According to Dana (1963), tetrahedrite is the most common member of the sulfosalt minerals, as it is very widespread in its occurrence and varied in its association. It may be an important ore of copper, in addition to silver. Tetrahedrite is found in many deposits usually as a primary mineral, associated with copper, lead, zinc and other silver compounds, sometimes with native gold. Boyle (1968) reports that investigations of tetrahedrite compositions in relation to distributions within several deposits, have discovered a marked decrease in silver content and cell edge with depth. The same investigations have also revealed that, in veins at the extreme border of the mineralized area, the tetrahedrite is high in mercury (5 to more than 10 per

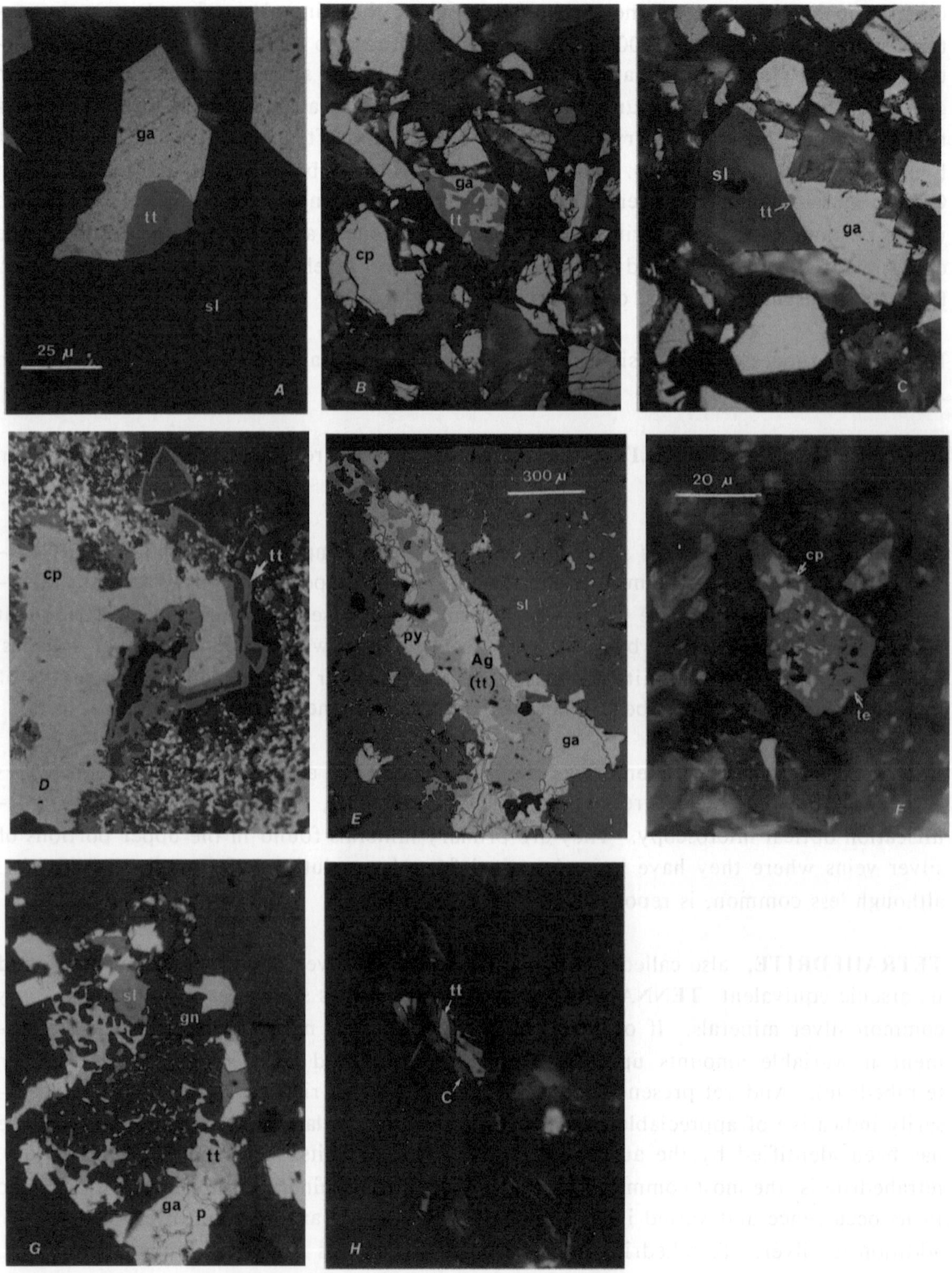

Fig. 38. *Occurrences of tetrahedrite. A, B and C: associations with galena, characterized by different grain sizes. D: coatings over chalcopyrite. E: composite grain with galena and pyrite, enclosed in sphalerite. F: relatively small grain with small inclusions of chalcopyrite. G: fine particles exsolving from and attached to sphalerite. H: fine particles associated with carbon.*

cent) while, in veins from the central part, the tetrahedrite is low in mercury or mercury-free. Figure 38 shows some common occurrences of the mineral.

POLYBASITE, PERCEITE, and **STEPHANITE** are other relatively common sulfosalts of antimony/arsenic with and without copper.

STROMEYERITE is a mineral frequently found in deposits rich in copper. It may be of hypogene and supergene origin. If supergene, it is usually associated with other secondary copper minerals.

CERARGYRITE, or **chlorargyrite** or *horn silver* **AND THE OTHER SILVER HALIDES (bromyrite, iodyrite, embolite, etc.)** are secondary products occurring in the altered or weathered portions of many silver deposits, particularly those in arid regions. Although common throughout the world, the silver halides are seldom found in Canadian localities.

ARGENTOJAROSITE AND ARGENTIAN PLUMBOJAROSITE are supergene silver minerals resulting from the oxidation under mildly acid conditions of deposits containing pyrite, along with the silver mineralization. Argentojarosite is therefore found in oxidized ores of simple mineralogical composition, where the predominant opaque minerals are oxides of iron secondary after pyrite. Argentian plumbojarosite is found in oxidized ores which originally consisted of pyrite and lead minerals.

When in this type of oxidized environments, minor amounts of silver are often detected within portions of the secondary iron oxide. Because usually variable, it is possible that such silver is not in solid solution in the iron oxide, but in discrete particles of a separate mineral, such as native silver for example, which is too fine to be resolved optically or by the average SEM magnifications.

MANGANESE MINERALS of which silver is an essential constituent, are **aurorite** (in oxidized environments) and **samsonite**, these however are relatively scarce when compared to those in which silver occurs in minor/variable amounts. Manganese minerals associated with silver enrichments, are more often represented by a large number of oxides, and lesser silicates, carbonates and chlorides, in whose lattice small quantities of silver occur distributed in solid solution. These minerals are complex to study as well as to treat metallurgically. Difficulties in the microanalytical determinations are related to their extremely large variety in chemistry, which makes the individual positive identifications sometimes impossible. Although a number of oxides/hydroxides, namely, pyrolusite, polianite, manganite, pyrochroite, backstromite, psilomelane, are defined, a much larger number of compounds can be recognized during the microanalytical studies. These are characterized by different reflectivities (from medium, to low), iso/anisotropism, and by a number of minor elements some of which are lead, zinc, copper chlorine, calcium and potassium (in addition to silver), all present in variable amounts. Presence and variation of any of these elements changes their appearance radically. The occurrence and distribution of many manganese minerals indicates a chemical deposition characterized by successive stages which manifest themselves in layered or concretionary structures. Variations in mineralogical compositions can sometimes be observed within less than 100 microns (figure 39). Silver solid solutions in these minerals are not immediately detected by a routine EDS-SEM analysis, because they are represented by amounts of 3 per cent or less. If a microprobe is not available, strong correlations between manganese and silver values (as obtained by wet chemical analyses) and absence of separate silver minerals, are usually sufficient to confirm the existence of solid solutions in this type of occurrence.

The term *wad* is used to indicate a mixture of a number of predominantly manganese oxides, with or without silver, and containing variable amounts of water.

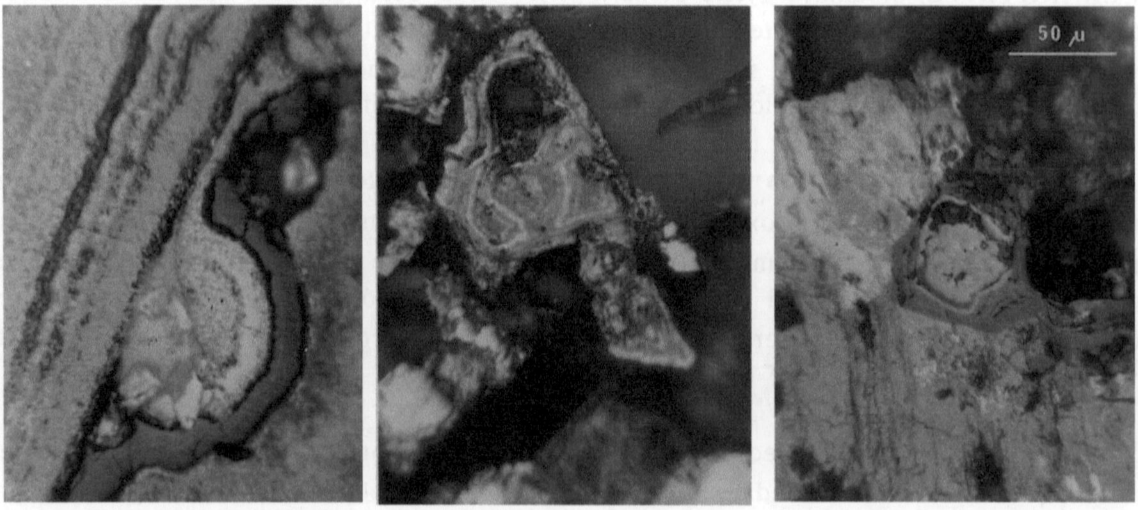

Fig. 39. *Silver-bearing manganese oxides showing concretionary structures and variations in optical properties, due to variations in minor element contents.*

Some of the less common silver-bearing minerals are shown in figure 40.

OTHER MINERALS IN WHICH SILVER IS A MINOR CONSTITUENT. While many of the minerals bearing silver in major or essential amounts are clearly defined in the literature, information on minerals bearing the element in minor/variable amounts is not always accurate. Electron microprobe and EDS-SEM determinations indicate that silver is definitely in solid solution replacing copper and gold in many sulfides, sulfosalts, tellurides and selenides. Secondary copper minerals such as bornite, chalcocite and covellite may contain silver in amounts up to 5 per cent. Silver probably also occurs in chalcopyrite, but usually in lower quantities and may be more difficult to detect. A type of chalcopyrite, characterized by advanced tarnishing, is reported as silver-bearing by several researchers including Riley (1974) and Knights (1982) in ores from Mount Isa in Australia and by Ilkka Tuoko (1982, personal communication) in ores from Finland. The chalcopyrite was as-

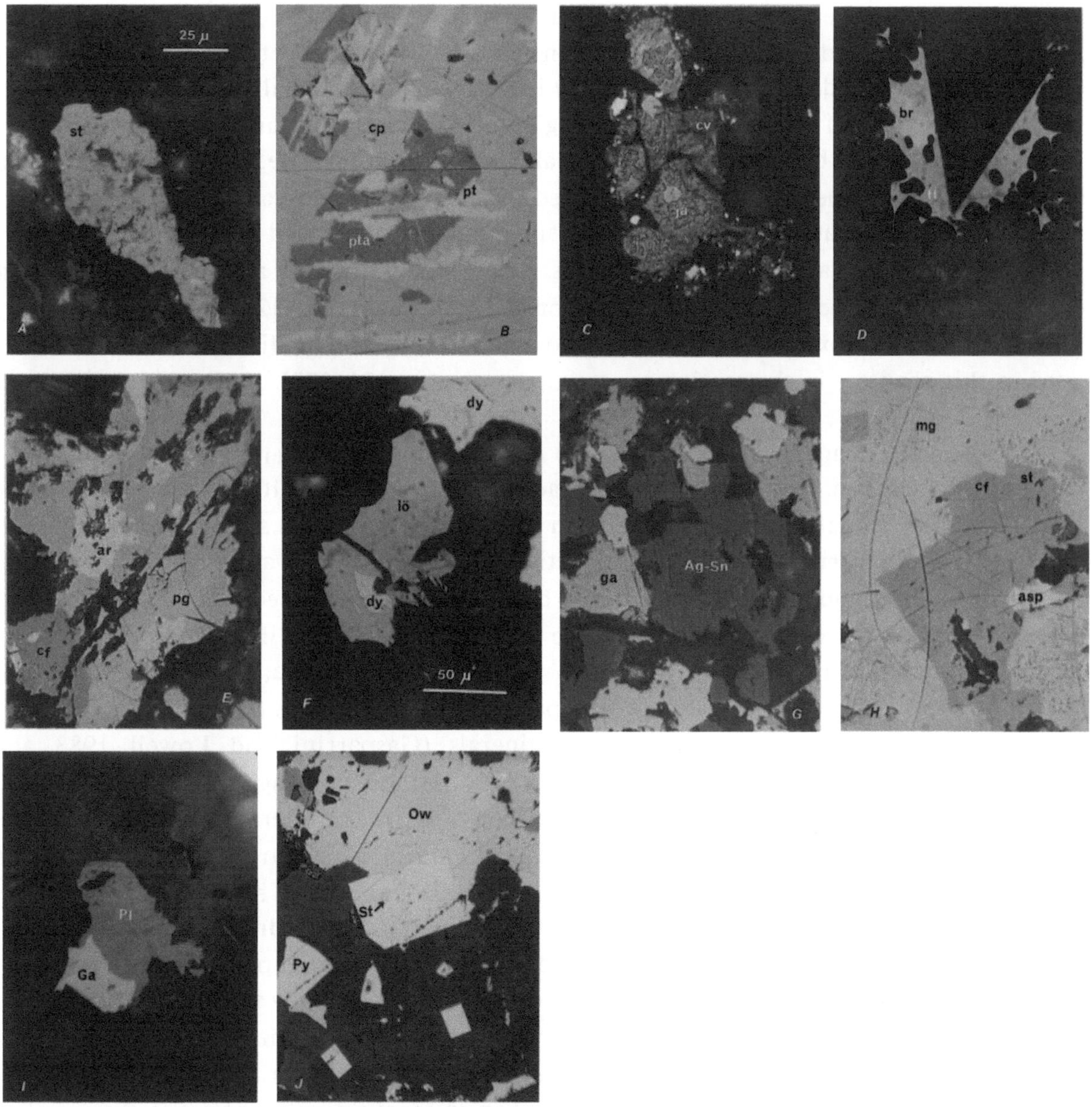

Fig. 40. *Some of the less common silver-bearing minerals. A: stephanite in silicate. B: argentian pentlandite and pentlandite exsolving from chalcopyrite. C: jalpaite coated bv covellite. D: bournonite in silicate. E: aramayoite and canfieldite with pyrargyrite. F: dycrasite with lollingite. G: stannite IV. H: miargyrite and canfieldite with arsenopyrite and stannite. I: polybasite with galena. J: owyheete and very minor stephanite with pyrite.*

sumed to be silver bearing because, when analyzed in the microprobe, gave silver counts above the background. Chen et al (1980) on the other hand, report that the tarnishing on some chalcopyrite and tennantite which they investigated, was actually a silver sulfide coating resulting from diffusion phenomena originated from adjacent **unstable** silver minerals. It was not produced by a silver-bearing chalcopyrite or tennantite. It is possible that the silver-bearing tarnished chalcopyrite reported by the other investigators is a similar silver-free chalcopyrite, coated by a thin film of silver sulfide developed from the adjacent silver minerals. The tarnishing and coatings reported by Chen et al are similar to those described for the electrum in the chapter on gold on pages 122-123. Silver sulfide tarnishing may affect extractions of silver and other metals and is discussed in more detail later in this chapter, in the section on metallurgy (pages 171-172).

As already noted (page 151), in addition to copper and gold, silver is reported to replace elements such as iron, manganese, zinc, palladium, nickel and cobalt. Silver is also reported in solid solution in galena, probably as often as gold is reported in solid solution in pyrite. In the author's experience, silver solid solutions in minerals such as sphalerite, pyrite, or even galena, in amounts within the limits of detection of the electron microprobe (approximately 0.01 per cent, when using some of the more accurate instruments and a competent operator) are unusual. Often, when associated with galena, a large proportion, if not all of the metal, occurs in discrete fine particles of tetrahedrite (figure 41), or of other minerals (Gasparrini and Lowell 1983 and 1985). Because average silver contents of galena ores are considerably higher than those of the gold of most pyrite ores (by factors of up to 100), one would expect easier detection of the separate tetrahedrite particles, by an accurate optical microscope study. Yet, although low concentrations may affect the study of gold occurrences, it is the small size and similar optical properties of the tetrahedrite to galena (when in finely divided particles that are less than 10 microns in diameter), which elude discovery. Particularly so, when using the lower microscope magnifications of 100x or 200x.

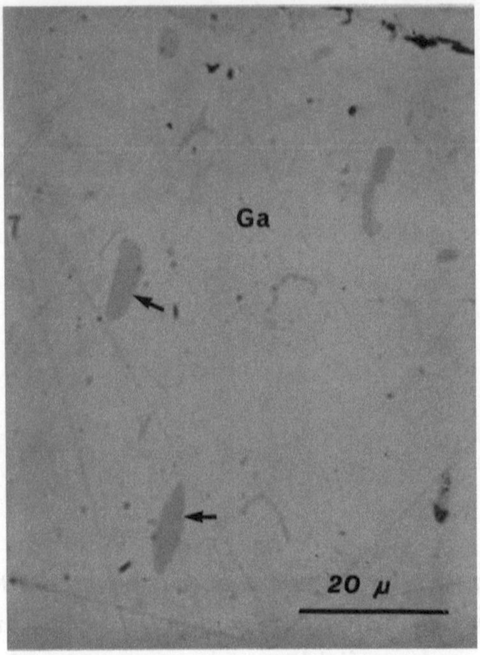

Similar considerations to those made for the gold solid solutions in sulfides, as opposed to finely disseminated particles (page 108), apply

Fig. 41. *Fine inclusions of tetrahedrite in galena photographed by using high power magnification (1250x). When observed at lower magnifications, these inclusions can easily be overlooked, particularly if the polished surface is poor.*

to the silver solid solutions in galena, for those ores where silver is exploited as a primary metal.

There is a limited amount of available information on the remaining minerals, which can be used for the understanding of the silver occurrence, when viewed in relation to the process of extraction. It would, for example, be useful to know any variations in silver mineralogy which can be related to the occurrence within the deposit (margins, depth, surface, etc.), or to the general ore chemistry. Theoretical data, such as crystallographic structures, optical properties and common world occurrences are available in most textbooks and tables, for example, Schouten 1962, Dana 1963, Picot and Johan 1977, Kuhnel, Prins and Roorda 1980, Ramdohr 1980. Because of no immediate interest for the purposes of this book, such data are not discussed in any more detail.

Descriptions of silver minerals, including many tellurides, as well as other less common compounds, are also provided by Boyle (1968).

Grain size

Due to the higher grades of many economic deposits, when compared to those of gold, silver-bearing minerals occur in particles of coarser average sizes than those of the native gold. These are usually in a range between 30 and 100 microns. Grades between a fraction of a percentage and 1-2 per cent are coupled with coarser grain sizes up to 1 mm or more. Finer sizes, in the range of 1 to 10 microns, occur, but except for special situations where they account for a very large portion of the total silver in the ore (see pages 168-169 and figure 43A), they are uneconomical and of no metallurgical interest.

Grain shapes are irregular, crystalline forms being unusual.

A relation is sometimes observed between grain size and silver content, in minerals which contain silver in variable amounts, for example, tetrahedrite.

Case study 16. Grain size/silver content of tetrahedrite. A microanalytical study conducted on samples from a silver-copper ore in British Columbia, revealed that the grain size of the tetrahedrite was inversely proportional to its silver content with the finer sizes bearing greater amounts of silver.

Boorman et al (1982) report an opposite trend in the New Brunswick lead-zinc-copper-silver ores where the larger sizes of the tetrahedrite are characterized by greater silver contents.

Host minerals

Silver is commonly associated with lead, copper, zinc and gold minerals, such as galena, chalcopyrite, secondary copper minerals, sphalerite, various copper and lead sulfosalts, complex sulfosalts with arsenic, antimony, lead and bismuth, electrum, native gold and gold tellurides. High silver concentrations may be found associated with nickel and cobalt arsenides.

Silver-bearing minerals also occur in predominantly pyritic ores, but in many such occurrences, their concentrations are too minor and liberation would be too expensive for the deposit to be of economic value.

Mineral associations

The type of association of the silver with the host mineral varies: galena-tetrahedrite associations are characterized by intergrowths and totally enclosed particles, depending on the galena to tetrahedrite ratio (figures 38A, B and C on page 156). Associations with other host minerals are largely dependent on the general textural features of the ore. For example, in a fractured ore, the silver minerals may be distributed in fractures, in a compact ore, the silver minerals may form totally enclosed particles. The basic associations, for the smaller size particles, are similar to those described for the gold minerals.

Mineral stability

As discussed on pages 42-44, many of the silver minerals are unstable, which means that they tend to decompose after the grinding process. Such property affects the recoveries of the silver itself, as well as those of the other metals. Unstable minerals are, in the author's experience, native silver, electrum, acanthite, and several less common others. Very little information is available on the degree of instability of each mineral and on the actual processes involved in these chemical changes.

METHOD OF STUDY

Microanalytical studies of gold ores are time consuming because of the scarcity and small size of the particles which need to be located on the surface of the polished mounts. Microanalytical studies of silver ores are complex, as well as time consuming, because of the large number of minerals in which the element is commonly concentrated, and which need to be identified. Particular attention and accuracy are required because:

- Silver occurs in major and minor amounts in several minerals within the same deposit. All these minerals must be correctly identified, during any study conducted with the purpose of developing an efficient recovery strategy.
- Silver minerals display optical properties which are not as distinctive as those of the gold minerals. Their studies therefore require a considerably higher amount of mental concentration.

- Minor amounts of silver (less than 5 per cent) are not readily detected by the ore microscope alone and are difficult to detect by the SEM, used for qualitative determinations. The microprobe is the best instrument for such determinations. Use of the microprobe takes additional time and requires more specialized expertise than use of the ore microscope and SEM only.

- In a particular ore, silver may occur in minerals which do not commonly contain the metal, for example, secondary copper minerals such as bornite, chalcocite and covellite. Direct microprobe analyses of all of the sulfide/sulfosalt and other ore minerals are the only means for the determination of the correct silver distributions. Again, more time and specialized expertise are necessary.

- A commonly silver-bearing mineral such as tetrahedrite does not always carry portion of the silver in the ore under study. Direct analyses provide the only positive answers. Again, the need for more time and specialized expertise.

Based on the above observations, a study of a silver ore may include the following:

- Detailed microscopic examinations of polished mounts of the ore, to locate and photograph all the grains with the optical properties of the silver-bearing minerals and of any minerals which are likely to contain silver in minor amounts. This study does not require the preparation of as many polished mounts as were recommended for the study of the gold ores (six per sample), because silver grades are usually 10 to 100 times as high as those of gold. Two to four mounts per sample are sufficient in most situations for a good indication of the general silver mineralogy. The study is not usually as lengthy, because, due to the higher silver concentrations, no tedious searches similar to those conducted during the examinations of gold ores, are necessary. However, the location and tentative identification of all the silver minerals, the determination of their grain size, identification of the host minerals and definition of the type of association, still need to be completed accurately.

- Qualitative determinations of the silver and host minerals located by the optical microscope. This study is best conducted by the SEM, rather than by the microprobe, because of the ease with which fine intergrowths and mineral inclusions are resolved and displayed on the oscilloscope screen. Determinations of the mineral chemistry take longer when studying silver ores than when studying gold ores, because of the larger number of mineral particles which need to be analyzed in each polished mount.

- Quantitative determinations of the silver content of the major sulfides and sulfosalts which are suspected to contain silver in minor amounts, such as copper minerals, sulfosalts of copper with lead, bismuth, mercury, antimony and arsenic. Quantitative chemical analyses of some of the silver-bearing minerals in which silver is present as an essential constituent, to confirm their identifications. This study is best conducted by an electron microprobe with crystal spectrometers, because of its lower limits of detection and higher accuracy of analysis. This phase of the study is not necessary when examining gold ores, unless accurate determinations of minor elements such as mercury, arsenic, etc., are needed (page 95).

- Identifications of all the silver-bearing and other minerals, based on their optical properties and chemistry.

Microprobe determinations of silver (like those of other elements) are traditionally performed by comparison of the counts emitted by the grain under study at the Ag L alpha emission wavelength (or peak counts), with the counts emitted at the same wavelength by a silver-bearing standard of known composition, and by subtracting background counts collected from the sample and standard. When dealing with minor amounts of silver, of 1 per cent or less, electron microprobe background determinations are a delicate operation, particularly so when analyzing minerals of large mean atomic number, such as galena or lead and bismuth sulfosalts: the background of such minerals is normally high, because of the matrix effect caused by the lead and bismuth. Consequently high background values may be interpreted as peak counts.

In addition to the problems associated with the high counts, if using an ADP diffracting crystal or other crystal with a similar wavelength range, the shape of the spectrum at the Ag L alpha emission wavelength is such that background determinations require the collection of counts on both sides of the peak. Unless accurately performed, these determinations may give incorrect values which again may suggest the existence of peak counts.

Due to the above reasons, accurate silver determinations, when dealing with values in the range of 1 per cent or lower in minerals of high mean atomic number, such as galena, may not always be successful by the standard techniques. An alternate technique could involve the collection of silver peak counts on the galena in question, and on a standard of similar composition, and then the use of the counts emitted by a silver-free galena as background for both sample and standard, without adjusting the spectrometer peak position.

METALLURGY

Only about 15 to 25 per cent of the silver produced in the world comes from deposits exploited primarily for silver. The remaining 75 to 85 per cent is recovered as a byproduct of gold and several base metals. If a byproduct of copper, silver is separated with the copper by smelting and then recovered as a silver (or silver-gold) anode mud after electrolysis. If a byproduct of lead, or lead-zinc ores, silver is recovered with the lead bullion. The pure metal is produced from the crude by-product either by electrolysis or by chemical methods (Smith 1988, also refer to the section on lead metallurgy on page 260).

Byproduct silver losses, or distributions with more than one product may occur during the beneficiation stages when preparing the copper, lead and zinc concentrates. Due to variations in the chemical composition of the prevalent silver-bearing minerals or in their grain sizes, the silver may report with more than one product, for example copper and zinc, (as in case study 5, pages 30-31), or copper and lead (as in case study 17, following page). Some silver may remain in the tails. In these situations, microanalytical studies are conducted to determine the reasons for the differential separation and to assist in the development of a more suitable approach to the preparation of the concentrates, for example, use of an agent capable of floating all of the silver minerals with one product, or use of finer grinding sizes.

Routine recovery techniques used for those ores where silver is either the primary metal, or a byproduct of gold, are presently the same as, or very similar to those applied for the recovery of gold, and include flotation with or without amalgamation, then cyanidation.

Cyanidation is carried out through very similar processes to the ones described for the gold ores (page 119). Where native silver is the most abundant silver carrier, the reactions with the cyanide and with the zinc dust are essentially the same. The silver precipitate is then filtered, melted and cast into bars. If silver is associated with gold, separation from the dore' is done by parting or by electrolysis.

Sterling silver, the highest silver product produced by these processes, consists of 92.5 per cent silver alloyed with another metal functioning as a hardening agent. Most common alloying metal is copper (Smith 1988).

When exploiting ores where silver is the primary product, microanalytical studies prior to and at the same time as the process of recovery are essential, much more so than when silver is a byproduct.

Of the different variables affecting silver recoveries, silver minerals and mineral instability represent the principal sources of complication. The problems derive mainly from the complexity of the silver mineralogy and from the ease with which many of the minerals alter to secondary compounds, following the grinding process.

Rock types hosting silver

Rock types hosting silver are critical during cyanidation, as they determine the choice of the leaching technique, for example heap leaching, or more accurate techniques, such as agitation, pressure leaching, etc. Similar considerations to those made when discussing gold (pages 120-121) apply to silver, keeping however in mind that the lower silver market value does not always justify the high expenditures involved in the exploitation of gold ores.

Silver-bearing minerals

The large number of existing silver minerals, and the silver distributions in several mineral phases within the same ore (Bajon et al 1982, Boorman et al 1982, Chen and Petruck 1980, Gasparrini 1984-A), sometimes as many as ten, are two important factors to keep in mind when planning or developing techniques for silver recovery. Different minerals may react differently to the same flotation or leaching agent. If silver is concentrated in similar amounts in six minerals all of which are exposed or liberated, and if three of these minerals do not respond to a particular reagent, the amount of silver separated by that reagent will only be 50 per cent of the total.

NATIVE SILVER, ELECTRUM, ACANTHITE, CERARGYRITE AND SECONDARY COPPER MINERALS. All of these minerals, when treated in laboratory tests, respond to the standard flotation techniques, if they form grains of large enough size for liberation by grinding. When in copper-lead-zinc ores, native silver, electrum and acanthite normally report with the copper concentrates, because of their similar flotation properties, and also because, when unliberated, they are more often associated with the chalcopyrite than with the galena and sphalerite.

These minerals, again in laboratory tests, also respond to similar leaching techniques to those used for native gold. If in grains of adequate size, native silver and electrum are recovered by amalgamation.

Although very responsive in laboratory tests, native silver, electrum and acanthite are not always as amenable to those treatments during routine extractions from ores, due to their instability which is responsible for chemical changes, even under normal pressure-temperature conditions. Mineral instability for electrum was discussed in the chapter on gold (pages 122-123), and is discussed again in this chapter on pages 171-172.

TETRAHEDRITE- TENNANTITE. These, and particularly so the **tetrahedrite**, are complex minerals because of their variable silver contents and grain sizes.

Case study 17. Products with which silver-bearing tetrahedrite reports. A microanalytical study of a silver ore rich in tetrahedrite found that although most of the mineral was associated with the galena, after grinding and flotation, the larger liberated particles were reporting with the copper concentrates (figure 42A). The finer ones intergrown with or totally/partially enclosed in galena were reporting with the lead concentrates (figure 42B). The few very fine particles associated with pyrite were being lost in the tails (figure 42C).

Boorman et al (1982) found a correlation between silver content and final product in which the silver was being recovered in the New Brunswick lead-zinc-copper-silver ores: the grains containing higher silver (11.37 per cent average) were reporting with the copper concentrates, the grains containing lower silver (5.25 per cent average) were reporting with the lead concentrates, while the grains with the lowest silver content were lost in the tails. The same study also revealed that floatability and cyanide solubility of the tetrahedrite increase with increased silver contents.

Case study 16 on page 161, the studies by Boorman et al, and some of the observations by Boyle (1968, page 155-157), indicate the existence of correlations between silver content of the tetrahedrite and:

- Occurrence within the deposit (surface, depth, margins, etc.).
- Grain size.
- Floatability/leachability.

Knowledge of these correlations could be used for the planning of efficient silver recoveries. For example, accurate microanalytical studies of the silver-bearing tetrahedrite could lead to the development of sophisticated techniques where the mineral is differentially floated or leached according to its occurrence within the deposit and to its grain size. All the attention could be focused on the deposit location(s) and tetrahedrite size bearing the highest silver, while the remainder would be ignored, not only because of the lower silver content, but because of the poor response to the recovery agents.

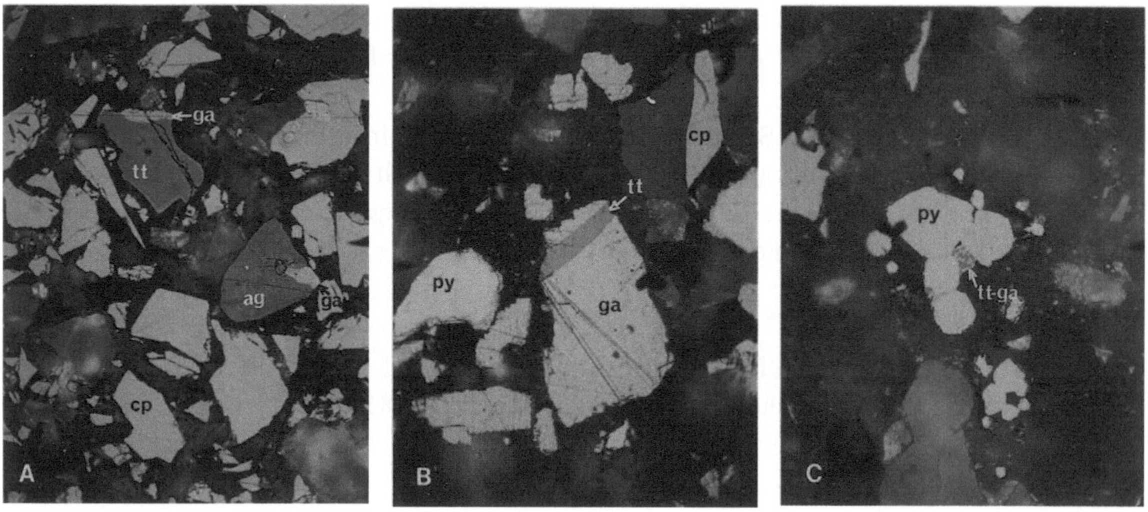

Fig. 42. *Tetrahedrite differentially distributed among the chalcopyrite concentrate (A), the galena concentrate (B) and the tails (C), as a result of a variation in grain size.*

MANGANESE MINERALS. According to Dorr and Bosqui (1950), ores consisting of the higher oxides of manganese, where the silver occurs either in solid solution or in separate minerals, are not generally amenable to the common treatments used for silver recoveries. Manganese fouls the mercury, if amalgamation is attempted. Insoluble compounds of manganese and silver are formed, if leaching is attempted. A way of controlling this problem is by use of the *Caron process* described by Clevenger and Caron (1917). When oxidized ores containing insoluble compounds of manganese and silver are heated in a reduced atmosphere, the higher manganese oxides are converted into manganous oxides. If cooled so as to prevent oxidation, such insoluble compounds become amenable to cyanidation.

Perhaps it should be remembered that, because manganese-silver concentrations tend to occur in the oxidized zone only of many ores (page 153), lack of response to the conventional techniques may often be limited to the silver occurring at the surface of the deposit.

A method developed by Scheiner et al (1973) to recover silver from ores consisting of manganese oxides and carbonates, iron oxides, silver sulfides and argentojarosite from the Candelaria District in Nevada and the Round Mountain District in Colorado, involves leaching with a sulfurous acid-sodium chloride system. According to the experimenters, the method improved recoveries by factors between 9 and 61 per cent, leading to a final silver extraction of 80 to 85 per cent.

Tables 10 and 11 summarize flotation and cyanidation data obtained by Dorr and Bosqui (1950) in laboratory tests on relatively pure samples of the minerals discussed above and of other common minerals such as proustite, pyrargyrite, stephanite and polybasite. The tests did not take into account the instability of some of the minerals and were only conducted on a small number of compounds. The data are far from comprehensive.

OTHER SILVER MINERALS. Unfortunately very little information is available on the solubility and other metallurgical properties of the majority of the silver minerals. Although less common than the ones discussed, many of these minerals occur and may be important constituents of some deposits. The necessity for systematic research in the field becomes apparent.

A compilation of the productive silver mines in Mexico, of their mineralogy and methods of treatment has been published by Marley and Hagni (1983)

Grain size and host minerals

Fine grain sizes are not as major a problem in silver metallurgy as they are in gold metallurgy, because, silver minerals occur in coarser grains which are easily liberated or exposed by grinding. Finer grain sizes, if present and if accounting for a small portion only of the total silver, can be left in the tails without major economic losses.

A case of difficult recovery could be the one where the silver-bearing minerals being floated show a wide range in grain sizes. If that is the case, liberated and unliberated particles may be distributed in more than one product, as we have seen for the tetrahedrite of case study 17 on page 166.

Another situation where grain sizes may complicate the process of extraction could be the one where the silver occurs predominantly in particles of fine sizes, for example exsolutions and emulsions, but in sufficiently high amounts to justify the process of recovery. Awareness of the presence of these exsolutions and identification of the host mineral(s) become then essential for the choice of an appropriate dissolving agent. Figure 43 shows fine inclusions and exsolutions of silver-bearing particles in silicate and pyrite hosts, some of which can be of economic value (for example those in 43A, because of their high fre-

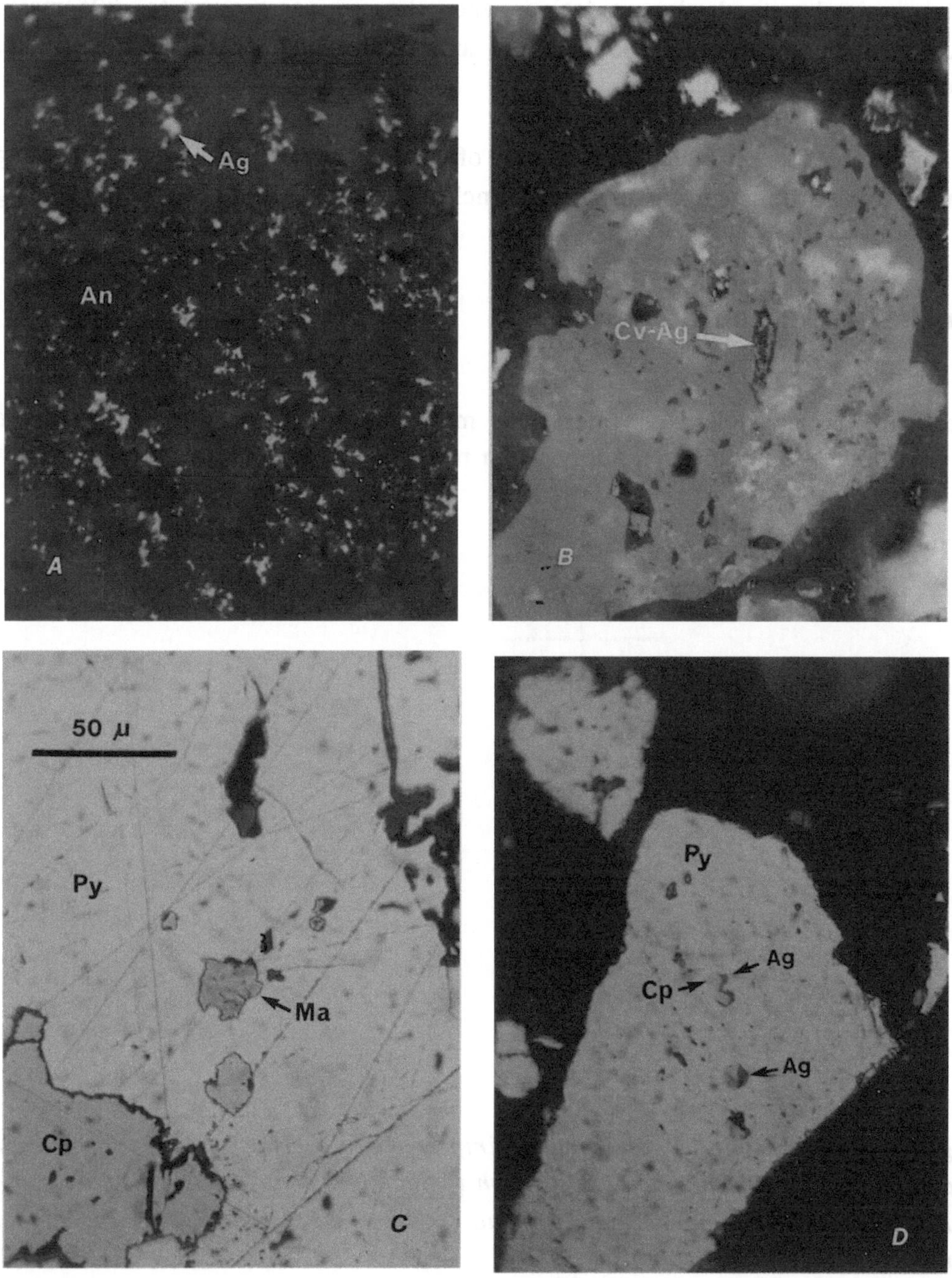

Fig. 43. *Fine particles of various silver minerals in silicate (top) and pyrite (bottom) hosts. While the silver in the top left photomicrograph is abundant and could be of economic value, the one in the other three photomicrographs is scarce and uneconomical. Host An in the top left photomicrograph is anorthite. Cv is covellite and Ma is matildite.*

quency). When dealing with economic silver of fine size, recoveries are not obtained by standard cyanidation, unless some preliminary treatment is applied to the host.

McQuiston (1985) reports that a silver mineralization from the Calico district in California and occurring in silica gangue (possibly similar to the one in figure 43A), gave 47 to

60 per cent recoveries by standard cyanidation, when the ore was ground to -325 mesh. However, recoveries improved to 75-85 per cent after *salt roasting* -10 mesh material, prior to regrinding and cyanidation.

In general, when leaching the product, occurrences of silver minerals in fractures and along grain boundaries are advantageous over totally enclosed particles, like when recovering gold.

Minerals other than the host mineral

The same problems discussed and considerations made regarding the presence of these minerals in gold ores (page 134-139), apply when recovering silver by cyanidation.

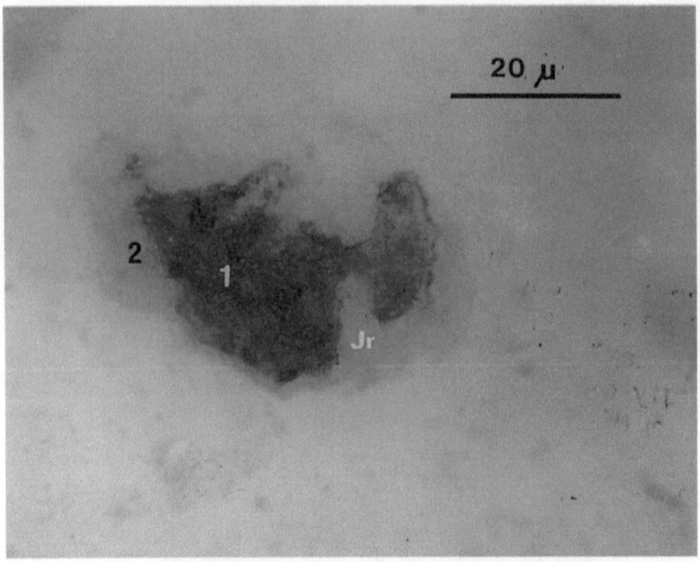

Fig. 44. *A grain of cerargyrite (1) showing an impermeable coating of clay/jarosite (2 and Jr). The coating shown in the photomicrograph is thick and clearly visible. Clay coatings are usually considerably thinner and difficult to detect, even when using the higher microscope and/or SEM magnifications.*

Case study 18. Silver minerals in ores rich in clay. A situation where the minerals other than the host mineral may affect silver recoveries in spite of the coarser sizes, is the one involving ores rich in clay. During the recoveries of these types of ores, coatings of clay over the surface of the silver minerals and possibly developed after the grinding process, prevent the cyanide solution from reaching the enclosed particle. If the coatings do develop after the grinding process, then further grinding does not help, because more coatings will quickly form over the resulting finer particles. Coatings of clay (as well as jarosite) over cerargyrite were found to be one of the causes for poor silver extractions in a gold-silver deposit in Nevada (figure 44).

These clay coatings represent a different problem from the one where the gold is invisible, because in this case the particles of cerargyrite are coarse and clearly observable. Yet, the presence of clay is just as troublesome.

Problems associated with clays are obviously varied and some of the most difficult to solve because of the often very fine sizes of the mineral of interest and/or of the coatings, which elude immediate detection and then control or elimination.

Mineral stability

Although the laboratory tests of native silver, electrum and acanthite give recoveries of nearly 100 per cent by the standard leaching or flotation techniques, routine treatments of ores containing these minerals are often accompanied by erratic results. This is a consequence of the dissolution and redeposition phenomena which cause the development of coatings over the different mineral particles in the product, through processes similar to those described by Chen et al (1980, discussed on pages 158-160), and in this book in the chapter on gold (pages 122-123). These processes may result in one of the following:

- Redeposition of a film of silver sulfide over the electrum or other silver minerals (figure 45). This does not affect the flotation or leaching properties of the enclosed mineral unless the film is mixed with some of the finer particles in the product, as described in the chapter on gold. If such mixing occurs, and if a sufficiently large number of particles is involved, the electrum or other silver mineral may acquire the flotation properties of the predominant coating particle and float with the wrong product. This results in silver losses from the final concentrates. If leaching the product, the particles may form an impermeable coating which prevents the cyanide solution from reaching the enclosed silver mineral, thus preventing its dissolution.

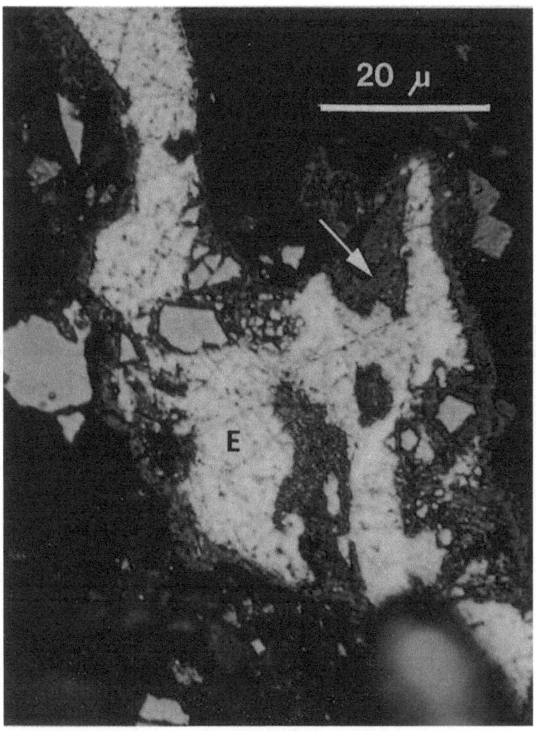

Fig. 45. *Particle of electrum coated by secondary silver sulfide.*

- Redeposition of the film of silver sulfide over particles other than those of the silver-bearing minerals, for example, galena, sphalerite or pyrite (figure 46). In such a situation, and if the silver coating is not mixed with other material, the enclosed particles may acquire the properties of the silver sulfide and float with the silver minerals, thus resulting in contaminations of the final silver concentrates.

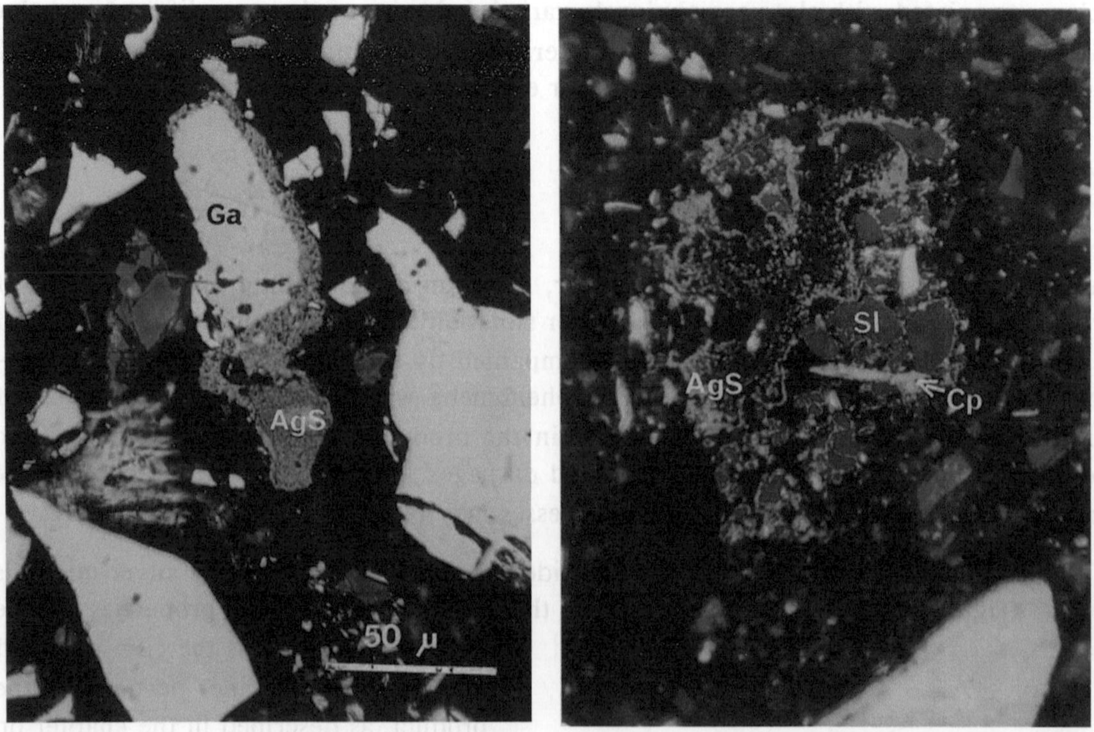

Fig. 46. *Particles of galena and sphalerite coated by secondary silver sulfide. Coatings such as these may modify the response to the flotation agent of the enclosed mineral, thereby resulting in contaminations of the final silver concentrates.*

DISCUSSION AND CONCLUSIONS

Silver ores, and consequently silver metallurgy, are complex because of:

- The relatively low concentrations of the silver.
- The large number of silver minerals existing in nature.
- The large number of silver minerals occurring within the same deposit.
- The substitutions of silver for copper and other elements in many minerals which may or may not be commonly silver-bearing.
- The instability of some of the silver minerals.

Because of such complexity, comprehensive *general* plans combining the microanalytical studies and test work, similar to the one suggested for gold in chart 11 (page 140) and per-

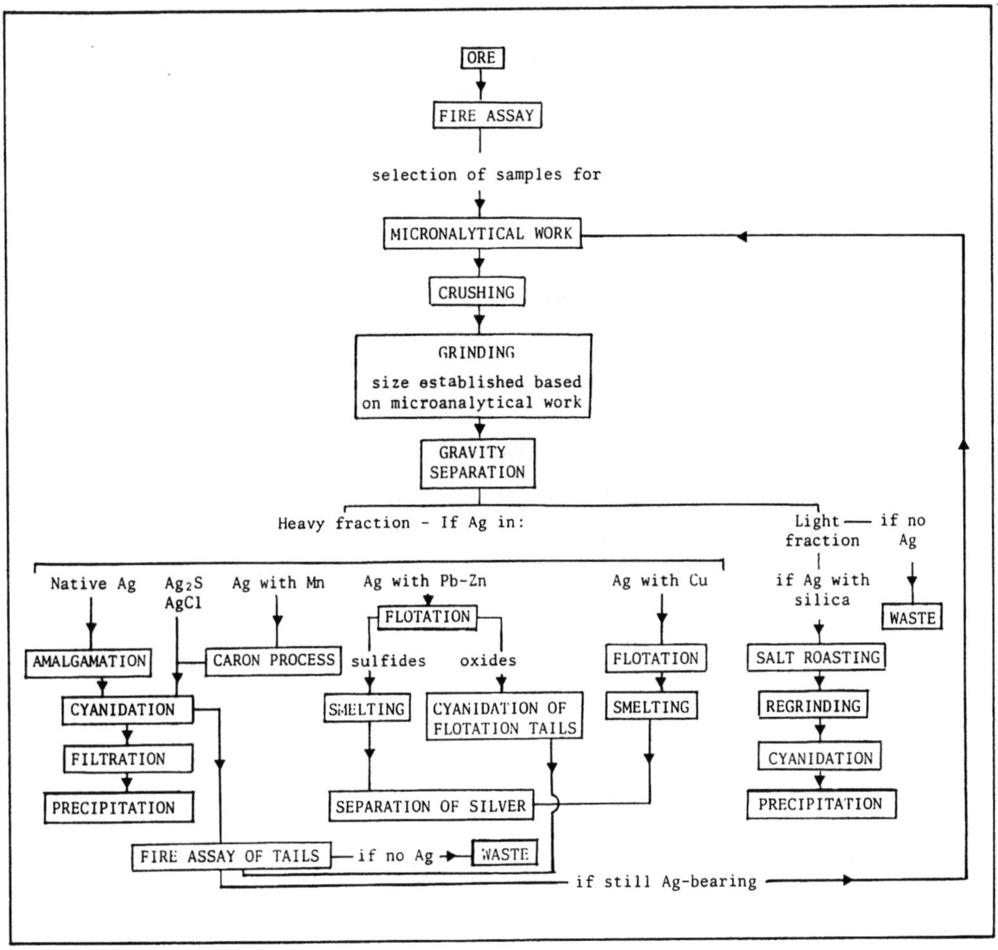

Chart 17. *Flowchart for silver recovery including microanalytical data for the more common silver minerals/occurrences.*

haps more so in chart 14 (page 143), are not as easily designed. Silver is probably the one metal for which complete microanalytical studies are necessary before any metallurgical test work can be planned or attempted, while no predictions as to its occurrence are usually possible. Such studies apply to the beneficiation stages of silver recoveries as a byproduct of the base metals, and to all the stages of silver recoveries as the primary metal or as a byproduct of gold.

Chart 17 is a *very general* flowchart outlining the different steps in the recovery of silver from the more common minerals (primary recovery) and ores (byproduct). The chart was constructed based on data from McQuiston (1985) and applies to a very broad classification of the silver ores, such as ores which are exploited primarily for silver, manganese ores, lead-zinc ores, copper ores, ores where silver is in the silicate gangue. The chart does not include all of the silver-bearing minerals because, as already observed (page 168), very little information on their metallurgical properties and treatments is available.

Case study 19. Exploitation of the high grade portions of a silver ore and suggestion of a plan, based on the data from the microanalytical study. This case study was conducted on an ore in the Western United States, which was giving poor recoveries when treated by leaching techniques. The study was conducted on two crushed diamond drill core feed composites, representative of the oxidized and unoxidized portions of the ore, and on their tailing products. Fire assay analyses for silver had indicated irregular distributions with high values up to 20 oz/ton or 684 ppm Ag, and average values, for the majority of the deposit, of less than 2 oz/ton or 68.4 ppm Ag.

The conclusions drawn from the study were:

- "The fire assay of the different portions of the deposit shows that nine intervals out of a total of 55 analyzed for silver, or 16.4 per cent, contain more than 4 oz/ton or 137 ppm Ag, thus accounting for about 61 per cent of the total silver in the deposit. An additional 7 intervals, or 12.7 per cent, contains between 2 and 4 oz/ton or between 68 and 137 ppm Ag, thus accounting for about 14 per cent of the total silver in the deposit. Therefore, about 75 per cent of the total silver is carried by less than 30 per cent of the analyzed ore. No relation is observed between silver content and degree of oxidation, silver distributions being irregular in both portions of the ore.
- Silver occurs in a large variety of minerals, more common of which are tetrahedrite, stromeyerite, secondary copper sulfides, acanthite, argentojarosite, and silver-bearing iron oxides.
- Silver content of the tetrahedrite varies between 0.5 and 15 per cent.
- Grain sizes range between a few and several 100 microns. Coarser sizes are associated with the quartz of the unoxidized portion of the ore.
- Host minerals are quartz, pyrite, secondary iron oxides, and, less commonly, sphalerite, galena and sericite.
- The unoxidized ore is generally compact. The oxidized ore shows more frequent fracturing often filled by iron oxide.
- A large amount of the silver-bearing and silver-free secondary iron oxide, supposedly soluble in cyanide, is still present in the tails.

Based on the above considerations, the following was recommended:

- Leaching of the high grade portions only, whether oxidized or unoxidized,.
- Finer grinding before leaching of the unoxidized material, for liberation of the silver-bearing particles. Therefore techniques other than heap leaching.
- Leaching of the oxidized material for longer periods without further grinding. Fracturing should promote access of the leaching agent and dissolution of the iron oxide first (barren and silver-bearing) and then of the enclosed argentojarosite and other silver minerals (assuming that these are all soluble).
- Study of the tails from the above products, if still silver bearing, to identify the insoluble silver minerals."

TABLE 7 - SILVER DEPOSITS

DEPOSIT	STRUCTURAL CHARACTER	TYPE LOCALITY	MAIN MINERALS	ASSOC. MINERALS	TYPICAL ELEMENTS	METALS MINED
Epigenetic depositions	Sedimentary rocks. Hypogene veins, lodes, stockworks, mantos	Kongsberg, Norway; Hartz Mountains and Freiberg, Germany; Jachymov, Czechoslovakia; Lurion, Greece; Keno and Galena Hills, Yukon Broken Hill and Mt.Isa, Australia; Santa Eulalia - Chihuahua, Mexico; Coer d'Alene, Idaho; Thunder Bay, Ontario; Ainsworth and Sullivan Mines, B.C.	Tetrahderite, tennantite, argentite-acanthite, pyrargyrite, proustite, (native Ag, dycrasite, electrum	Pyrite, sphalerite, chalcopyrite, Ni-Co arsenides, bismuthinite, native Bi and Au. Pitchblende. (Cassiterite and worlframite). Siderite, calcite, dolomite, quartz and barite. Fluorite and tourmaline.	Pb, Zn, Cd, Cu, Fe, In, Mn, Sb, As, S, (Sn, (Co, Ni, Bi, Se, Te)	Ag, Pb, Zn, sometimes Au
Ag-Au veins and lodes	Associated with volcanic flow-rocks	Transylvania, Rumania; Kalgoorlie, Australia; Hawaki, New Zealand; El Oro and Guanajuelo, Mexico; Black Mountains, Arizona; De Lamar, Idaho, Rawhide and Comstock, Nevada; Cripple Creek, Colorado; Yellowknife, N.W.T.	Acanthite, proustite, pyrargyrite, miargyrite, stephanite, polybasite, tetrahedrite, tellurides	Pyrite, arsenopyrite galena, sphalerite, chalcopyrite. Sometimes pyrrhotite, alabandite, molybdenite, scheelite. Quartz, calcite, dolomite, ankerite, barite, some fluorite, rodochrosite, and rhodonite	Ag, Au, Te, Sb, As, Cu, Pb, Zn, Ca, Mn,Fe, Mn	Ag, Au
		Red Lake, Porcupine, Kirkland Lake, Ontario	Native Ag as an alteration product	Chlorite, sericite	K, SO2, CO2, S less frequent Mo, W, Se, Ba, F	
Polymetallic veins	Cu-Pb-Zn-Fe sulfides	In mineral belts all over the world	Tetrahedrite, Ag sulfides-sulfosalts, native Ag	Pyrite, chalcopyrite, galena, sphalerite	Fe, Cu, Pb, Zn	Cu, Pb, Zn. Ag from metall. treatment
Complex geological environments	Sediments, volcanics and igneous granitized rocks	Potosi, Bolivia	Native Ag, cerargyrite, argentite, pyrargyrite-proustite	Smaltite-chloanthite, Ag, Sn loellingite, safflorite, rammelsbergite, skutterudite,	Ag, Sn	Ag, Sn

TABLE 7 - CONT.

DEPOSIT	STRUCTURAL CHARACTER	TYPE LOCALITY	MAIN MINERALS	ASSOC. MINERALS	TYPICAL ELEMENTS	METALS MINED
				arsenopyrite		
Complex geological environments		Cobalt, Ontario	Native Ag , dycrasite, argentite, tetrahedrite, stephanite, pyrargyrite-proustite, stromeyerite	Gersdorffite, cobaltite, glaucodot, niccolite, brauthauptite, cassiterite. Pyrite, chalcopyrite, sphalerite, galena. Native Bi, pentlandite, bismuthinite	Co, Ni, Cu, Zn Pb, Bi, As, S	Ag, Co, Cu, Ni, As, Bi
		Grat Bear Lake, N.W.T	Native Ag	Erytrite, galena, annabergite, scorodite, Mn oxides (wad), chalcopyrite, limonite. Quartz, hematite, calcite, pitchblende, secondary U, azurite, malachite	Ni, Co, As, S, Bi, Cu, Zn, Pb	Ag, U
Lead-zinc deposits	Carbonate rocks	Pine Point-Mississippi Valley type	Tetrahdrite (?)	Galena, sphalerite	Pb, Zn, Ag	Pb, Zn, (Ag)
Native copper deposits	Basalts, conglomerates, sandstones, etc.	Michigan Copper, Michigan	Native Ag , native copper	Native copper	Cu, Ag	Cu, Ag
Porphyry copper deposits	Porphyry coppers	Butte, Montana; Bingham, Utah; Ely, Nevada; Ajo, San Manuel and Ray, Arizona	Tetrahdrite-tennantite, stromeyerite, secondary copper minerals, enargite(?)	Chalcopyrite, bornite, covellite, enargite	Cu, Ag	Cu, Ag
Gold placers	Any gold placer	Any gold placer (if silver bearing)	Native Au,	Monazite, ilmenite,	Au, Ag, RE's	Au, Ag, RE's

TABLE 7 - CONT.

DEPOSIT	STRUCTURAL CHARACTER	TYPE LOCALITY	MAIN MINERALS	ASSOC. MINERALS	TYPICAL ELEMENTS	METALS MINED
			electrum	rutile, zircon, quartz, other silicates, hematite-magnetite	Ti, Fe	
Red bed sandstones	Cu-Ag sandstones Carbonaceous sandstones	Sandstones of Nova Scotia and New Brunswick; Corocoro, Bolivia; Kazakhstan, USSR	Acanthite, tetrahedrite-tennantite, stromeyerite, native Ag, betechtinite, pyrargyrite-proustite	Chalcopyrite, bornite, chalcocite, domeyckite, native Cu, azurite, malachite	Cu,Ag,Pb,Zn,Cd,Fe,Mn, Co,Ni,Cr,P,Mo,RE's, Bi,As,Se,S,Ba,Cl,F	Cu, Ag
Red Bed sandstones	U-Cu-V sandstones	U-V deposits of S.W. USA	Cerargyrite and the ones above	Uraninite, carnotite, vanadinite and those above	U, V and those above	U, V, Cu, Ag
Red Bed Sandstones	Pb-Zn sandstones	Laisvall, Sweden	Tetrahderite (?)	Galena, sphalerite, cerussite, pyromorphite, anglesite, Barite, calcite, fluorite, gypsum, quartz, pyrite, siderite, marcasite. Celestite may occur in all three types of deposits	Pb, Zn,	Pb, Zn, sometimes Ag
Shale deposits	Shales	Kupferschier, Germany-Poland; Zambian Copperbelt; El Boleo, Mexico (rich in Cl); White Pine, Michigan; Creta Area, Oklahoma; Redstone River, N.W.T.	Native Ag, argentite-acanthite, tetrahedrite, cerargyrite, secondary, Cu minerals	Pyrite, chalcopyrite, bornite, chalcocite, galena, sphalerite, native Cu, molybdenite. Barite, calcite, minor quartz, feldspar	Cu, Pb, Zn, Cd, Mo, Co, Ni, Fe, Mn, As, Sb, S	Cu, sometimes Ag, Pb and Zn
Ag bearing skarns	Skarns	Ammeberg, Sweden;	Tetrahedrite, argentite-	Pyrite, pyrrhotite,	Pb, Zn, Cu, Fe	Pb, Zn, sometimes Ag

TABLE 7 - CONT.

DEPOSIT	STRUCTURAL CHARACTER	TYPE LOCALITY	MAIN MINERALS	ASSOC. MINERALS	TYPICAL ELEMENTS	METALS MINED
		Chihuahua and Zactecas, Mexico; Western USA; Tetreault, Quebec	acanthite, polybasite, electrum	sphalerite, galena, chalcopyrite, bornite, molybdenite. Hematite-magnetite. Calcite, feldspar, tremolite, garnet, epidote, vesuvianite, dipside, scapolite, wallastonite, dolomite		
Eluvial and alluvial deposits	Appropriately mineralized areas	Guadalcazar, Mexico	Oxidized products of Ag	Galena, anglesite, bindheimite, cassiterite, cinnabar	Pb, Au, Sn, Hg	Pb, Ag, Au. Sn and Hg could exploited in the future
Sudbury type deposits	Astrobleme or layered intrusion	Sudbury, Ontario	Ag pentlandite	Pyrite, chalcopyrite, pentlandite, millerite, heazlewoodite	Cu, Ni	Cu, Ni, Ag
Fe-Cu-Zn deposits	Massive sulfides	Rammelsberg, Germany; Cerro de Pasco, Peru; Noranda, Quebec; Flin, Flon, Manitoba; Geco, Ontario; Willroy, Ontario; Bathurst, N.B.	Tetrahdrite, argentite-acanthite, native Ag, electrum, native Au, sylvanite, altaite,	Pyrite, pyrrhotite, enargyte, galena, marcasite, arseno-pyrite, bournonite, gratonite, emplectite, bismuthinite, realgar. Quartz, barite, sericite, alunite, clay minerals	Cu, Pb, Zn, Au Fe, As, Bi, Te	Cu, Pb, Zn, Ag
			Stromeyerite, Ag tellurides, aramoyite, polybasite. Supergene Ag, cerargyrite, argentojarosite	Secondary Cu minerals		

TABLE 8 - SILVER-BEARING MINERALS

Silver essential or major constituent

MINERAL	FORMULA	COLOR	REFL.	ISO/ANISO
Acanthite	**Ag_2S**	**gray**	**low**	**aniso**
(silver glance, orthorhombic, stable below 175°C)				
Aguilarite	Ag_4SeS	dull gray	low	iso
(has light etching properties)				
Alaskaite	$Pb(Ag,Cu)_2Bi_4S_8$	gray-white	M	aniso
Allargentum	Ag_6Sb			
Andorite	$PbAgSb_3S_6$	gray-white	M	aniso
Antimonial silver	(Ag,Sb)			
Aramayoite	$Ag(Bi,Sb)S_2$	gray-white	M	aniso
Arcubisite	Ag_6CuBiS_4			aniso
Ag plumbojarosite	$(Pb,Ag)Fe_{3-6}(SO_4)_{2-4}(OH)_{6-12}$			
Argentian pentlandite	$(Fe,Ni)_8AgS_8$	yellow	M	iso
Argentite	Ag_2S	gray	low	iso
(isometric, stable above 175°C)				
Argentojarosite	**$AgFe_3(SO_4)_2(OH)_6$**	**dull gray**	**low**	**iso**
Argentopyrite	$AgFe_2S_3$	pale brown	low	aniso
Argyrodite	Ag_8GeS_6	gray	low	iso
Argyropyrite	$Ag_3Fe_7S_{11}$	pale brown	low	aniso
Aurorite	$(Ag,Ba,Ca,Mn..)Mn_3O_7\ 3H_2O$			
Balkanite	$Cu_9Ag_5HgS_9$			
Benjaminite	$Pb_2(Ag,Cu)_2Bi_4S_9$			
Berryite	$Pb_2(Cu,Ag)_3Bi_5S_{11}$			
Betechtinite	$(Cu,Fe)_{11}(Pb,Ag)S_7$			
Bezsmertnovite	$Au_4Cu(Te,Pb)$			
Bideauxite	$Pb_2AgCl_3(F,OH)_2$			
Billingsleyite	$Ag_7(As,Sb)S_6$			
Boleite	$Pb(Cu,Ag)Cl_2(OH)_2\ H_2O$			
Bromyrite	AgBr	dull gray	low	iso
(bromargyrite)				
Canfieldite	$Ag_8(Sn,Ge)S_6$	gray	M	iso
Cerargyrite	**AgCl**	**gray**	**low**	**iso**
(chlorargyrite or horn silver)				
Chilenite	Bi bearing Ag			
Cocinerite	Cu_4AgS			
Criddleite	$Ag_2Au_3TlSb_{10}S_{10}$			
Crookesite	$(Cu,Tl,Ag)_2Se$			
Diaphorite	$Ag_3Pb_2Sb_3S_8$	gray-white	M	aniso
Dycrasite	Ag_3Sb	white	high	iso
Electrum	**(Au,Ag)**	**pale yellow**	**high**	**iso**
Embolite	Ag(Br,Cl)			
Empressite	AgTe	gray	M/low	aniso
Eskimoite	$Ag_{1.75}Pb_{1.75}Bi_{3.75}S_9$			
Eucairite	AgCuSe	pale yellow	M	aniso
Fischesserite	Ag_3AuSe_2			
Fizelyite	$Ag_2Pb_5Sb_8S_{18}$	gray-white	M	aniso
Freibergite	$Cu_3(Ag,Sb)S_3$	gray	low	iso
(Ag-bearing tetrahedrite)				
Freieslebenite	$Ag_5Pb_3Sb_5S_{12}$	white	M	aniso
Friesite	Fe,Ag,S mineral. No more info available			
Gustavite	$Bi_{11}Pb_5Ag_3S_{24}$			
Hessite	Ag_2Te	gray-white	M	aniso
Hocartite	Ag_2SnFeS_4			
Huantajayite	(Na,Ag)Cl			
Hutchinsonite	$(Pb,Tl)(Cu,Ag)As_5S_{10}$	gray-white	M	aniso
Incaite	$(Pb,Ag)_4FeSn_4Sb_2S_{13}$			
Iodyrite	AgI	gray	low	iso
(iodargyrite)				
Jalpaite	Ag_3CuS_2	blue-gray	low	aniso
Kittlite	Hg-Ag-S-Cu-Se mineral. No more info available			
Krennerite	$AuAgTe_4$	white	M-high	aniso
Kustelite	(Ag,Au)	pale yellow		
Kutinaite	Cu_2AgAs			

TABLE 8 - CONT.

MINERAL	FORMULA	COLOR	REFL.	ISO/ANISO
Laffittite	$AgHgAsS_3$			
Larosite	$(Cu,Ag)_{21}(Pb,Bi)_2S_{13}$			
Lengenbachite	$Pb_6(Ag,Cu)_2As_4S_{13}$	white	M-high	iso
Liujinynite	Ag_3AuS_2			
Marrite	$PbAgAsS_3$			
Matildite	$AgBiS_2$	white	M-high	aniso
McKinstryite	$Cu_{.8+x}Ag_{1.2-x}S$			
Miargyrite	$AgSbS_2$	blue-white	M	aniso
Miersite	(Ag,Cu)I			
Moschellandsbergite	Ag_2Hg_3			
Muthmannite	(Ag,Au)Te			
Native silver	**Ag**	**white**	**high**	**iso**
Naumannite	Ag_2Se	white	M	aniso
Neyite	$Pb_7(Cu,Ag)_2Bi_6S_{17}$			
Novakite	$(Cu,Ag)_4As_3$			
Ostwaldite	colloidal-like AgCl			
Ouryite	$Ag_{12.5}Pb_{15}Bi_{20.5}S_{52}$			
Owyeeite	$Pb_5Ag_2Sb_6S_{15}$	gray-white	M	aniso
Para-schachnerite	Ag_3Hg_2			
Pavonite	$AgSb_3S_5$	gray-white	M	aniso
Penzhinite	$(Ag,Cu)_4Au(S,Se)_4$			
Perceite	**$(Ag,Cu)_{16}As_2S_{11}$**	**gray-white**	**M**	**aniso**
Petrovskaite	AuAg(S,Se)			
Petzite	Ag_3AuTe_2	white	M	iso
Polyargyrite	$Ag_{24}Sb_2S_{15}$			
Polybasite	**$(Ag,Cu)_{16}Sb_2S_{11}$**	**gray-white**	**M**	**aniso**
Proustite	**Ag_3AsS_3**	**pale blue**	**low**	**aniso**
(ruby silver, rhombohedral-hemimorphic))				
Pyrargyrite	**Ag_3SbS_3**	**pale blue**	**low**	**aniso**
(ruby silver, rhombohedral-hemimorphic)				
Pyrostilpnite (monoclinic)	Ag_3SbS_3	pale blue	low	aniso
Ramdohrite	$Ag_2Pb_3Sb_6S_{13}$			
Rathite-I	$(Pb,Tl)_3As_4(As,Ag)S_{10}$			
Sakuraiite	$(Cu,Ag)_2(In,Sn)(Zn,Fe)S_4$			
Samsonite	$Ag_4MnSb_2S_6$	blue-gray	low	iso
Schachnerite	$Ag_{1.1}Hg_{0.9}$			
Schapbachite	$AgBiS_2$	white	M-high	aniso
Schirmerite	$PbAg_4Bi_4S_9$	white	M-high	iso
Schwazite	Hg-bearing tetrahedrite			
Silver amalgam	(Ag,Hg)			
Silver jarosite	$Ag_2Fe_5(OH)_{12}(SO_4)_4$			
Smithite	$AgAsS_2$			
Stannite(?)III	$CuAgSn_2S_4$	gray	low	iso
Stannite(?)IV	Sn-Ag-Zn-bearing tetrahedrite			
Stephanite	Ag_5SbS_4	gray-white	M-low	aniso
Stetefeldtite	Ag CuO FeO Sb_2O_5 S			
Stylotypite	$(Ag,Cu,Fe)_3SbS_3$			
Stuetzite	$Ag_{5-x}Te_3$			
Sternbergite	$AgFe_2S_3$	brown	low	aniso
Stromeyerite	$Cu_{1+x}Ag_{1-x}S$	gray-white	low	aniso
Sylvanite	$(Au,Ag)Te_4$	white	high	aniso
Telargpalite	$(Pd,Ag,Pb,Bi)_{4+x}(Te,Se)$			
Tennantite	**$(Cu,Ag,Fe)_{12}As_4S_{13}$**	**gray**	**low**	**iso**
Tetrahedrite (gray copper)	**$(Cu,Ag,Fe)_{12}Sb_4S_{13}$**	**gray**	**low**	**iso**
Trechmanite	$AgAsS_2$			
Uytenbogaardtite	Ag_3AuS_2			
Vikingite	$Ag_{1.25}Pb_2Bi_{3.25}S_{7.5}$			
Volinskite	$AgBi_{1.6}Te_2$			
Weishanite	$(Au,Ag)_3Hg_2$			
Xanthoconite (monoclinic)	Ag_3AsS_3	pale blue	low	aniso

The bold characters are used to indicate the more common minerals

TABLE 9 - SILVER-BEARING MINERALS

Ag minor constituent *

MINERAL	FORMULA
Altaite	$PbTe$
Baumhauerite	$Pb_{12}As_{16}S_{36}$
Berzelianite	Cu_2Se
Bismuthinite	Bi_2S_3
Bornite	Cu_5FeS_4
Boulangerite	$Pb_{2-5}Sb_{2-4}S_{5-11}$
Bournonite	$PbCuSbS_3$
Calaverite	$AuTe_2$
Chalcocite	Cu_2S
Chalcopyrite	$CuFeS_2$
Chalcostibnite	$CuScS_2$
Cinnabar	HgS
Clausthalite	$PbSe$
Cooperite	PtS
Cosalite	$Pb_2Bi_2S_5$
Covellite	CuS
Cuprobismuthinite	$CuBiS_2$ (?)
Cylindrite	$Pb_3Sn_4Sb_2S_{14}$
Digenite	$Cu_{2-x}S$
Enargite	Cu_3AsS_4
Famatimite	$Cu_3(Sb,As)S_4$
Frankeite	$Pb_5Sn_3Sb_2S_{14}$
Galena	PbS
Geochronite	$Pb_5(Sb,As)_2S_8$
Kobellite	$Pb_2(Bi,Sb)_2S_5$
Krennerite	$AuTe_2$
Klockmannite	$CuSe$
Jamesonite	$Pb_4FeSb_6S_{14}$
Lautite	$CuAsS$
Melonite	$NiTe_2$
Meneghinite	$CuPb_{18}Sb_7S_{24}$
Montbrayite	$(Au,Sb)_2Te_3$
Nagyagite	$Pb_5Au(Te,Sb)_4S_{5-8}$
Native bismuth	Bi
Native copper	Cu
Native gold	Au
Native tellurium	Te
Palladoarsenide	Pd_2As
Penroseite	$(Ni,Cu,Pb)Se_2$
Plagionite	$Pb_5Sb_8S_{17}$
Plumbopalladinite	Pd_3Pb_2
Rozhkovite	$(Cu,Pd)_3Au_2$
Selenocosalite	$Pb_2Bi_2(S,Se)_5$
Selenokobellite	$Pb_2(Bi,Sb)_2(S,Se)_5$
Semseyite	$Pb_9Sb_8S_{21}$
Stannite	Cu_2FeSnS_4
Stibnite	Sb_2S_3
Teallite	$PbSnS_2$
Telargpalite	$(Pd,Ag,Pb,Bi)_{4+x}(Te,Se)$
Tennantite	$(Cu,Fe)_{12}As_4S_{13}$
Tetrahedrite	$(Cu,Fe)_{12}Sb_4S_{13}$
Umangite	Cu_3Se_2
Wehrlite	$Bi_{2+x}Te_{3-x}$ (?)
Wittichenite	Cu_3BiS_3
Wittite	$Bi_6Pb_5(Se,S)_{14}$
Zinkenite	$Pb_6Sb_{14}S_{27}$

* The above minerals are normally silver-free, but have all been reported to sometimes contain the element in minor amounts, in the range of a few per cent or lower.

TABLE 10 - FLOTATION CHARACTERISTICS OF THE MORE COMMON SILVER MINERALS

MINERAL	FLOTATION CHARACTERISTICS	RATIO OF CONC.	BEST RECOVERY
Native silver	Normal flotation	-	-
Electrum	Normal flotation	-	-
Argentite-acanthite	Normal flotation. Lime has little effect. Recovery is lowered by Fe oxides. Starch helps concentrate grade.	107.4:1	98.5%
Cu minerals	Normal flotation	-	-
Ag halides (cerargyrite, bromyrite, embolite, etc.)	Normal flotation. Lime has little effect. Concentrate grade is lowered by slimes, but improved by starch	25.5:1	98.8%
Proustite	Lime is deleterious. Grade but not recovery is lowered by talcose material. Starch should not be used.	12.4:1	94.5%
Pyrargyrite	Lime is deleterious. Both recovery and grade are lowered by slimes. Starch should not be used. The mineral is very sensitive to changes in flotation conditions. Na sulfide is very harmful in all cases	68.7:1	97.0%
Stephanite	Lime is deleterious. Grade is lowered by talcose slimes, but starch is effective in correcting the problem	87.5:1	94.4%

TABLE 10 - CONT.

MINERAL	FLOTATION CHARACTERISTICS	RATIO OF CONC.	BEST RECOVERY
Polybasite	Lime has little effect. Grade is lowered by talcose slimes, but starch is effective in correcting the problem	67.5:1	98.7%
Tetrahedrite-tennantite	Lime has little effect. Grade, but not recovery is lowered by talcose slimes. Starch is effective in correcting the problem. Recovery improves with the silver content of the minerals (Boorman et al.1982)	243:1	99.1%
Pb-Zn minerals enclosing fine particles of silver minerals or or containing Ag in solid solution	Lime has little effect. Grade, but not recovery is lowered by talcose slimes. Starch is effective in correcting the problem.	-	-

TABLE 11 - CYANIDATION CHARACTERISTICS OF THE MORE COMMON SILVER MINERALS

MINERAL	CYANIDATION CHARACTERISTICS	BEST RECOVERY AND DISSOLUTION TIME
Native silver Electrum Argentite-acanthite Cu minerals Silver halides	All these minerals readily dissolve in NaCN solution particularly when finely divided (15 microns in size nd finer). Coarser grains of native silver and electrum dissolve slowly and may consume too much cyanide. If the minerals are finely mixed with insoluble hosts such as pyrite, sphalerite or quartz gangue, roasting hour using a temperature up to 460°C may be necessary for a 75% extraction. Temperature control is important (particularly for argentite-acanthite): high temperature (600°C and more) may lead to the formation of insoluble products.	87 to 99 % in 72 hours for all the minerals
Proustite	These minerals do not dissolve readily in the NaCN solution. Roasting may improve recovery	42.5% in 72 hours by using a warm solution. 91% after roasting and changing solution.
Pyrargyrite		67% in 72 hours by using a warm solution. 88% after roasting and changing solution.

TABLE 11 - CONT.

MINERAL	CYANIDATION CHARACTERISTICS	BEST RECOVERY AND DISSOLUTION TIME
Stephanite		90% in 72 hours. Recovery is not improved by roasting
Polybasite		80% in 72 hours by using a warm solution. 90% after roasting
Tetrahedrite-tennantite	Neither mineral dissolves well Solubility appears to be proportional to the Ag content (Boorman et al. 1982). Solubility also improves after roasting, up to 70% from an ore giving 25% recovery when un-roasted	Variable depending on the compos-ition of the mineral
Pb-Zn minerals enclosing fine particles of Ag minerals or containing Ag in solid solution	These minerals dissolve with great difficulty. Silver is best recovered as a byproduct by smelting of the Pb-Zn ore	Not determined
Mn-Ag minerals (for example aurorite, samsonite and) Mn oxides)	These minerals dissolve with great difficulty. Ag is leached with H_2O_2 in acid solution or after reduction by roasting (see text for details on the Caron process)	Not determined

Most of the above data were obtained from laboratory tests on relatively pure samples (modified after Dorr and Bosqui 1950)

CHAPTER 8

DIFFERENCES BETWEEN GOLD AND SILVER AND THEIR EFFECTS ON RECOVERIES

There are many differences between gold and silver and several affect the decisions regarding ore exploitability first, then the choice of recovery agents and techniques. In order to understand these differences (listed in table 12), we need to go back to the variables discussed in chapter 3.

We have seen that the first and most important variable affecting the decision as to exploitability of a certain ore, is the **market value** of the metal of interest. The higher the market value of our metal, the lower the minimum economic grade of the deposit. Market value of gold is higher than that of silver by a factor of about 60. This means that lower grades can be exploited and expensive techniques are more than justified for its recovery. Consequently, economic quantities for gold can be as low as 0.02 oz/ton or 0.684 ppm Au, where the ore responds to the less costly methods of extraction. 2-3 oz/ton or 68.4-102.6 ppm Ag, unless readily recoverable, may not represent a sufficiently high amount to initiate mine production. The common **lower** gold **quantities** therefore are balanced by the higher market value.

The other differences are largely related to the microcomposition of the ore. They are important when selecting the most appropriate recovery technique for the two metals in any given deposit.

TABLE 12. DIFFERENCES BETWEEN GOLD AND SILVER

GOLD	SILVER
High market value	Lower market value
Low quantities in ores	Higher quantities in ores
Limited number of minerals	Very large number of minerals
Native gold/electrum common	Many very common
Usually only one mineral per deposit	At least 5 or 6 minerals per deposit
Fine mineral grain sizes	Coarser mineral grain sizes
Stable minerals	Unstable minerals

Gold and silver, when occurring together, are often treated by using the same techniques, that is amalgamation or cyanidation, applied to the ore for the extraction of both metals simultaneously. Grinding sizes and other ore preparation prior to the actual recovery step are the same or very similar. Recoveries however are often different, in many instances better for gold than they are for silver. Metallurgists are confused because variations in the laboratory conditions seldom help.

In the light of our discussions so far, the reasons for the metallurgists' confusion become obvious. They try to improve their recoveries while unaware of the differences between gold and silver which require a differential approach to the recovery process.

When extracting **GOLD**, we are dealing with a smaller number of **minerals** and with frequent occurrences in native gold/electrum. This means that a limited number of reagents and techniques for its separation is required, for all practical purposes only two, amalgamation, for the coarser sizes and cyanidation, for the finer ones.

Problems during gold recoveries are seldom the result of insolubility of the gold itself. They are rather a consequence of the very fine gold **grain sizes** which do not always become exposed or liberated at the normal grinding sizes. If fine gold grain sizes are the reason for poor recoveries in many ores, then what is necessary for improved results is not an increase in cyanide strength, nor a change in the pH of the solution, nor the addition of chemicals, all recommendations made by *the overbearing consultant in gold*. Certainly not as much as the determination of the correct gold grain sizes and the **positive** identification of the **host mineral(s)**. It is the host mineral which requires all the attention, not the gold itself, because it needs to be broken to similar sizes to those of the native gold. Only after the gold has become exposed, by adequately breaking the host mineral, can the cyanidation step take place, along with the most appropriate conditions for that particular ore, **including** correct cyanide strength, high or low pH, addition of the correct chemicals, etc. Again, decisions regarding laboratory conditions will be made, based on the knowledge of the microcomposition of the ore.

We have seen that grinding sizes can only be brought down to 10-15 microns for most ores. If such fine sizes are not sufficient to liberate or expose the gold, which can be as fine as 1 or 2 microns, then host breakdown by chemical methods becomes necessary. Roasting and other techniques used to convert sulfides such as pyrite and arsenopyrite into less stable phases, are costly, yet the higher gold **market value** justifies the additional expenses.

When extracting **SILVER**, we are dealing with **many minerals** and with several mineral species all occurring within the same ore. We have seen (page 166) that if six different minerals account for the total silver in the ore, and three of them do not respond to the reagent used (for example cyanide or mercury), a partial loss is experienced. Consequently, additional reagents may be necessary for optimum recoveries. Again, these reagents will be chosen only after the silver-bearing minerals have all been identified and appropriately tested. Variations in laboratory techniques without a prior knowledge of the minerals which do not react and why, will only cause unnecessary wastes in time and expenditures.

Contrary to gold recoveries, fine grain sizes are not generally a problem during silver recoveries because the **higher** silver **quantities** are often coupled with **coarser average grain sizes** which are easily liberated by the grinding process. Silver extractions as a byproduct, provided that all of the silver is separated with the primary mineral, are not affected by fine grain sizes. If some of the finer particles remain enclosed in one or more of the gangue minerals, then the lower silver **market value**, which does not justify as expensive techniques as those used for gold recoveries, allows for larger portions of the metal to be left in the tails without significant economic losses.

Mineral stability is not a problem during gold recoveries, unless dealing with a native gold which is very rich in silver (electrum) or with a gold-silver ore containing unstable silver minerals. Several silver minerals are **unstable** and cause the impermeable coatings preventing silver dissolutions, and the other complications discussed in the previous chapter. Poor mineral stability represents the other important hindrance during silver recoveries.

We can therefore conclude that, during gold recoveries, main causes of problem are fine grain sizes and difficulty to break the host minerals to sizes comparable to those of the gold. During silver recoveries main causes of problem are variety in mineralogy and mineral instability. The necessity for the use of differential techniques for efficient recoveries of the two metals, even when occurring together in the same ore, should be apparent.

As for the other variables:

Minerals other than the host mineral affect gold and silver recoveries in a similar manner. We have seen on pages 134-139 and 170-171 that the presence of clays can be particularly troublesome, because it lowers the amounts of both gold and silver recovered in several ways. Clays also complicate processes such as the actual mining, ore handling, disposal of tails, etc.

Host rocks affect gold and silver extractions in a similar manner.

Confined higher concentrations are probably more critical to detect during gold recoveries because of the higher market value and the costs involved for optimum extractions. Their correct definition is however important for the recovery of silver as well, to avoid the leaching of those portions of the ore containing the metal in amounts which are too low to be of economic significance.

Case study 20. Microanalytical study of a mineralization requiring differential treatments for gold and silver, as well as for the different portions of the ore. This case study provides a practical example of data obtained from microanalytical studies of a suite of samples representative of different sections of a deposit in the Western United States. In these samples, the gold and silver occurred in different minerals of variable grain sizes, depending on the section of the ore where they were collected, therefore they required treatments which needed to be differentially applied to the two metals, as well as to the various portions of the ore. The following are the conclusions drawn in the report.

"The samples studied are characterized by several types of gold and silver mineralizations, each of which requires a different metallurgical approach for optimum recovery.

Section A. Gold and silver occur together in fine invisible high silver native gold (electrum). Sericite/clay is abundant. Fine and invisible grain sizes, presence of sericite and clay and presence of electrum are all factors which contribute to poor recoveries. This portion of the ore will be a difficult one metallurgically, unless special treatments which take into consideration all of the above potential sources of problem are applied.

Section B. Native gold is relatively coarse and contains only 15 to 20 per cent of silver. A large portion of the silver minerals identified, such as acanthite, embolite, native gold, are soluble in cyanide. Gold and silver occur in cavities and fractures of the quartz, which allow cyanide penetration. An ore with similar mineralogy should give easier recoveries because both the nature and type of occurrence of most of the gold and silver lend themselves to extractions by the standard leaching techniques. Proustite, also identified in this section, may require special attention. There is some sericite/clay. A good approach to the treatment of this type of ore might involve a preliminary test run. Then and only if poor recoveries are experienced, the leached tails could be examined by microanalytical techniques and decisions could be made as to the use of possible specialized treatments.

Section C. The mineralization from this section, characterized by high manganese and a very close silver-manganese association, needs special treatments applied to the Mn-Ag minerals for successful extractions. Again, studies of leached tails, following a preliminary test work, should help in the choice of the most appropriate treatment.

Perhaps we should remember that manganese enrichments in silver deposits are often only superficial. If that is the case, the presence of manganese may not be a problem, when exploiting the deeper unoxidized levels of the deposit.

Section D. Silver is probably with the manganese minerals. Although the relation between silver and manganese is not as clear here as in the previous section, samples from this portion of the ore might be treated in a similar manner to that employed for the other manganese oxide portions.

Section E. This section contains acanthite and silver-manganese mineralization. Consequently, two processes may be necessary, standard cyanidation for acanthite, and special treatments for silver-manganese minerals. These processes may need to be applied following the differential concentration of the acanthite and of the silver-manganese minerals. Acanthite-iron oxide associations, frequently observed here, should pose no problem during cyanidation because of the porosity and high solubility of the iron oxide which will promote penetration of the leaching solution. The presence of secondary iron oxides however, also observed in the other sections of the ore, may consume a portion of the cyanide.

The writer has no knowledge regarding the solubility of argentian plumbojarosite or of plumbojarosite containing minor amounts of silver, both identified in several of the above sections, neither is she aware of any special treatments for these minerals. Again, the observation of leached tails may provide some answers."

CHAPTER 9

PLATINUM GROUP ELEMENTS

INTRODUCTION

Platinum group elements are ruthenium, rhodium and palladium (of similar weight as, but lighter than silver, the next element in the periodic table by atomic number) and osmium, iridium and platinum (of similar weight as, but lighter than gold, the next element in the periodic table by atomic number). As opposed to gold and silver, known probably since the stone age, the first certain discovery of platinum, in the alluvial deposits of the Rio Pinto - Columbia, is as recent as 1741. Platinum was introduced to Europe by Sir Charles Wood who brought it to England during that same year. Osmium and iridium were isolated together from the residues of platinum ores in aqua regia in 1803 or 1804 by English chemist Smithson Tennant. The existence of rhodium and palladium was established in 1803 by English doctor W. Hyde Wollaston. Ruthenium was recognized as an element in 1844 or 1845 by Russian chemist C.E. Claus.

Until very recently, platinum and palladium have been far more important commercially than the other four members, the potential uses of them being not completely developed (Wise 1988). It is only during the late eighties that the importance of rhodium in the automotive industry was recognized. Applications of the platinum group elements, as provided by Mohide (1979) and the **Encyclopaedia Britannica** (1988), are summarized below.

RUTHENIUM. Catalyst in the chemical, motor vehicle and petroleum refining industries, and hardener in platinum alloys. Also important because of its refractoriness and chemical inertness.

RHODIUM. Catalyst in the chemical, pharmaceutical and petroleum refining industries. Also important because of its refractoriness and chemical inertness. Demand for rhodium increased dramatically during the early nineties, when the potential of the element as an emission control catalyst was realized and put to the services of the automotive industry.

PALLADIUM. Catalyst in the pharmaceutical, motor vehicle, petrochemical and petroleum refining industries. Alloys used for high voltage regulators, electrical and electronic apparatus, thermostats, resistors, gas turbine engines and atomic reactors; in spinnerets to produce rayon fiber, in dentistry and in the jewelry industry (alloyed with gold, page 119).

OSMIUM. Component of electrical contacts, phonograph needles, and (formerly) gold fountain pen nibs, when alloyed with other platinum group elements. Also used in the jewelry industry; in medicine and as a detector of finger prints. Osmium wires were used for filaments of early incandescent lamps, before the introduction of tungsten.

IRIDIUM. Catalyst in the chemical, motor vehicle and petroleum refining industries. Also important because of its refractoriness and chemical inertness.

PLATINUM. Catalyst in the chemical, motor vehicle, and petroleum refining industries. Also used in the jewelry industry, often alloyed with other platinum group elements and with gold; in electrical contacts; in dentistry and medicine. Important because of its refractoriness and chemical inertness. The international prototype standard kilogram of mass is made of an alloy containing 90 per cent of platinum and 10 per cent of iridium.

* * * * * *

The world's first most important deposits of platinum were discovered in 1822 in the Ural Mountains of Russia. Other platinum localities are in the Transvaal - South Africa, in the Sudbury district and other localities in Northern Ontario - Canada, in the Stillwater Complex, Montana and in Northern Minnesota - United States, in Finland, Ireland, Borneo, New South Wales, New Zealand, Brazil, Peru and Madagascar.

Recently, announcements have been made regarding the presence of several platinum group element deposits in Canada very near the United States (Runck 1988). These deposits include significant concentrations at Lac des Iles in Southern Ontario and at Grasshopper-Tulameen and Platinum Blonde in Southern British Columbia. The platinum group element deposit of the Stillwater Complex, and the appreciable values reported in Northern Minnesota, together with the newly announced deposits in Ontario and British Columbia, suggest a distribution of those elements along the Canadian-United States border.

World's largest producers of the six platinum group elements are South Africa (Merensky Reef in the Bushveld Complex) and the USSR (Ural Mountains and Noril'sk in Siberia). During 1987, South Africa produced 131 tons of platinum, the Soviet Union produced 121 tons, thus together accounting for 94 per cent of the total world production of the six metals. Canada was third with approximately 5 per cent, the remaining 1 per cent being produced by the rest of the world. If we were to compare the amount of platinum group elements produced by one locality in South Africa to the amount produced by the entire North American continent, where numerous important deposits occur, perhaps the same considerations regarding the exploitation of gold deposits (page 89), also apply to the exploitation of the platinum group element deposits.

MARKET VALUE

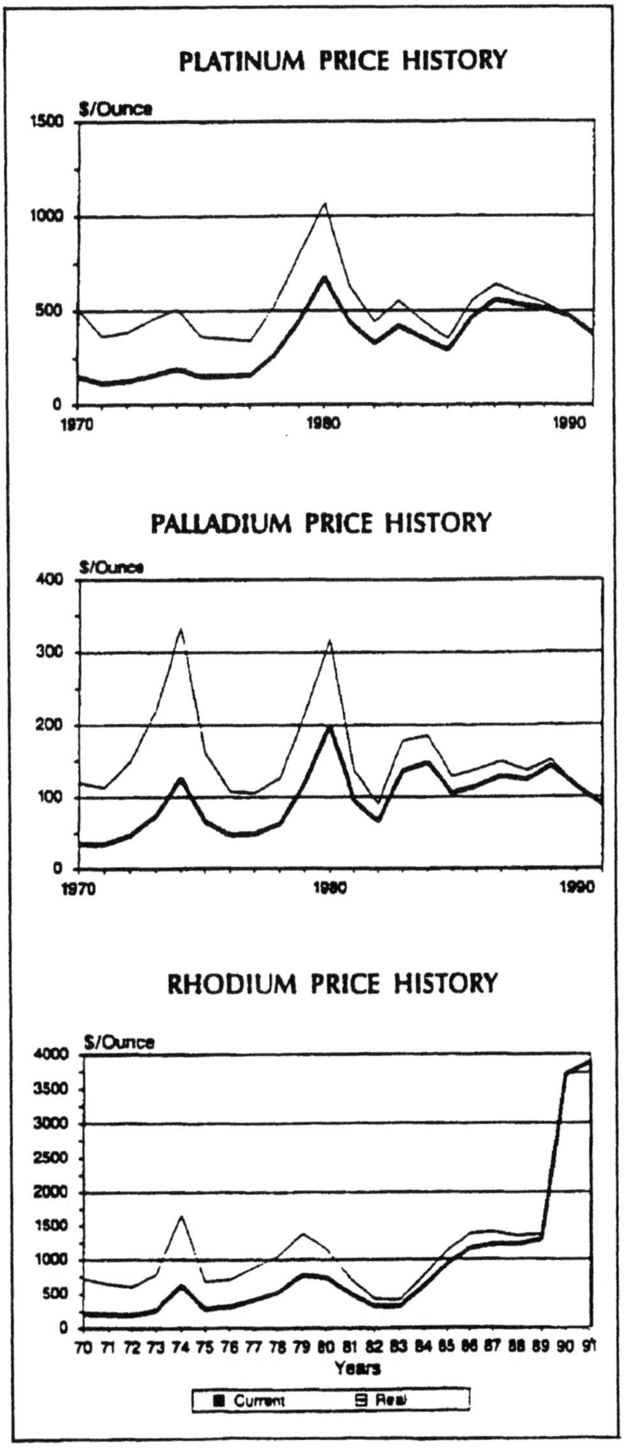

Chart 18. *Fluctuations in the price of platinum, palladium and rhodium during the 70's, 80's and very early 90's. (From Maule 1992).*

Over the past decade, two of the six platinum group elements have been more expensive than gold, namely rhodium (by factors of up to 20) and platinum itself (by much smaller factors). Over the same period, the market price of the other platinum group elements has been either lower but within the same range as gold (for example, iridium), or lower by factors of one half (for example, osmium) to one quarter or one fifth (for example, palladium and ruthenium).

Prices per ounce, reported in the June 5th 1992 issue of **American Metal Market**, were: platinum, US$ 373; palladium, US$ 81-81.50; iridium, US$ 315; ruthenium, US$ 64; rhodium, US$ 2600. The price of osmium was not provided, but was probably in the range of US$ 150. The price of gold, reported in the same issue, was US$ 338.50-340.80.

Chart 18 (modified after Maule 1992) shows the fluctuations in the price of platinum, palladium and rhodium during the past 20 years. Comparison of the curves shows that all three elements reached peaks at approximately the same time as gold (mid-seventies and very early eighties - see chart 9 on page 90). The same comparison also shows that rhodium, relatively constant and more stable than gold, silver or platinum at just under US$ 500 in 1983, peaked for a third time in 1990. The price of rhodium in fact started to increase steadily after 1983. Between that year and 1986, it more than doubled to US$ 1200 per ounce, then resumed its climb in late 1989 to reach its highest peak ever at US$ 7,000 the following July 1990. It declined to US$ 1600 towards the end of 1991 and was volatile during the following months, with fluctuations as wide as 30 per cent within a few days, sometimes within even one day. It came back up to US$ 2600 towards the middle of 1992. US new laws controlling the emission of car exhaust gases into the air, and a resulting greater demand for rhodium as an auto-catalyst, are considered responsible for the meteoric rise of the price of the metal during the past decade. Consumer resistance to the purchase of the element, after the price started to decline, and a weak world economy, particularly so in the auto industry sectors, might explain the extreme volatility that followed such rise (Kaeser 1992).

GEOCHEMISTRY SYMBOLS: Ru, Rh, Pd, Os, Ir, Pt ATOMIC #: 44, 45, 46, 76, 77, 78

The six platinum group elements belong to the same group VIII (transition metals) as iron, cobalt and nickel. For that reason, according to Westland (1981) these nine elements tend to be concentrated together in various geological formations. Platinum group elements, along with iron, cobalt and nickel, are assumed to be concentrated in the mantle in amounts which are 5 to 30 times as high as in the crust. They are found on shallower levels in the economic deposits, as a consequence of processes of interchange of matter between the mantle and the overlying crust.

Abundance of the six platinum group elements in the Earth's crust is too low to be recorded (Brownlow 1979). Platinum abundance, for economic deposits exploited primarily for platinum group elements, would probably be similar to that of gold. Because the platinum group elements are presently recovered largely as byproducts of copper and nickel, lower values are or can be extracted.

Chemical determinations of the platinum group elements by fire assay (pages 91-93) are difficult, particularly when gold and/or silver are present. Problems and *some* means of controlling them are described by Bugbee (1981). Presently not all fire assay laboratories in North America will provide analyses for low quantities of the platinum group elements (in the range detected in ores). Those that do, may not be able to analyze for all of them, rhodium, palladium, iridium and platinum being the easier ones to handle. The author's experience suggests that even some of the more reputable laboratories sometimes misinterpret the experimental results and report values representative of platinum **and** gold as representative of gold only, or of platinum only. Publications by the **International Precious Metals Institute**, mostly in the form of proceedings of conferences and seminars, may help in the understanding of the analytical techniques for the platinum group elements.

ROCK TYPES HOSTING THE PLATINUM GROUP ELEMENTS

More common host rocks for the platinum group elements are mafic and ultramafic formations with or without nickel-copper or chromium. Other associations do however occur. Besides in compact ores, the platinum group elements are found in alluvial or placer deposits where, following weathering of the host rock, their higher density and resistance to corrosion have lead to their concentration in river beds. In some locations, gold is also present in the same placers, although not usually originated from the same mother lode. (Wise 1988). Some platinum group elements have been detected in manganese sea nodules (T.P. Mohide 1988, personal communication).

Two attempts at classifications of the platinum deposits, one by Cabri (1981-A) and one by Naldrett (1981) are in the CIM publication **Platinum group elements: mineralogy, geology and recovery.** Both classifications are incomplete, and contain data which are not always in agreement with one another. They are however as comprehensive as the author has been able to find in the literature. Table 13 at the end of the chapter, listing some of the major platinum group element deposits, was prepared based on the information in the article by Naldrett.

ORE MICROCOMPOSITION

The published number of platinum group minerals, listed in table 14, is 75 (Cabri 1981-B). The author's experience, acquired largely while conducting microanalytical research in South Africa, suggests that there are many more unnamed minerals which are recognized as new or different from the ones reported; the available analytical data however, is not sufficient for their official classification as new minerals. Cabri (1972) estimates that altogether the actual number of minerals bearing the platinum group elements is possibly close to at least 150.

The existence of minerals in which the platinum group elements occur in minor to trace amounts (like those discussed for gold and silver), should be expected. Information on these minerals is again limited, however substitutions are probably possible with iron, cobalt and nickel, the three elements within the same group VIII of the periodic table. Palladium and platinum are the most likely to replace nickel in its sulfides, in a similar fashion as gold and (more so) silver replace copper. In consideration of the fact that minor gold (and very likely silver) are detected in some of the platinum group minerals, for example, **atokite**, **atheneite**, **palladoarsenide**, and **zvyagintsevite** (pages 99-100), minor platinum group elements probably also occur in the gold and silver minerals.

The author's experience with the platinum group elements, far more limited than the one with gold and silver, indicates that there are basic differences in the general appearance

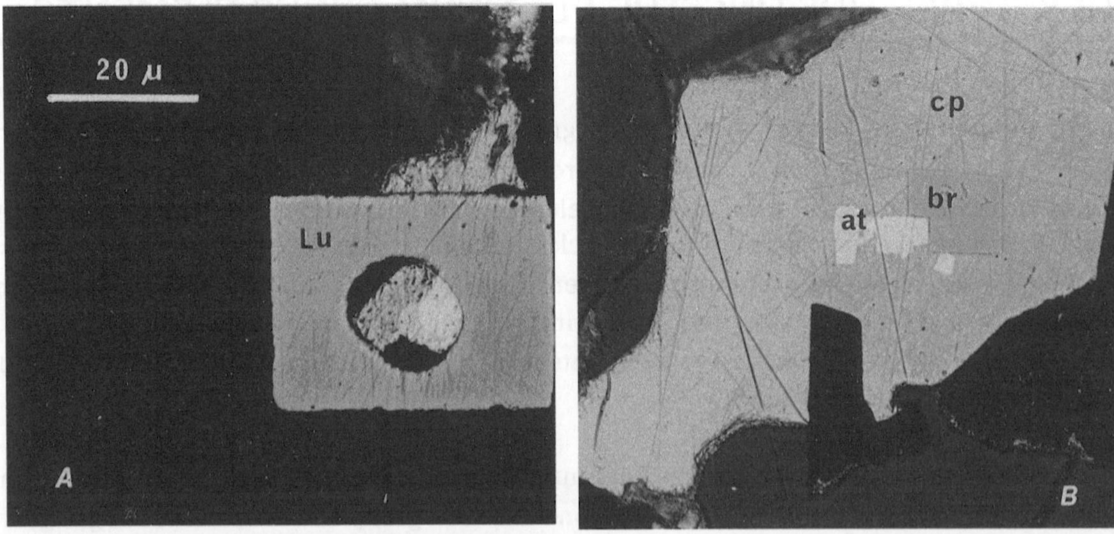

Fig. 47. *Platinum group minerals from the Bushveld complex in South Africa. A: laurite. B: atokite and braggite in chalcopyrite. (From Schwellnus, Hiemstra and Gasparrini 1976).*

of the minerals in the two groups. Native gold is yellow with high reflectivity, and soft, therefore seldom scratch-free when observed in the polished mounts. It usually forms anhedral particles. Fine sizes of less than 10 microns are very common. Silver minerals show a very wide range in optical properties, grain size and hardness, but anhedral shapes are common to most grains. The platinum group minerals, on the other hand, although not always euhedral with sharp contours, like those represented in figure 47, tend to occur in more crystalline shapes. Many of them are white with medium to high reflectivity and high hardness, thereby always taking a good polish. The average grain sizes, at least for the deposits studied by the author, are larger (5 to 30 microns), and the mineral grains *relatively* easy to locate during the microanalytical studies of samples from both South Africa and North America.

Case study 21. Platinum and palladium in sulfide ore. The assay of a nickel sulfide ore had shown the presence of platinum and palladium in amounts of 0.005 per cent for each element. Because the deposit was not being exploited for nickel at the time, it was necessary to determine if a concentrate could be obtained, consisting of the platinum group elements only. This involved finding whether the elements were in solid solution in the nickel sulfide(s) or were present as separate minerals. If so, it would be necessary to identify those minerals and determine their grain size and host mineral(s). The microanalytical study showed that the sulfides were pyrrhotite (predominant), containing minor amounts of nickel, and violarite. Most abundant platinum-palladium mineral was merenskyite occurring enclosed in the pyrrhotite with grains 20-30 microns in diameter. No platinum or palladium were detected in solid solution replacing nickel in the pyrrhotite or violarite. It was concluded that the platinum group elements were of economic value, because they all occurred in discrete grains of adequate size for flotation by the standard techniques.

METHOD OF STUDY

The study of the distribution of the platinum group elements is probably the most time consuming of the precious metals (refer to chart 5 on page 60) because their ores are characterized by both lower concentrations-fine grain sizes (typical of gold) and by large numbers of minerals (typical of silver). The optical properties, not always as distinctive as those of the native gold-electrum, contribute additional difficulty.

An efficient method of study might include:

- The location of any possible platinum-group mineral by use of the optical microscope.
- The identification of those minerals as *platinum group element-bearing* by use of the SEM.
- The determination of their quantitative chemical composition by the electron microprobe with crystal spectrometers (WDA).
- Their positive identification based on optical properties and **quantitative** chemistry.

The use of the EDS-SEM for positive identifications, without the microprobe, is not recommended, unless the investigator is very familiar and is dealing with some of the more common or more obvious of the platinum group minerals, for example, the native metals and those with simpler chemical compositions, such as sperrylite, cooperite, braggite, etc..

Some researchers prefer the use of the SEM without the ore microscope for the location of the platinum group minerals, as well as for the determination of their chemistry. This method requires less skilled a person because, due to their high mean atomic number, platinum-group minerals are easily recognized on the SEM oscilloscope screen. The method however is expensive, particularly if the SEM time is being purchased from outside laboratories. It does not provide as complete information, because no data becomes available on optical properties, nor on quantitative chemistry. Positive identifications again may only be possible of the more common minerals.

METALLURGY

Platinum was long known to exist in the arsenide form (sperrylite) in Canadian copper-iron-nickel sulfides from which some recovery was made, beginning in 1900. Not until the adoption of the electrolytic methods for refining copper and nickel however, did substantial production of the platinum group elements take place.

Platinum and palladium dominate in many deposits, but all of the platinum group elements are usually present and are recovered.

Little information on ore microcompositions which can be of practical value for the process of extraction is available in the literature. The author's experience on the subject is also limited, because during the twelve years that she has acted as a consultant, there has been very little demand for data on platinum group elements. Although the author's literature search has not been overly extensive, the CIM publication **Platinum group elements: mineralogy, geology and recovery** (1981), of more academic than applied significance, could very well be the most comprehensive reference on the subject.

Some of the reasons for the limited amount of published data on the occurrence of the platinum group elements in their ores and on their relation to the process of extraction could be summarized as follows:

- Approximately half the volume of the total platinum group elements produced in the world has been coming from Russia. The attitude of the Russians towards the publication of their techniques and findings, was not as liberal as in the Western World, under communist regime.
- Although a large amount of research is being conducted in South Africa, this is done predominantly by mining companies, most of which prefer to keep the information confidential.
- There are very few mines in the Western World being exploited primarily for platinum group elements and few universities seem interested in conducting systematic research on the subject of platinum group element metallurgy.

The **SME Mineral Processing Handbook** (1985) gives a good indication of the amount of interest that there is in North America in the different precious metals. Of the 35 pages on the subject, 29 are dedicated to gold, 3 to silver and 2 and 1/2 to all six of the platinum group elements.

Wise (1988) describes approaches to platinum group element recoveries in Canada and South Africa, which follow the usual beneficiation processes represented by gravity separation and/or flotation. These have been summarized as follows:

Canadian recoveries. The platinum group elements, in addition to gold and silver, are obtained as byproducts of copper-nickel ores, treated by the International Nickel company in the Sudbury-Copper Cliff district in Ontario. The ore is selectively floated to yield fractions containing copper and nickel. The nickel fraction, converted into nickel-rich matte by blast furnace and converter treatment, is cooled, crushed, ground and passed over a magnetic separator to recover nickel-rich metallics which contain most of the platinum group elements. The copper fraction is given further treatment and again selectively floated to yield a low copper and a high copper sulfide. Nickel and copper products are then refined by the usual electrolytic methods. The platinum metals are recovered from the nickel electrolysis and then shipped to Acton, England, where they are treated to isolate the individual elements. Gold and silver are recovered from the copper electrolysis.

South African recoveries. Recoveries are mostly from the pyroxenite deposits of the Merensky Reef, today the largest single source of platinum in the world. The coarser particles of the platinum group minerals are separated by corduroy tables, further concentrated and shipped to England for additional processing. The tailings from the table concentration are sent through a flotation cycle which separates a concentrate of nickel, copper and iron, plus the remaining platinum group minerals. This concentrate is pelletized and converted into a copper-nickel matte containing substantial amounts of platinum group elements. The matte is then treated to separate the copper and nickel sulfides and ultimately to produce electrolytic copper and nickel. The platinum group elements are recovered from the anode slimes produced during the nickel electrolysis.

The steps followed during the recovery of the platinum group elements by Rustenburg Platinum Mines in South Africa and including the microanalytical studies (added by the author) are summarized in chart 19 (from data in Carlisle 1985, pp. 18-34 and 18-35).

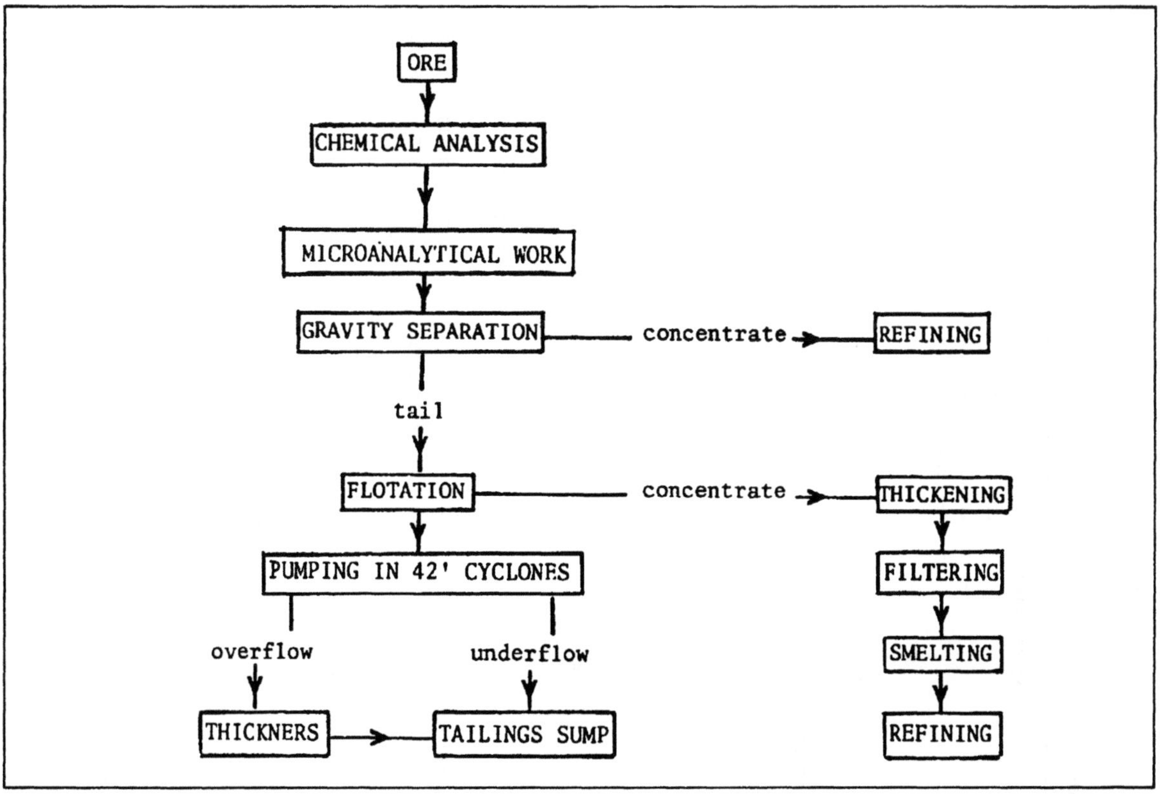

Chart 19. *Flowchart for platinum group element recovery.*

A review of the classical, as well as modern methods for the refining of the platinum group elements, such as solvent extraction and ion exchange, is given by Demopoulos (1989), who also draws attention to the necessity for newer more efficient recovery techniques, given the high demand for these elements and their high tech applications. The IPMI volume **Precious Metals Science and Technology** (1991) also gives a comprehensive review of the techniques for extraction of the platinum group metals (along with those used for gold and silver). Solvent extraction of the platinum group metals is discussed by Cote' et al. (1992).

TABLE 13 - PLATINUM GROUP ELEMENT DEPOSITS

DEPOSIT	STRUCTURAL CHARACTER	AGE	TYPE LOCALITY	MAIN MINERALS	ASSOC. MINERALS	TYPICAL ELEMENTS
Pge's predominant - Merensky type	Large bodies of basaltic magma intruded into stable continental rocks	Precambrian	Bushveld Complex, South Africa; Great Dyke, Zimbabwe; Stillwater, Montana	Laurite, cooperite, braggite, Fe-Pt, vysotskite, monkeite	Scattered sulfides, chromite	Fe, Cr,....
Pge's predominant - Hydrothermal deposits	Irregylar pods in hydrothermally decomposed metadiorite and metagabbro	na	New Rambler, Wyoming	na	Pyrite, chalcopyrite, pyrrhotite, covellite, marcasite, minor pentlandite, and electrum	Fe, Cu, Ni, (Au)
Pge's predominant - Placer deposits	Eluvial and alluvial deposits	Triassic	Nizhnii-Tagil, Urals; Noril'sk, Siberia	na	na	na
Ni-Cu predominant	Noritic rock associated with an astrobleme	na	Sudbury, Ont.	Froodite, sizwaite, sudburyite kotulskite, merenkyite, mertierite II, michenerite, moncheite, niggliite, sperrylite	Cu and Ni sulfides, cobaltite, gersdorffite, melonite, pyrite, pyrrhotite, magnetite	Fe, Cu, Ni...
Ni-Cu predominant	Intrusive equivalents of flood basalts	Triassic	Noril'sk, Siberia; Duluth, Minnesota	Native metals, alloys with Sn and Pb, compounds of Pd with Pb, Cu, Ni, S, Bi, As	Pyrrhotite, pentlandite, chalcopyrite	Sn, Pb, Cu Ni, As, Bi, Sb
Ni-Cu predominant	Tholeitic intrusions	na	Pechenga, Kola Peninsula; Lynn Lake, Manitoba	na	na	na
Ni-Cu predominant	Komatiite lavas and intrusions	Precambrian	Kambalda, W. Australia; Thompson, Manitoba; Ungava Peninsula, Quebec	Sperrylite, Pd melonite, Pd-Bi tellurides	Pyrrhotite, pentlandite pyrite, magnetite, minor chalcopyrite	Fe, Cu, Ni...

na: not available

TABLE 14 - PLATINUM GROUP MINERALS

Modified after Cabri (1981) *

MINERAL	FORMULA	MINERAL	FORMULA
Anduoite	$(Ru,Os,Ir)As_2$	Palarstanide	$(Pd,Pt)_8(Sn,As,Pb)_8$
Arsenopalladinite	$Pd_8(As,Sb)_3$	Palladium	Pd
Atheneite	$(Pd,Hg)_3As$	Palladoarsenide	$(Pd,Ag,Au)_2As$
Atokite	Pd_3Sn	Pallado-bismutharsenide	$Pd_{10}As_4Bi$
Borovskite	$(Pd,Pt)_3(Sb,Bi)Te_4$	Palladseite	$Pd_{17}Se_{15}$
Braggite	$(Pd,Pt,Ni)S$	Paolovite	Pd_2Sn
Cooperite	PtS	Platarsite	$Pt(As,S)_2$
Daomanite	$(Pt,Cu)AsS_2$	Platiniridium	(Ir,Pt); Ir=+50-80 at.%
Erlichmanite	$(Os,Rh,Ir)S_2$	Platinum	(Pt,Fe,Cu,Pd)
Froodite	$PdBi_2$	Plumbopalladinite	$(Pd,Ag)_3(Pb,Bi)_2$
Genkinite	$(Pt,Pd,Rh,Ni)_4Sb_3$	Polarite	Pd(Bi,Pb)
Gervesite	$PtSb_2$	Potarite	PdHg
Guanglinite	Pd_3As	Prassoite	$Rh_{17}S_{15}$
Hollingworthite	RhAsS	Rhodium	(Rh,Pt)
Hongshiite	PtCu	Ruarsite	RuAsS
Isizwaite	$PtBi_2$	Rustenburgite	Pt_3Sn
Irarsite	IrAsS	Ruthenarsenite	RuAs
Iradarsenite	$IrAs_2$	Rutheniridosmine	(Os,Ir,Ru,Rh,Pt,Pd,Cu)
Iridium	(Ir,Pt,Os..)	Ruthenium	(Ru,Pt,Ir,Rh..)
Iridosmine	(Os,Ir..)	Ruthenosmiridium	(Ir,Os,Ru,Pt,Rh,Fe,Ni,Pd)
Isoferroplatinum	Pt_3Fe	Sobolevskite	PdBi
Isomertieite	$Pd_{11}Sb_2As_2$	Sperrylite	$PtAs_2$
Keithconnite	$Pd_{3-x}Te$ ($0.14<x<0.43$)	Stannopalladinite	Pd_3Sn_2
Kotulskite	Pd(Te,Bi)	Stibiopalladinite	$Pd_{5+x}Sb_{2-x}$
Laurite	$(Ru,Os)S_2$	Stillwaterite	Pd_8As_3
Majakite	PdNiAs	Stumpflite	PtSb
Malanite	$(Pt,Ir)_2CuS_4$	Sudburyite	PdSb
Maslovite	PtBiTe	Telargpalaite	$(Pd,Ag)_{4+x}Te$
Merenskyite	$PdTe_2$	Telluropalladinite	$Pd_9(Te,As,Bi)_4$
Mertieite	Pd_8Sb_3	Temagamite	Pd_3HgTe_3
Michenerite	Pd(Bi,Sb)Te	Testibiopalladinite	PdSbTe
Moncheite	$PtTe_2$	Tetraferroplatinum	PtFe
Niggliite	PtSn	Tulameenite	Pt_2FeCu
Omeiite	$(Os,Ru,Ir)As_2$	Urvansevite	Pd(Bi,Pb)
Oosterboschite	$(Pd,Cu)_7Se_5$	Vysotskite	PdS
Osarsite	OsAsS	Xingzhongite	(Ir,Cu,Rh)(S,Pb)
Osmiridium	(Ir,Os,Pt,Ru,Rh,Fe,Cu)	Zvyagintsevite	Pd_3Pb
Osmium	(Os,...) Os>80%		

* Additional minerals have been described since the table was compiled (for example, Desborough and Criddle 1984, Corrivaux and Laflamme 1990, Nixon et al 1990, Cabri and Genkin 1991, Harris and Cabri 1991), but by the time the new data was published, the manuscript had already gone to the printers.

CHAPTER 10

URANIUM

INTRODUCTION

Uranium, a dense, hard, radioactive nickel-white metal with the heaviest atomic weight of any naturally occurring element, was discovered in the oxide form in 1789 by German chemist M.H. Klaproth, who named it after the newly observed planet. It was isolated as a metal in 1841 by the reduction of uranium tetrachloride with potassium. "It was not until much later, in 1896, that the French physicist Henry Becquerel discovered the phenomenon of radioactivity through its ability to expose photographic film, a technique that continues to be used for the location and characterization of radioactive minerals in rock and ore samples". (Steacy and Kaiman 1978, p. 107).

Early uses of uranium were for small-scale technical and industrial applications, particularly in coloring glass and ceramics. Following the discovery in 1938 that the uranium nucleus could be made to fission or split when bombarded with neutrons, uranium became critically important for military applications and, more recently, as a nuclear fuel. The discovery opened the dramatic possibility that the uranium atom could be made to give up its energy in a sustained chain reaction. Recoverable uranium ores, as sources of nuclear fuels, contain many times more energy than all the known recoverable deposits of fossil fuels, as it is estimated that one pound of uranium yields as much energy as three million pounds of coal. As a consequence, intense studies of the element, together with extensive exploration resulted in the first non-laboratory application of the 1938 discovery, in the form of the original atomic bomb dropped on Hiroshima in 1945. After world war II, exploitation of nuclear energy as a power source and for other peaceful purposes was developed in the United States, Great Britain, the USSR, France and other countries. By the 1970's uranium had become a metal of vast economic significance. Estimates of future United States and world requirements indicate a continuous need to develop substantial additional resources. (King and Adler 1985, pp. 24-2 and 24-3, and Williams 1988, p. 433). According to Shade (1986), approximately 20 per cent of the present combined electrical power consumed in the United States, Europe and Japan is generated by nuclear stations.

Major world uranium producers are the Mesozoic (Jurassic) and Tertiary (Eocene) sandstones in the Western United States, and the quartz-pebble conglomerates of Canada (Great Bear Lake and Ontario) and of the Witwatersrand in South Africa. In the Witwatersrand, uranium is extracted as a byproduct of gold mining. Other major uranium locations are in Zaire, France, Czechoslovakia, Australia, South America. The USSR and China are also believed to hold substantial deposits (Peterson 1985, and Williams 1988).

MARKET VALUE

The market value of uranium has oscillated between high and low prices. During the late sixties and then late seventies, it reached above US$ 50 per pound of U_3O_8. At the time the book was being completed it was in the range of US$ 15 per pound of U_3O_8 (US$ 13.75 on September 30 1988).

GEOCHEMISTRY SYMBOL: U ATOMIC #: 92

"Uranium is the most abundant element of the uranium decay series, with an average crustal abundance of 2.6 ppm. Its high mobility in the surficial environment is due to oxidation to the uranyl ion which forms soluble carbonate complexes. Uranium-decay products generally do not form soluble compounds and so secular disequilibrium of the series is commonly observed. Disequilibrium is almost invariably due to removal of uranium under oxidizing conditions or uranium addition under reducing conditions. Re-establishment of radioactive equilibrium requires time spans of several half lives of the daughter products. The stable end-products of uranium, i.e., lead and helium, represent the extremes of mobility of the decay products: lead is relatively immobile and helium is able to escape to outer space." (Dyck 1978, p. 57).

Uranium abundance in the Earth's crust, as expressed per 10^6 atoms of silicon, is 1.1 (Brownlow 1979). Uranium economic abundance in the ore is in the range of 0.5 to 2 per cent or higher.

ROCK TYPES HOSTING URANIUM

"The classification of mineral deposits is always a difficult and contentious exercise because most natural phenomena are not readily separable into distinct subdivisions. For uranium deposits, this problem is compounded because the deposits are commonly polygenetic, and may have undergone a complex history of deposition, concentration and remobilization prior to preservation as an ore deposit....." (McMillan 1978, p. 187).

"Uranium of igneous origin has been found in several alkaline and carbonatite complexes in the Canadian Shield and the Cordillera. These complexes occur in circular and radial structures near large-scale rift zones and range in age from Proterozoic to Cretaceous.... Deposits of metamorphic origin include those which are found in fold belts where metamorphism has reached upper amphibolite to granulite grade. The deposits are of two main types; (a) felsic segregations consisting of pegmatitic or alaskitic sills and dikes and (b) skarns. Both rock types are believed to have formed from some precursor, generally of syngenetic origin..... Detrital deposits include the paleoplacers in the Elliot Lake district, one of the most important reserves of uranium in the world...Hydrogenic deposits *(which include hydrothermal, syngenetic, epigenetic and supergene in table 15 at the end of the chapter* - sentence in italics added by the author) are deposited from aqueous solutions by precipitation of uranium minerals or by sorption of uranium onto clay minerals, carbonaceous material, phosphates, or hydrous oxides. Most hydrogenic deposits are

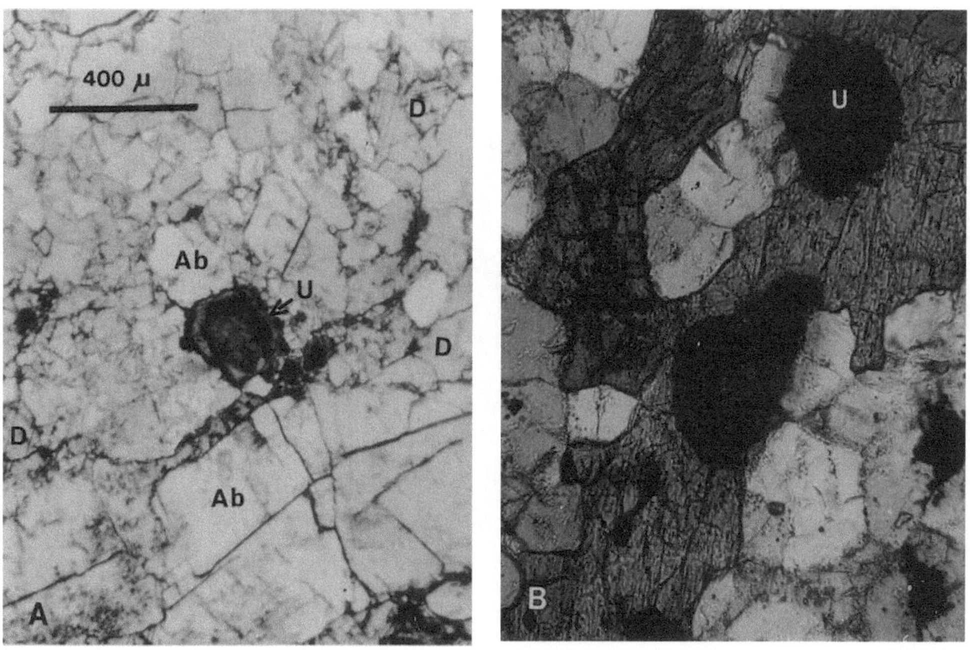

Fig. 48. *Uraninite in albite (A) and in amphibole-quartz (B) hosts, showing subhedral crystalline shapes. Note the brown radioactive halos in the photomicrograph to the right.*

precipitated at low temperatures and are characterized by uranium in the form of pitchblende." (McMillan 1978, pp. 190-192). The classification presented in table 15 is from McMillan (1978). The table has been modified to a format similar to the one of the other tables of metal deposits in this book.

Because of their relatively high density and hardness, mechanical enrichment of uranium ores is possible (Williams 1988).

ORE MICROCOMPOSITION

Uranium-bearing minerals

Uranium occurs in nature in a very large number of minerals, (over 150 are reported, Steacy and Kaiman 1978), associated with a variety of elements in both its tetravalent and hexavalent states. Table 16 is a compilation of all the recognized uranium minerals as reported by Steacy and Kaiman (1978). Many of them are secondary. Many uranium minerals are transparent, hence they are studied by using transmitted light microscopy.

More common uranium minerals are uraninite-pitchblende, the many secondary products after uraninite-pitchblende, coffinite and a number of compounds of the element with

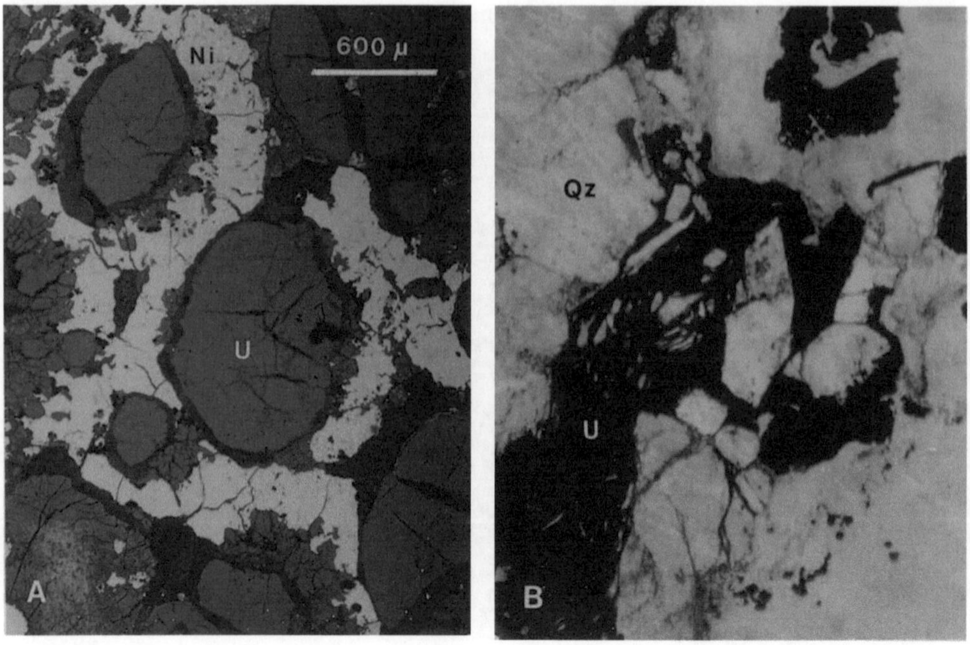

Fig. 49. *Pitchblende forming spheroids associated with niccolite (A) and irregular veins distributed through the quartz (B).*

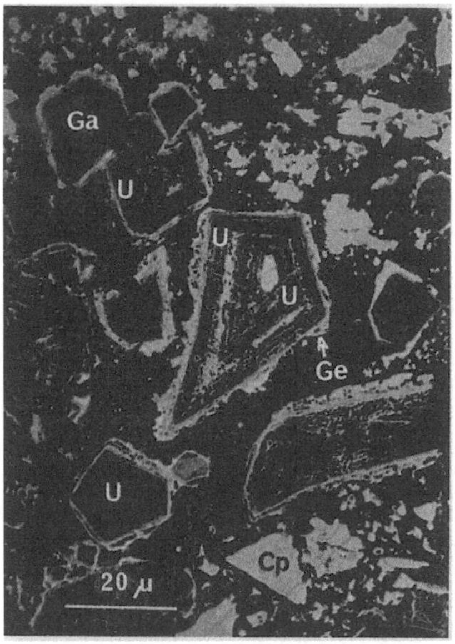

Fig. 50. *Pitchblende finely intergrown with gersdorffite, galena and chalcopyrite.*

titanium, niobium, tantalum, rare earths, and thorium commonly referred to as the complex multiple oxides. Uranothorite and thucholite are also important uranium carriers. Minor amounts of uranium are sometimes detected in some of the accessory minerals within the host rock.

URANINITE-PITCHBLENDE. The terms uraninite and pitchblende refer to uranium oxides the ideal formula of which is UO_2, and in which the uranium is usually oxidized to some extent from the tetravalent to the hexavalent state. The reported limit of oxidation is $UO_{2.6}$ (Steacy and Kaiman) or U_3O_8. Although the two minerals have the same crystal structure, in North America the term **uraninite** is *traditionally* used to indicate specimens characterized by sub to euhedral crystalline shapes (figure 48), by the presence in their lattice of minor amounts of thorium and rare earths, and by their association with high temperature environments such as granites and pegmatites and their metamorphic facies. An example of this type of occurrence are the pegmatite ores of Bancroft, Ontario.

The term **pitchblende** is used to characterize specimens with no observable crystal form (in the hand sample or under the microscope), but rather showing botroidal-colloform textures, or forming spheroids (figure 49A) which may be coalescing, or simply distributed as veins or other anhedral masses in the host rock (figure 49B). Pitchblende commonly occurs in low temperature environments in simple mineral assemblages, for example quartz-pitchblende, like in figure 49B, or in complex nickel-cobalt associations, like in figures 49A and 50. The occurrences of the Athabaska basin in Saskatchewan are an example of the latter.

Both uraninite and pitchblende are dull gray with low reflectivity, when observed in reflected light microscopy, and opaque when observed in transmitted light microscopy.

The terms uraninite and pitchblende **are not** indicative of two different chemical compositions, nor of two different oxidation states of uranium, as some metallurgists believe.

SECONDARY PRODUCTS AFTER URANINITE-PITCHBLENDE. Uraninite and pitchblende become readily oxidized and decomposed especially when they occur in acid environments. Their alteration products show an extremely large variety of chemical compositions and optical properties, as they include hydrous oxides, silicates and, when in the chemically viable environments, phosphates, arsenates, sulfates, vanadates, etc. These secondary deposits may reach sizes of economic value. Secondary alteration products after uraninite and pitchblende commonly have bright colors in the hand sample and under the microscope such as yellow, red, orange, green and brown and, because their alteration

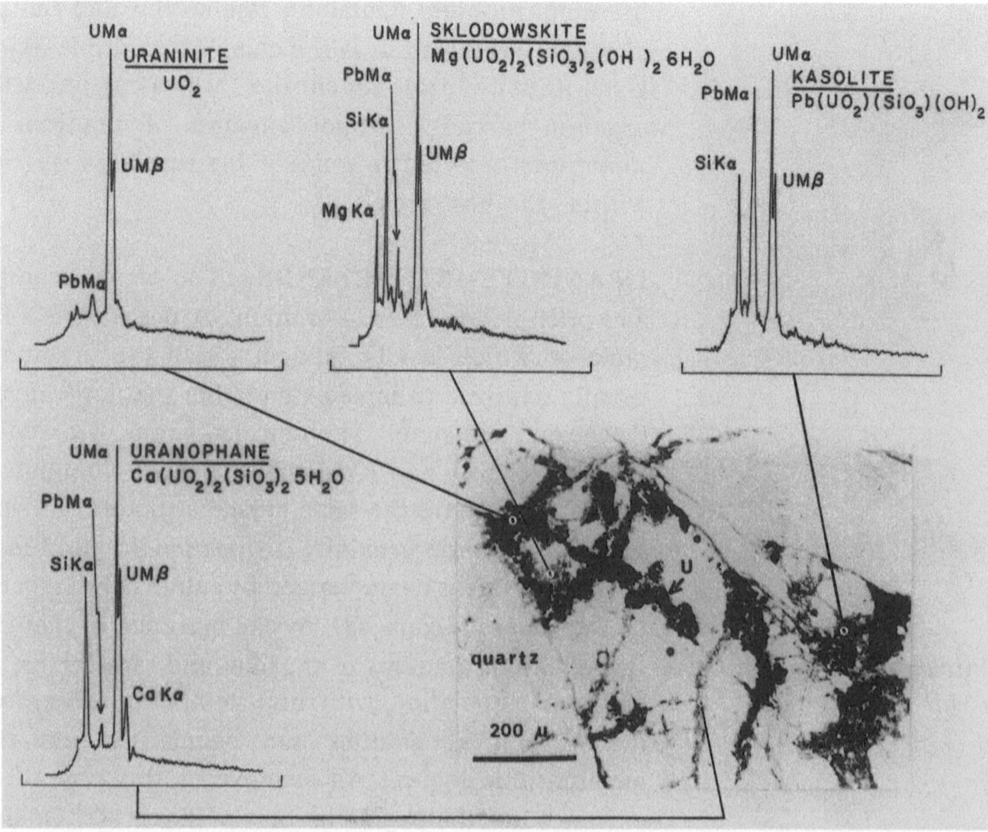

Fig. 51. *Secondary uranium minerals after uraninite showing a wide variety of chemical compositions within very short intervals.*

develops gradually, they show no definite boundaries, but rather replace one another through subtle changes in color and chemical composition (figure 51). A large number of these products, formerly all known by the general term of **gummite**, have now been identified and named. In the author's experience, the changes in chemistry are so great and so subtle that probably many unreported species still exist. Some of the more common of the secondary uranium minerals are **uranophane, weeksite, kasolite, sklodowskite, autunite and carnotite.**

COFFINITE. Coffinite is a silicate of uranium the ideal formula of which is $USiO_4$, and which may contain variable amounts of hydroxyl to give the formula $U(SiO_4)_{1-x}(OH)_{4x}$. The mineral is gray with low reflectivity, when observed in reflected light, and rather similar to pitchblende with which it may be confused. It may also be confused with carbonaceous material with which is sometimes associated.

COMPLEX MULTIPLE OXIDES. Brannerite is the most important of the complex multiple oxides. When occurring in discrete grains that are coarse enough to be observed under a moderate microscope magnification, brannerite is reddish-brown to completely opaque (depending on the titanium to uranium ratio) in transmitted light, and gray with low reflectivity in reflected light (figure 52A) Besides in single crystals, brannerite also occurs in fine intergrowths with rutile with or without quartz and sericite as shown in figure 52B.

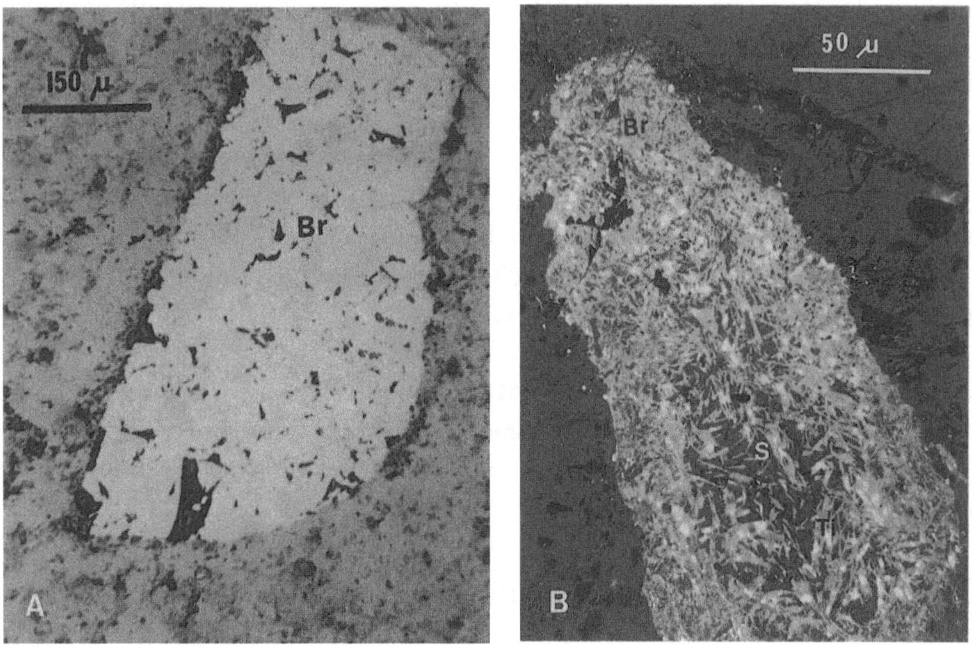

Fig. 52. *A homogeneous grain of brannerite (A) and an intergrowth of brannerite-rutile-sericite (B).*

This latter occurrence is typical of much of the brannerite at Elliot Lake, Ontario where uraninite is the other major uranium mineral.

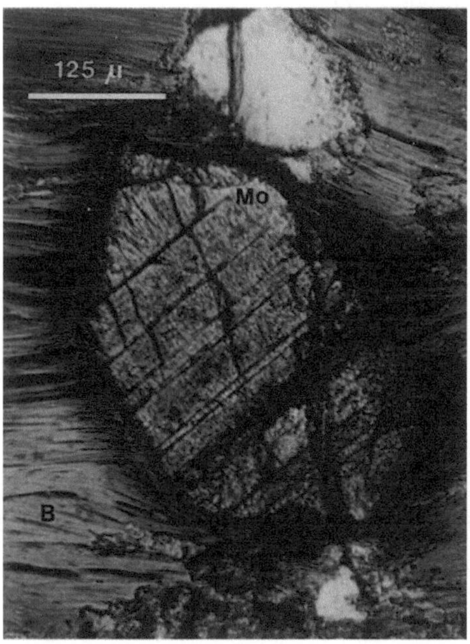

Fig. 53. *Uranium-bearing monazite in biotite. Note the radioactive halo.*

The other complex multiple oxides are relatively common, but do not generally occur in deposits of large proportions. These oxides are **davidite, betafite, euxenite, pyrochlore** and **samarskite.**

Unlike uraninite-pitchblende, the complex multiple oxides are stable under most environment conditions and there are very few secondary minerals resulting from their alteration.

URANOTHORITE. Uranothorite is another uranium mineral which may be of economic value, as it contains the metal in amounts of up to 10 per cent. Uranothorite is often associated with uraninite.

THUCHOLITE. The term thucholite, derived from the chemical symbols of its major elements (Th, U, C, H and O), was originally used by Ellsworth (1928) to describe a radioactive solid hydrocarbon from a pegmatite dike in the vicinity of Parry Sound, in Ontario. Today the name is used to denote certain natural solid hydrocarbonaceous materials that contain uranium and/or other radioactive elements, especially thorium. The best example of a thucholite deposit is provided by the extensive uraniferous carbonaceous materials in the Witwatersrand conglomerates of South Africa. In more recent years the organic fractions in thucholite were identified as being primarily terrestrial plant derivatives or coal (Vassiliou and Kerr 1973), while the uranium and related urano-organic material are primarily in the form of highly disseminated mineral grains such as uraninite and coffinite (Vassiliou 1980). Therefore, according to some authors, thucholite is not a true mineral, but a mixture of uranium minerals and natural organic derivatives.

MISCELLANEOUS ACCESSORY MINERALS. Accessory minerals such as **monazite** (figure 53), **apatite, zircon, sphene and allanite**, sometimes contain uranium in minor or trace amounts. Because of the small amounts of uranium which they carry, these minerals are seldom of economic value.

Grain size, host minerals and mineral associations

Economic amounts of uranium are represented by concentrations which are considerably larger than those of the precious metals, therefore these three variables affect extractions in a limited manner. Grain sizes are coarse, and liberation and/or exposure are usually achieved without excessive effort. An exception may be the fine intergrowths of uranium minerals with complex nickel-cobalt sulfide/sulfosalts, like those of the Athabaska basin shown in figure 50 (page 209), which need special laboratory conditions during recovery. (See case study 22 on page 217 and the discussion on page 219).

Minerals other than the host mineral

These are all the minerals which dissolve in the leaching agents, or which in other ways interfere with the leaching processes used for uranium recovery. They may include carbonates, sulfides, oxides and clays, and are discussed in the section on metallurgy.

Mineral stability

Although uraninite-pitchblende alter easily in the natural environment, mineral instability poses no problem during the uranium recoveries. The oxidation process which produces the secondary minerals, results in greater uranium solubility and therefore eases and expedites the process of extraction.

METHOD OF STUDY

The presence, abundance and distribution of uranium in ore samples are rather simple operations when compared to those applied to the determination of most other economically important elements.

Due to its radioactive nature, the presence of uranium is easily detected in the field by geiger counters and, if confirmed by chemical analysis (which also provides the correct abundance), its distribution within the individual samples is just as easily determined by autoradiographs. Autoradiographs may give an indication of the minerals with which the element is associated and, in some cases, may suggest the origin of the mineralization, for example, formed at the same time as the host rock (if in a compact formation) or secondary (if distributed in fractures, figure 54).

Fig. 54. *Rock sample (A) and autoradiograph (B) showing that the radioactivity is distributed in a fracture and therefore of secondary origin.*

Autoradiographs are obtained by exposing a smoothed surface of the sample to a photographic plate for a length of time which varies according to the intensity of the radioactivity and the speed of the photographic plate. If the speed of the plate is known, the time of exposure can be estimated roughly by using a geiger counter. The quality of the autoradiograph and the time involved in its preparation depend on the film used. Both x-ray and ordinary photographic films give images, although they seem to be sensitive to different radiations and, in some cases, give different images for the same sample. Polaroid films give very fast impressions and they are especially adapted to field exploration due to the minimal amount of extra equipment which they require for development. Radioluxography, involving the use of a screen of silver-activated zinc sulfide between the sample and the photographic plate (Dooley 1958, Dooley et al 1977, Ragland 1964) provides im-

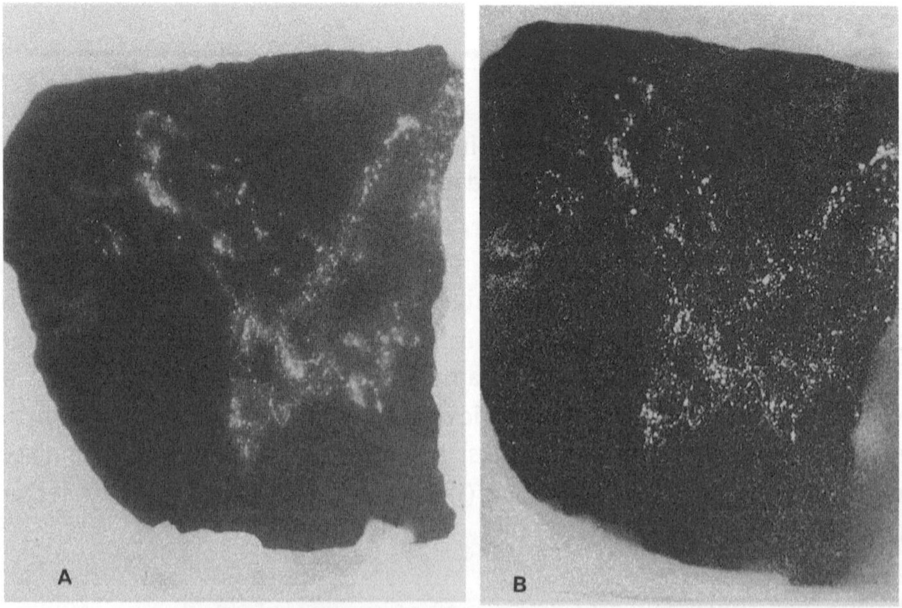

Fig. 55. *Autoradiographs of the same rock sample taken by using a regular (orthochromatic) photographic film (A) and a polaroid film with a screen of silver activated zinc sulfide (B). The time of exposure was 10 days for A and 24 hours for B. The combined use of the zinc sulfide screen and polaroid film improves the resolution of the image and increases the speed of determination (Photographs provided by Martha Sizgoric, Inco Metals).*

proved resolution of the uranium distribution image, supposedly over a shorter period of time (figure 55).

Autoradiographs of polished chips or mounts and polished-thin sections (figure 56) can be invaluable to the uranium researcher because they indicate the exact location of the radioactive minerals. This is particularly important when the uranium is distributed in fine particles and/or intergrown with non-radioactive minerals. In such instances, the uranium-bearing phases may not be coarse enough to show distinctive properties for location and identification by the optical microscope alone. Autoradiographs of polished and polished-thin sections may also prove useful in those situations where the uranium is present in minor amounts in minerals which do not normally contain the metal (for example, zircon, apatite, monazite, etc).

Whereas the presence, abundance and distribution of uranium are determined relatively easily and inexpensively by geiger counters-chemical analysis-autoradiographs, the next step in the study of uranium ores, that is the identification and description of the uranium mineralization, is rather complex and can be time consuming. This is due to:

- The large number of existing uranium minerals, particularly secondary ones, representing many alteration stages within very limited often microscopic areas.
- The lack of distinctive optical properties of some of the more common ones, such as uraninite and coffinite.

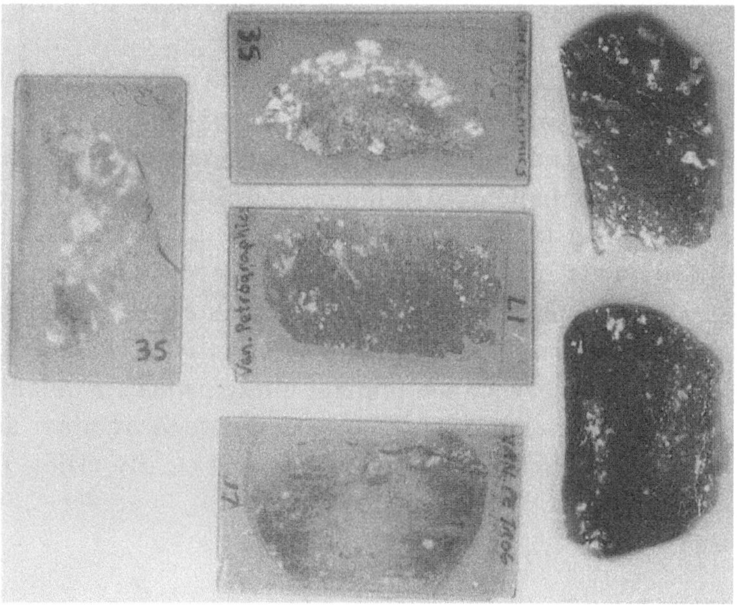

Fig. 56. *Autoradiographs of polished-thin sections and polished rock chips.*

- The fine intergrowths which they sometimes form with other non-radioactive minerals.
- The metamict state of some of them. (The term *metamict* is defined in chapter 4, page 54).

Both EDS-SEM and electron microprobe techniques, as well as x-ray diffraction methods (Morton 1978) are used for the identification of uranium minerals. X-ray diffraction methods are, or at least have been used successfully for the study of many uranium mineral deposits, due to their higher abundance and coarser grain sizes. Yet, the EDS-SEM and microprobe give in many (although not all) instances more reliable results.

Identification of uraninite and pitchblende is possible by x-ray diffraction methods and by EDS-SEM/electron microprobe techniques with equally satisfactory results. Identification of the secondary products, on the other hand, is best accomplished by the EDS-SEM or microprobe, due to their often amorphous state, fine grains sizes and to their subtle changes in chemistry within very short intervals of the polished or polished-thin section.

Identification of coffinite is obtained by x-ray diffraction methods and EDS-SEM/electron microprobe techniques. Identification may be particularly difficult where coffinite is associated with pitchblende and finely intergrown with silicate minerals. In many such instances, the EDS-SEM or electron microprobe will not resolve the coffinite-pitchblende from the silicate, due to the difficulty to decide whether the silicon peak is emitted by the coffinite or by the intergrown silicate. If sufficient material is available, x-ray diffraction techniques may be more successful, if a competent scientist capable of resolving the various lines, is in charge of the identification.

Identification of the complex-multiple oxides is best obtained by EDS-SEM or electron microprobe techniques. Due to their metamict state, their identification by x-ray diffraction methods may be difficult and time consuming. Particularly difficult may be the identification of brannerite by the microprobe, where the mineral is finely intergrown with rutile or other titanium rich phases. In such a situation, it may not become immediately apparent whether we are dealing with (i) a titanium-rich brannerite of variable composition, or with (ii) an intergrowth of brannerite and rutile, or with (iii) an intergrowth of pitchblende and rutile. Fine associations of uranium and titanium minerals are common, but the uranium phase is not necessarily brannerite. Intergrowths such as these may also be difficult to resolve optically, due to the poor polish which they often take. Best results in the resolution and identification of uranium-titanium mineral intergrowths are obtained by the EDS-SEM: due to the large difference in atomic number between uranium and titanium, the two phases, when two phases are present, are resolved and clearly visible for analysis on the display screen (figure 57).

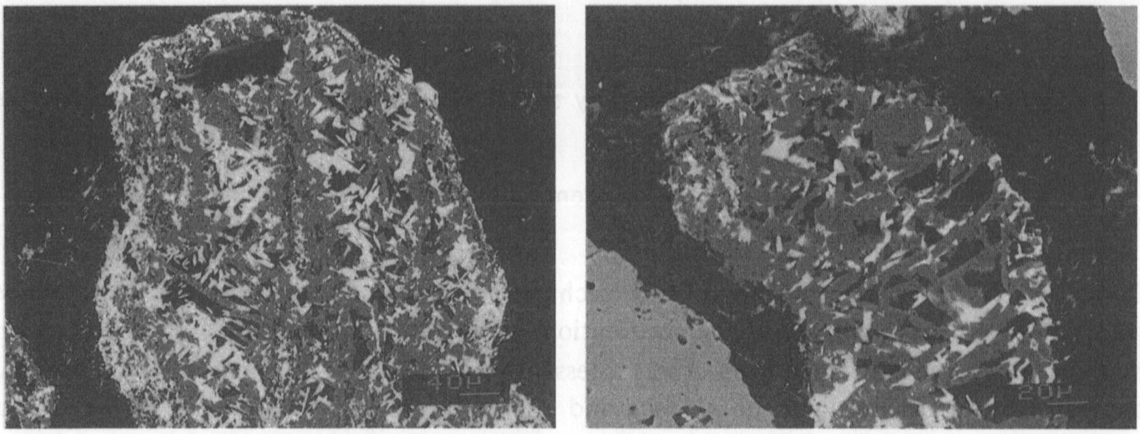

Fig. 57. *Scanning electron microscope photographs of two grains consisting of intergrowths of brannerite (white) and rutile. (Photographs provided by Martha Sizgoric, Inco Metals).*

Minor amounts of uranium in minerals which are not usually uranium-bearing are best detected by the microprobe, because x-ray diffraction methods do not give information on minor element distributions. And because, due to the poor limits of detection, EDS-SEM determinations would only be valid for higher uranium contents (above 2 or 3 per cent).

Case study 22. Uranium-nickel mineralization consisting of fine intergrowths. Part of the uranium associated with a nickel-arsenic mineralization from the Athabaska basin in Saskatchewan, was being lost in the nickel-arsenic concentrate. X-ray diffraction studies on some of the radioactive grains, isolated by autoradiographing the concentrates, were not helpful because they only showed the patterns of gersdorffite and niccolite. The microanalytical study revealed that pitchblende and secondary uranium minerals were finely intergrown with nickel minerals and, less commonly, with galena like in figure 50 on page 209. X-ray diffraction studies had not been successful because the patterns were either masked by the stronger arsenide and sulfoarsenide lines or were just not obtained because the minerals (or some of the minerals) were amorphous. In addition to the identification of the different uranium minerals, the microanalytical studies contributed the description of the fine intergrowths responsible for the impermeability of much of the uranium. Such intergrowths had not been discovered before. Conclusions were that, because the uranium and nickel minerals were so finely associated, a leaching agent or stronger leaching conditions to dissolve all the phases simultaneously was necessary before separation of the individual components.

METALLURGY

Numerous specific and general papers are available on both the theory and practice of uranium extractions. The most comprehensive of these is probably the one by Merrit (1971).

Uranium is recovered from its ores by leaching in either acid or alkaline treatment. Most commercial operations utilize an acid circuit with sulfuric acid as lixiviant. Nitric and hydrochloric acids are less widely used for several reasons: both acids are more expensive than sulfuric acid and both acids form uranyl ions in dilute solutions which are incompatible with anion exchange resins. Hydrochloric acid is extremely corrosive and cause severe material handling problems. There are strict environmental regulations on the permissible concentrations of nitrate and chloride in tailing effluents. The following discussion is restricted to sulfuric acid leaching.

Before proceeding, it would be informative to provide a background on the behavior of sulfuric acid in aqueous solution and the dissolution of uranium in sulfuric acid. The ionic composition of the leach liquor is important in leaching studies and also in subsequent purification steps.

In aqueous solution, sulfuric acid ionizes to form sulfate, bisulfate and hydrogen ions as shown in the following reactions:

$$H_2SO_4 + H_2O = H_3O^+ + HSO4^-$$

$$HSO_4^- + H_2O = H_3O^+ + SO_4^{=}$$

At high acid concentrations, the bisulfate ion is the main anionic species, whereas at low acid concentrations, the sulfate ion is the main species.

The dissolution of uranium in sulfuric acid produces uranyl sulfate and complex uranyl sulfate anions. The actual uranyl sulfate species present in the leach solution is determined by the acid concentration, by the uranium concentration and by other factors such as temperature. The proposed reactions are as follows:

$$UO_3 + 2H^+ = UO_2^{++} + H_2O$$

$$UO_2^{++} + SO_4^{=} = UO_2SO_4$$

$$UO_2SO_4 + SO_4^{=} = UO_2(SO4)_2^{=}$$

$$UO_2(SO4)_2^{=} + SO_4^{=} = UO_2(SO_4)_3^{4-}$$

The mineralogy of the uranium ore will suggest the initial conditions with which to examine the leaching characteristics of that particular ore and in this regard, it is important to remember that the chemistry of the leaching reaction of pure uranium minerals must give way to the compromise of the multi-component systems treated in practice.

The main process variables controlling the leaching reaction are grinding size, acid concentration, oxidant addition, temperature, retention time and pulp density. Understanding the mechanism of the leaching reaction is essential for optimization of the process variables which, because of the variation and complexity of the uranium ores, require a comprehensive test work program for each individual case.

There are two fundamental criteria in the uranium leaching reaction. First there must be sufficient free acid to dissolve the uranium mineral, and second, an oxidizing environment must be provided to assure that all the uranium is in the hexavalent state.

The free acid concentration of the leach solution is the difference between the acid addition and the acid consumption. The acid consumption of the uranium ore is a function of the leaching conditions required to extract the uranium and of the composition of the host rock. Common acid consumers in the host rock are calcium, magnesium, and iron carbonates, iron sulfides and oxides, and finely divided weathered or metamorphosed silicate minerals such as those in clays and gneisses, which may dissolve to form aluminum sulfate, silicic acid and colloidal silica.

The second criterion, that of maintaining or producing an oxidizing environment in the leach solution, has been established because, while uranium occurs naturally in either the tetravalent or hexavalent state, only the latter is soluble in acid solution. Tetravalent uranium is insoluble in non-oxidizing systems such as dilute sulfuric acid and consequently the first step in the leaching reaction is an oxidation step from tetravalent to hexavalent uranium prior to dissolution.

Ferric iron is considered to be the oxidizing species. The criterion is to maintain a Fe^{+3}/Fe^{+2} concentration ratio such that the oxidation-reduction potential in the leach pulp is greater than 425 mV as measured with a platinum saturated calomel electrode. The oxidants most commonly employed to maintain this potential are sodium chlorate, manganese dioxide and air or oxygen. The following equations are used to describe the reaction:

$$U^{+4} + 2Fe^{+3} = U^{+6} + 2Fe^{+2}$$

$$6Fe^{+2} + NaClO_3 + 6H^{+} = 6Fe^{+3} + NaCl + 3H_2O$$

$$2Fe^{+2} + MnO_2 + 4H^{+} = 2Fe^{+3} + Mn^{+2} + 2H_2O$$

$$4Fe^{+2} + O_2 + 4H^{+} = 4Fe^{+3} + 2H_2O$$

* * * * * *

The Olympic Dam project in Australia, which came into production in 1988, represents a deposit of some significance in the production of uranium. The deposit is one of the few which have the distinction of consisting of an ore exploited for copper and uranium. In such deposit, the byproduct sulfuric acid from the copper smelter is used to partially satisfy the requirements of the leach circuit for uranium. (J.C. Taylor 1988, personal communication by letter)

THE PRIMARY URANIUM MINERALS contain variable proportions of tetravalent uranium (100 per cent ideally). As already mentioned, uranium in uraninite is essentially tetravalent, whereas hexavalent anions result from oxidation either through weathering or liberation of oxygen with radioactive disintegration to lead:

$$UO_2 \longrightarrow PbO + 1/2O_2 + (6He)$$

When in simple mineral assemblages, the primary minerals uraninite-pitchblende, coffinite and uranothorite, require only moderate leaching conditions of temperature, acid concentration and oxidant addition to obtain high metal extraction. For example, the pegmatite ores from Bancroft, Ontario, containing uraninite and uranothorite are leached at ambient temperature with 20 to 40 Kg/ton of sulfuric acid and 0.5 Kg/ton of chlorate.

The Athabaska basin matasediments in Saskatchewan contain considerable amounts of nickel arsenides and nickel sulfides (gersdorffite, millerite, niccolite, etc.) in conjunction with pitchblende and coffinite. As shown in figure 50 (page 209), some of the pitchblende/coffinite grains are totally enclosed in the nickel arsenide/sulfide matrix and vice versa. Because of the complex mineral association, it is necessary to oxidize and dissolve the nickel arsenide/sulfide to obtain satisfactory uranium recovery. Consequently, the leaching conditions are more severe than what they are for a simple pitchblende (uraninite)/coffinite ore. Typical leaching conditions are 50°C, 20-40 gpl free sulfuric acid under oxygen pressure in an autoclave.

THE SECONDARY PRODUCTS contain uranium mainly in the hexavalent state. These minerals are readily soluble in acid solution and oxidants are added only to prevent stray reducing reactions from allowing the uranium to reprecipitate.

THE COMPLEX MULTIPLE OXIDES. **Brannerite,** whether intergrown with rutile or forming individual discrete grains, requires a stronger acid attack for its dissolution. The Elliot Lake, Ontario, uranium ores, which consist of quartz conglomerates containing uraninite and brannerite in the pebble matrix, are leached at 70-80°C, with sulfuric acid and 40-60 hours retention time.

Even stronger leaching conditions in the order of 150°C and 1000 gpl sulfuric acid are required to dissolve the metal values from some of the other complex multiple oxides such as **euxenite, betafite** and **pyrochlore.**

"After leaching, the crude solution is separated from the pulp by filtration or settling with the raw mineral solution gradually concentrated as it moves along the processing line. The solution is then purified by various chemical techniques. Precipitation from the purified solution, separation and drying constitute the final stage of the extraction of uranium concentrates. The solution is neutralized with caustic soda, ammonia or magnesia. By controlling the acidity, the iron is precipitated out first and then the uranium, as a uranate, usually sodium diuranate, the principal component of *yellow cake*. After thickening, the uranium precipitate passes through automatic filter presses or vacuum filter drums and is allowed to dry. This crude uranium compound concentrate is in the form a bright clay-like material. It contains about 60 per cent by weight of uranium with a variety of impurities." (Williams 1988, p. 435).

The final purification stages where pure uranium is produced, are represented by hydrometallurgical techniques such as solvent extraction or ion exchange.

A method of extracting uranium from sulfide ores is by **bacterial leaching** (pages 77-78) which uses biological agents to liberate the uranium from its gangue. The technique is employed successfully at The Denison Mine in Elliot Lake.

Heap leaching techniques (page 75-76) similar to those used for gold recoveries, are applied to low grade uranium deposits.

Finally, microanalytical investigations show that uranium leach residues frequently contain a disproportionate amount of uranium in the -200 mesh size fraction (Gasparrini and Williamson 1981). This phenomenon has been attributed to (i) reprecipitation, (ii) absorption, or (iii) insoluble uranium minerals. Studies on the leach residues from several South African mines by Laxen (1964) indicate that insoluble minerals did not themselves account for the uranium content of the leach residues. However, to Williamson's knowledge, little has been published since such work. Like in several other cases mentioned in this book, the field is open to the metallurgist and microanalyst.

TABLE 15 - URANIUM DEPOSITS

DEPOSIT	STRUCTURAL CHARACTER	TYPE LOCALITY	MAIN MINERALS	ASSOC. MINERALS	TYPICAL ELEMENTS
Igneous	Carbonatite, Alkaline syenite	Prairie Lake, Ont.; Nova Beaucage, Ont.; Pocos de Caldas, Brazil	Uranothorianite, pyrochlore, betafite, perovskite	Niccolite, ilmenite, apatite, zircon	Nb, Th, Cu, P, Ti, Zr, RE's
Metamorphic	Pegmatite (alaskite), Skarn	Bancroft, Ont.; Rossing, Namibia	Uraninite, uranothorite, betafite	Molybdenite, fluorite, zircon	Th, Mo, RE's Nb, Ti
Detrital(placer)	Pyritic quartz-pebble conglomerate	Elliot Lake, Ont.; Witwatersrand, South Africa	Uraninite, brannerite	Monazite, pyrite, native Au	Th, Ti, RE's Au, Zr, C
Hydrothermal	Volcanogenic	Rexpar, British Columbia	Uraninite, uranothorite	Fluorite, celestite, pyrite	Th, RE's, Mo Cu, F, Sr
Hydrothermal	Hydrothermal veins	Bokan Mountain, Alaska	Uraninite, Brannerite, Thorite Allanite	Quartz,fluorite, carbonates, sulfides	Th, RE's, Be Nb, Zr
Syngenetyic	Shale, phosphorite, evaporitic limstone, duricrusts (calcrete)	Ranstad, Sweden; Kitts, Labrador; Todilto Limestone, New Mexico; Yeelirrie, W. Australia;	Pitchblende, carnotite	Apatite, gypsum, carbon	P, V, Cu, Co, Ni, As, Ag, C
Epigenetic	Sandstone, tabular roll, etc.	Colorado Plateau; Wyoming and Texas basins	Pitchblende, coffinite, carnotite	Pyrite	Cu, V, Se, Mo, C

TABLE 15 - CONT.

DEPOSIT	STRUCTURAL CHARACTER	TYPE LOCALITY	MAIN MINERALS	ASSOC. MINERALS	TYPICAL ELEMENTS
Epigenetic	Channel conglomerate	Kelowna-Beaverdell district, B.C.; Ninge Toge, Japan	Pitchblende, coffinite, autunite	Marcasite	C
Epigenetic	Lignite	Cypress Hills, Saskatchewan; Dakotas	na	na	Mo, V, C
Epigenetic	Classical veins	Beaverlodge, Sask.; Port Radium, N.W.T.; Schwartzwalder, Colorado	Pitchblende	Pyrite, chalcopyrite, Ni-Co arsenides, native Au and Ag	Cu, Co, Au, Ag, V, Ni, As, Mo, Bi, Se
Syngenetic	Unconformity veins	Wallaston Key and Cluff Districts, Sask.; Midnite, Washington	Pitchblende	Pyrite, Ni-Co arsenides, chalco-pyrite, native Au and Ag	Cu, Co, Au, Ag, Ni, As, Se, Mo, V, C
Supergene	Cappings, enrichments	Bolger and Eldorado, Sask.; Pocos de Caldas, Brazil; Rossing, Namibia	Gummite, uranophane, carnotite, coffinite	na	Cu, Ni, Ag, As, Si, Ca

Page 2

TABLE 16 - URANIUM MINERALS

(Modified after Steacy and Kaiman 1978)

MINERAL	FORMULA
Abernathyite	$K(UO)_2(AsO_4)\ 4H_2O$
Absite	(see brannerite)
Aeschynite-Niobo-aeschynite-Aeschinite-(Y) series	AB_2O_6

Where A = REE's, Fe^{+2}, Mn,Ca, Th, Pb; B = Nb, Ta, Ti, Fe^{+3}

Aeschynite	Ce predominant REE, Ti>Nb
Lyndochite	Th, Nd-rich variety of aeschynite
Niobo-aeschynite	Ce predominant REE, Nb>Ti
Aeschynite -(Y)	Y predominant REE, Ti>Nb

MINERAL	FORMULA
Agrinierite	$(K_2,Ca,Sr)U_3O_{10}\ 4H_2O$
Andersonite	$Na_2Ca(UO_2)(CO_3)_3\ 6H_2O$
Arrenuranospathite	$(H,Al)_{0.5}(UO_2)_2(AsO_4)_2\ 20H_2O$
Arsenuranylite	$Ca(UO_2)_4(AsO_4)_2(OH)_4\ 6H_2O$
Autunite	$Ca(UO_2)_2(PO_4)_2\ 10\text{-}12H_2O$
Autunite group	$A(UO_2)_2(XO_4)_2\ 8\text{-}12H_2O$

Where A = Ca, Na_2, Ba, Mg, Fe^{+2}, Cu, Mn; X = P, As, V

Autunite	Ca,U,P	Saleeite	Mg,U,P
Fritzscheite	Mn,U,V	Na autunite	Na,U,P
Heinrichite	Ba,U,P	Torbenite	Cu,U,P
Kahlerite	Fe,U,As	Uranocircite	Ba,U,P
Novacekite	Mg,U,As	Uranospinite	Ca,U,P
Sabugalite	(H,Al),U,P	Zeunerite	Cu,U,P

MINERAL	FORMULA
Bassetite	$Fe(UO_2)_2(PO_4)_2\ 8H_2O$
Bauranoite	$BaU_2O_7\ 4\text{-}5H_2O$
Bayleyite	$Mg_2(UO_2)(CO_3)_3\ 18H_2O$
Becquerelite	$CaU_6O_{19}\ 11H_2O$
Bergenite	$Ba(UO_2)_4(PO_4)_2(OH)_4\ 8H_2O$
Betafite	see pyrochlore group
Beta-uranophane	$Ca(UO_2)_2Si_2O_7\ 6H_2O$
Billietite	$BaU_6O_{19}\ 11H_2O$
Boltwoodite	$K_2(UO_2)_2(SiO_3)_2(OH)_2\ 5H_2O$
Brannerite	$(U,Ca,Ce)(Ti,Fe)_2O_6$
Absite	Th rich variety
Calciosamarskite	see samarskite
Calciouranoite	$(Ca,Ba,Pb)U_2O_7\ 5H_2O$
Calcurmolite	$Ca(UO_2)_3(MO_4)_3(OH)_2\ 11H_2O$
Carnotite	$K_2(UO_2)_2(VO_4)_2\ 3H_2O$
Clarkeite	$(Na,Ca,Pb)_2U_2(O,OH)_7$
Cliffordite	UTe_3O_9
Co-zippeite	$Co(UO_2)_6(SO_4)_3(OH)_{10}\ 16H_2O$
Coconinoite	$Fe_2^{+3}Al_2(UO_2)_2(PO_4)_4(SO_4)(OH)_2\ 20H_2O$
Coffinite	$U(SiO_4)_{1-x}(OH)_{4x}$
Compreignacite	$K_2U_6O_{19}\ 11H_2O$
Cousinite	$MgU_2Mo_2O_{13}\ 6H_2O$
Cuprosklodowskite	$Cu(UO_2)_2Si_2O_7\ 6H_2O$
Curienite	$Pb(UO_2)_2(VO_4)_2\ 5H_2O$
Curite	$Pb_2U_5O_{17}\ 4H_2O$
Davidite	$(Fe^{+2},La,U,Ca)_6(Ti,Fe^{+3})_{15}(O,OH)_{36}$
Demesmaekerite	$Pb_2Cu_5(UO_2)_2(SeO_3)_6(OH)_6\ 2H_2O$
Derriksite	$Cu_4(UO_2)(SeO_3)_2(OH)_6\ H_2O$
Dewindtite	$Pb(UO_2)_2(PO_4)_2\ 3H_2O$

TABLE 16 - CONT.

MINERAL	FORMULA
Euxenite-polycrase series	AB_2O_6
Where A = REE's, Fe^{+2}, Mn, Ca, Th, U, Pb; B = Nb, Ta, Ti, Fe^{+3}	
Euxenite	$TiO_2:Nb_2O_5+Ta_2O_5$<or=3:166
Polycrase	$TiO_2:Nb_2O_5+Ta_2O_5>3:1$
Dumontite	$Pb_2(UO_2)_3(PO_4)_2(OH)_4\ 3H_2O$
Ekanite	$(Th,U)(Ca,Fe,Pb)_2Si_8O_{20}$
Ellsworthite	(synonym for uranpyrochlore)
Fergusonite-Formatite series	ABO_4
Where A = REE's, U, Th, Zr, Fe^{+2}, Ca; B = Nb, Ta, Ti	
Fergusonite	essentially $YNbO_4$
Formanite	essentially $YTaO_4$
Furongite	$Al_2(UO_2)(PO_4)_2(OH)_2\ 8H_2O$
Fourmarierite	$PbU_4O_{13}\ 4H_2O$
Francevillite	$Ba(UO_2)_2(VO_4)_2\ 5H_2O$
Fritzscheite	$Mn(UO_2)_2(VO_4)_2\ 10H_2O$
Grimselite	$K_3Na(UO_2)(CO_3)_3\ H_2O$
Guilleminite	$Ba(UO_2)_3(SeO_3)_2(OH)_4\ 3H_2O$
Gummite	generic term for alteration prods. of uraninite and pitchblende
Haiweeite	$Ca(UO_2)_2Si_6O_{15}\ 5H_2O$
Hallimondite	$Pb_2(UO_2)(AsO_4)_2$
Hatchettolite	synonim for uranpyrochlore
Heinrichite	$Ba(UO_2)_2(AsO_4)_2\ 10\text{-}12H_2O$
Hjelmite	$(Y,Fe,U)(Nb,Ta,Sn)_2O_6$
Huegelite	$Pb_2(UO_2)_3(AsO_4)_2(OH)_4\ 3H_2O$
Ianthinite	$UO_2\ 5UO_3\ 10H_2O$
Iriginite	$(UO_2)Mo_2O_7\ 3H_2O$
Ishikawaite	$(U,Fe,Y,Ca)(Nb,Ta)O_4$
Johannite	$Cu(UO_2)_2(SO_4)_2(OH)_2\ 6H_2O$
Joliotite	$(UO_2)CO_3\ nH_2O$
Kahlerite	$Fe(UO_2)_2(AsO_4)_2\ nH_2O$
Kasolite	$Pb(UO_2)SiO_4\ H_2O$
Kobeite	$(Y,U)(Ti,Nb)_2$90,$(O,OH)_6$
Lermontovite	$(U,Ca,Ce)_3(PO_4)_4\ 6H_2O$
Liebigite	$Ca_2U(CO_3)_4\ 10H_2O$
Lyndochite	see aeschynite series
Mg zippeite	$Mg_2(UO_2)_6(SO_4)_3(OH)_{10}\ 16H_2O$
Marthozite	$Cu(UO_2)_3(SeO_3)_3(OH)_2\ 7H_2O$
Masuyite	oxide of Pb and U
Mckelveyite	$(Na,Ca)(Ba,Y,U)_2(CO_3)_3\ 1\text{-}2\ H_2O$
Meta-ankoleite	$K_2(UO_2)_2(PO_4)_2\ 6H_2O$
Meta-autunite	$Ca(UO_2)_2(PO_4)_2\ 2\text{-}6\ H_2O$
Meta-autunite group	$A(UO_2)_2(XO_4)_2\ 4\text{-}8H_2O$

Where A = K_2, $(NH_4)_2$, Ca, Ba, Mg, Fe^{+2}, Co, Cu, Zn;
X = P, As

Abernathyite	K,U,As	Metanovaoekite	Mg,U,As
Bassetite	Fe,U,P	Metatorbernite	Cu,U,P
Meta-ankoleite	K,U,P	Meta-uranocircite	Ba,U,P
Meta-autunite	Ca,U,P	Meta-uranospinite	Ca,U,As
Metaheinrichite	Ba,U,As	Metazeunerite	Cu,U,As
Metakahlerite	Fe,U,As	Na-uranospinite	(Na,Ca),U,As
Metakirchheimerite	Co,U,As	Uramphite	NH_4,U,P
Metalodevite	Zn,U,As		

Metacalciouranoite	$(Ca,Na,Ba)U_2O_7\ 2H_2O$
Metaheinrichite	$Ba(UO_2)_2(AsO_4)_2\ 8H_2O$
Metakahlerite	$Fe^{+2}(UO_2)_2(AsO_4)_2\ 8H_2O$
Metakirchheimerite	$Co(UO_2)_2(AsO_4)_2\ 8H_2O$
Metalodevite	$Zn(UO_2)_2(AsO_4)_2\ 10H_2O$
Metanovacekite	$Mg(UO_2)_2(AsO_4)_2\ 4\text{-}8H_2O$
Metaschoepite	$UO_3\ nH_2O$
Metatorbernite	$Cu(UO_2)_2(PO_4)_2\ 8H_2O$

TABLE 16 - CONT.

MINERAL	FORMULA
Metatyuyamunite	$Ca(UO_2)_2(VO_4)_2$ 3-5H_2O
Meta-uranocircite	$Ba(UO_2)_2(PO_4)_2$ 8H_2O
Meta-uranopilite	$(UO_2)_6(SO_4)(OH)_{10}$ 5H_2O
Meta-uranospinite	$Ca(UO_2)_2(AsO_4)_2$ 8H_2O
Metavandendriesscheite	PbU_7O_{22} nH_2O (n<12)
Metavanuralite	$Al(UO_2)_2(VO_4)_2(OH)$ 8H_2O
Metazellerite	$Ca(UO_2)(CO_3)_2$ 3H_2O
Metazeunerite	$Cu(UO_2)_2(AsO_4)_2$ 8H_2O
Microlite	see pyrochlore group
Moctezumite	$Pb(UO_2)(TeO_3)_2$
Moluranite	$H_4U^{+4}(UO_2)_3(MoO_4)_7$ 18H_2O
Ni Zippeite	$Ni_2(UO_2)_6(SO_4)_3(OH)_{10}$ 16H_2O
Ningyoite	$(U,Ca,Ce)_2(PO_4)_2$
Nb aeschynite	see aeschynite series
Novacekite	$Mg(UO_2)_2(AsO_4)_2$ 12H_2O
Obruchevite	synonim for yttropyrochlore
Paraschoepite	UO_3 2H_2O
Parsonsite	$Pb_2(UO_2)(PO_4)_2$ 2H_2O
Phosphuranylite	$Ca(UO_2)_3(PO_4)_2$ 2H_2O
Phurcalite	$Ca_2(UO_2)_3(PO_4)_2(OH)_4$ 4H_2O
Pitchblende	see uraninite-thorianite-cerianite
Plumbopyrochlore	see pyrochlore group
Polycrase	see euxenite-polycrase series
Priorite	synonim for aeschynite-(Y)
Przhevalskite	$Pb(UO_2)_2(PO_4)_2$ 4H_2O
Pseudo-autunite	$(H_3O)_4Ca_2(UO_2)_2(PO_4)_4$ 5H_2O
Pyrochlore group	$A_{2-m}B_2O_6(O,OH,F)_{1-n}$ pH_2O
Where A = Na, Ca, K, Sn, Ba, REE's, Pb, Bi, U; B = Nb, Ta, Ti	
Pyrochlore	Nb+Ta>2Ti and Nb>Ta
Uranopyrochlore	high in U
Plumbopyrochlore	high in Pb
Yttriopyrochlore	high in Y
Microlite	Nb+Ta>2Ti and Ta> or = Nb
Betafite	2Ti> or = Nb+Ta
Rabbittite	$Ca_3Mg_3(UO_2)_2(CO_3)_6(OH)_4$ 18H_2O
Rameauite	$K_2CaU_6O_{20}$ 9H_2O
Rauvite	$Ca(UO_2)_2V_{10}O_{28}$ 16H_2O
Renardite	$Pb(UO_2)_4(PO_4)_2(OH)_4$ 7H_2O
Richetite	oxide of Pb and U
Roubaultite	$Cu_2(UO_2)_3(OH)_{10}$ 5H_2O
Rutherfordine	$(UO_2)CO_3$
Sabugalite	$HAl(UO_2)_4(PO_4)_4$ 16H_2O
Saleeite	$Mg(UO_2)_2(PO_4)_2$ 8H_2O
Samarskite	$(Y,Ce,U,Ca,Pb)(Nb,Ta,Ti,Sn)_2O_6$
Ca-samarskite	relatively high in Ca
Schmitterite	$(UO_2)TeO_3$
Schoepite	UO_3 2H_2O
Schroeckingerite	$NaCa_3(UO_2)(CO_3)_3(SO_4)F$ 10H_2O
Sedovite	$U(MoO_4)_2$
Sengierite	$Cu(UO_2)_2(VO_4)_2$ 8-10 H_2O
Sharpite	$(UO_2)CO_3$ H_2O
Sklodowskite	$Mg(UO_2)_2Si_2O_7$ 6H_2O
Soddyite	$(UO_2)_5Si_2O_9$ 6H_2O
Na autunite	$Na_2(UO_2)_2(PO_4)_2$ 8H_2O
Na boltwoodite	$(H_3O)(Na,K)(UO_2)(SiO_4)$ H_2O
Na uranospinite	$(Na_2,Ca)(UO_2)_2(AsO_4)_2$ 5H_2O
Na zippeite	$Na_4(UO_2)_6(SO_4)_3(OH)_{10}$ 4H_2O
Strelkinite	$Na_2(UO_2)_2(VO_4)_2$ 6H_2O
Studtite	UO_4 4H_2O
Swartzite	$CaMg(UO_2)(CO_3)_2$ 12H_2O
Thorianite	see uraninite-thorianite-cerianite
Thorite	$ThSiO_4$ usually with U
Uranothorite	U rich variety
Thorogummite	$Th(SiO_4)_{1-x}(OH)_{4x}$

TABLE 16 - CONT.

MINERAL	FORMULA
Thucholite	U hydrocarbon
Torbernite	$Cu(UO_2)_2(PO_4)_2\ 8\text{-}12H_2O$
Troegerite	$(UO_2)_3(AsO_4)_2\ 12H_2O$
Tyuyamunite	$Ca(UO_2)_2(VO_4)_2\ 5\text{-}8H_2O$
Umohoite	$(UO_2)MoO_4\ 4H_2O$
Uramphite	$(NH_4)(UO_2)(PO_4)\ 3H_2O$
Uraninite-Thorianite-Cerianite series	
Uraninite	Essentially UO_2, normally with Th and REE's
Th uraninite	Th rich uraninite
Pitchblende	Essentially UO_2 with little or no Th or REE's
Thorianite	Essentially ThO_2 normally with U and REE's
U thorianite	U rich thorianite
Cerianite	CeO_2
Uranite	General term for phosphates and arsenates of the autunite and and meta-autunite groups
U thorianite	see uraninite-thorianite-cerianite
Uranocircite	$Ba(UO_2)_2(PO_4)_2\ 12H_2O$
Uranophane	$Ca(UO_2)_2Si_2O_7\ 6H_2O$
Uranopilite	$(UO_2)_6(SO_4)(OH)_{10}\ 12H_2O$
Uranosphaerite	$Bi_2U_2O_9\ 3H_2O$
Uranospinite	$Ca(UO_2)_2(AsO_4)_2\ 10H_2O$
Uranopyrochlore	see pyrochlore group
Uvanite	$U_2V_6O_{21}\ 15H_2O$
Uranothorite	see thorite
Vandenbrandeite	$CuUO_4\ 2H_2O$
Vandendriesscheite	$PbU_7O_{22}\ 12H_2O$
Vanuralite	$(UO_2)_2Al(VO_4)_2(OH)\ 11H_2O$
Vanuranylite	$(H_3O,Ba,Ca,K)_{1.6}(UO_2)_2(VO_4)_2\ 4H_2O$
Voglite	$Ca_2Cu(UO_2)(CO_3)_4\ 6H_2O$
Walpurgite	$(BiO)_4(UO_2)(AsO_4)_2\ 3H_2O$
Weeksite	$K_2(UO_2)_2Si_6O_{15}\ 4H_2O$
Widenmannite	$Pb_2(UO_2)(CO_3)_3$
Woelsendorfite	$(Pb,Ca)U_2O_7\ 2H_2O$
Wyartite	$Ca_3U^{+4}(UO_2)_6(CO_3)_2(OH)_{18}\ 3\text{-}5H_2O$
Yttrotantalite	$(Y,U,Fe)(Ta,Nb)O_4$
Yttrocolumbite	$(Y,U,Fe)(Nb,Ta)O_4$
Yttrocrasite	$(Y,Th,U,Ca)Ti_2(O,OH)_6$
Zellerite	$Ca(UO_2)(CO_3)_2\ 5H_2O$
Zeunerite	$Cu(UO_2)_2(AsO_4)_2\ 10\text{-}16H_2O$
Zn zippeite	$Zn_2(UO_2)_6(SO_4)_3(OH)_{10}\ 16H_2O$
Zippeite	$K_2(UO_2)_6(SO_4)_3(OH)_{10}\ 4H_2O$
Zippeite group	Hydrous uranium sulfates (containing Co, Mg, Ni, Na or Zn in the place of K)

✱ There is presently limited agreement regarding the correct full definition of the non-precious metals. One definition by a metallurgist was given on page 4. The following classification is provided by A.M. Evans (1987) in **AN INTRODUCTION TO ORE GEOLOGY**: (a) non-ferrous metals (copper, lead, zinc, tin, aluminum); (b) iron and ferro-alloy metals (iron, manganese, nickel, chromium, molybdenum, tungsten, vanadium, cobalt); (c) minor metals and related non-metals (antimony, arsenic, beryllium, bismuth, cadmium, magnesium, mercury, rare earths, selenium, tantalum, tellurium, titanium, zirconium). (G.C. Amstutz, personal communication). In this book, all non-precious metals (other than uranium) are referred to as base metals, consistent with the definition in **GLOSSARY OF PRECIOUS METAL INDUSTRY TERMS** (1987) published by the International Precious Metals Institute. Industrial minerals (some of which are included in this chapter, for example, barite, some forms of carbon and compounds of beryllium, boron, magnesium) are non-metallic minerals used often in their natural state in industrial and manufacturing processes.

CHAPTER 11

BASE METALS *

INTRODUCTION

This chapter conveys a far more concise amount of information than the one provided for the precious metals and uranium. As discussed in the introduction (pages 3 and 7), data on base metals (whether in this chapter or in the rest of the book) were not included to give assistance in the understanding or planning of recovery processes, only to emphasize the differences in the basic approach and techniques, when compared to the precious metals. The data in this chapter should be regarded only as a handy reference listing the uses, basic mineralogy and more common extraction techniques applied to the various base metals. Such information is seldom found together in one book.

The base metals are listed in order of increasing atomic number, except for lead/zinc and niobium/tantalum which are listed together. Their descriptions have been taken from the SME Mineral **Processing Handbook** (1985), and from the **Encyclopaedia Britannica - Micropaedia and Macropaedia** (1988). Although much of the data from the SME volume are quoted "as printed", information from the Encyclopaedia Britannica was interpreted and rephrased. Authors and/or editors from the latter reference are Clark (mining), Elliott (iron), Lorig (metallurgy), Mackintosh (tin), Miller (tungsten), Minarcik (lead), Ogden (titanium), Richards (zinc), Rizley (magnesium), Schlechten (copper), Smith (silver), Van Horn (aluminum), Wise (nickel), and Young (cobalt). Complete titles for their articles are in the **references** section at the end of the book.

Table 17, at the end of the chapter, lists the same metals in alphabetical order, along with the year in which they were first isolated, the market value in US dollars for some of them, as reported in the October 24 1988 issue of **The Norther Miner,** their abundance when compared to 10^6 atoms of silicon (Brownlow 1979), the major deposits, the main economic minerals and the more common extraction processes.

Additional descriptions of metallurgical processes for copper, iron, nickel, lead, zinc and uranium can be found in Habashi (1986). The TMS symposium volume **World Survey of Nonferrous Smelters** (1988) provides a coverage of smelters, used for the recovery of copper, lead and zinc in many world localities.

When quoting from the TMS volume, some mineral names and formulae, in those rare occasions where they were missing or incorrect, were inserted or corrected to maintain the uniformity of the text.

METHOD OF STUDY

As emphasized in several occasions throughout the book, the correct approach to the studies of the occurrence and distribution of the precious and base metals in their ores is different. Studies of precious metals are complex, difficult and time consuming. They require specialized techniques, equipment and expertise. Studies of base metals are relatively simple, they require less specialized techniques. Detailed microanalytical studies are not usually necessary for base metal identifications and descriptions, neither are experienced microanalysts. We are back in the field of mineralogy, as intended in the more traditional meaning of the word, and that is the study of minerals which can be readily observed and identified by some of the older techniques. An experienced mineralogist with a good ore microscope, even if unfamiliar with the use of the microprobe and/or the scanning electron microscope, will most of the time provide the answers needed for the understanding of the ore, or for the planning of a successful recovery strategy. Minerals of many base metals, such as galena, chalcopyrite, sphalerite, molybdenite, etc., are easily recognized and positively identified by the ore microscope alone. Because of their high concentrations and coarse grain sizes, base metal minerals are often also identified directly in the original sample, when observed by the naked eye or with the aid of a hand lens.

Ease of characterization of many base metal ores when compared to the precious metal ores, is a consequence of:

- The considerably higher abundance of many base metals (usually by a factor of one million or more), which makes location of their main minerals fast and simple.

- The occurrence of several of them in limited numbers of minerals of economic value. Metals such as titanium, chromium, lead, zinc, molybdenum, tin, barium, tungsten, mercury, are, in many deposits, extracted from only one or from very few minerals. Such type of occurrence dramatically simplifies the process of identification.

- The coarse mineral grain sizes which only require average microscopes and magnifications. Use of hand picking or hand sorting, for physical separation (see descriptive section and table 17 in this chapter), gives an indication of the ease with which the minerals of some of these metals (for example, antimony, beryllium, bismuth, niobium and tantalum) are located for observation and study.
- The lower market value which makes any fine sizes and difficult occurrences uneconomical, if accounting for minor portions of the total in the deposit.

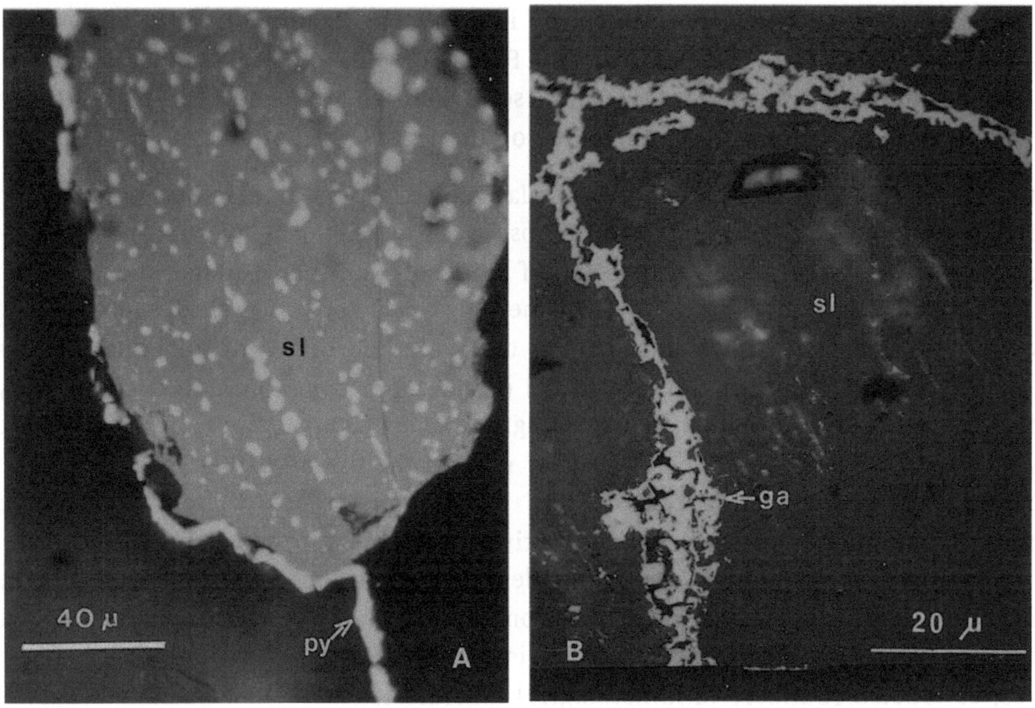

Fig. 58. *Sphalerite showing fine coatings of pyrite (A) and galena (B).*

In many ores (those exploited by pyrometallurgical techniques), detailed mineralogical observations may not be needed at all, although studies of the ore are sometimes necessary during the beneficiation stages or during hydrometallurgical recoveries. Some mineralogical features which may affect the base metal recoveries, and which may need the observation of polished mounts of representative samples, are:

- Intergrowths which prevent particle liberation, and therefore separation of clean concentrates. For example, chalcopyrite and sphalerite, galena and sphalerite, galena and chalcopyrite, galena and tetrahedrite (figures 11C on page 40 and 38B on page 156).
- Coatings over the surface of the particle of interest by minerals or by secondary products, which may produce changes in flotation properties and/or impermeability to the leaching solution. For example, pyrite, galena, and secondary

copper minerals over sphalerite (figures 58A and B, and figure 11F on page 40), causing zinc losses and contaminations of the lead and copper concentrates.

- Occurrences of the metal of interest in more than one mineral, due to variations in the ore chemistry and conditions of crystallization, or to secondary alteration; and lack of response of some of these minerals to the recovery agent. For example, nickel occurrences in pentlandite/millerite, amenable to the standard flotation techniques, and in violarite not amenable to the same techniques.
- Fine inclusions/exsolutions and solid solutions of an unwanted mineral or element in the lattice of the mineral of interest. For example, galena and chalcopyrite exsolutions (figures 11E and F on page 40) and iron solid solutions in sphalerite (figure 8 on page 37), iron solid solutions in cassiterite (figure 65 on page 266), all causing contaminations of the final concentrates.

Any of the above problems are pinned down relatively easily by a person with a training in the earth sciences, provided that an ore microscope is available. A microprobe or EDS-SEM, may be necessary for the determination of solid solutions and for the identification of the finer or unusual mineral particles. Yet the long and tedious searches, such as those for native gold, the identifications and analyses of five or six different minerals, such as those of silver, are not necessary. Intergrowths of the more common minerals are easily observed in the optical microscope, when using average magnifications. Coatings are readily visible, although, if very fine, they may require the use of higher magnifications. Violarite and millerite are identified in the optical microscope by an experienced mineralogist. If in doubt, a microprobe quantitative analysis will provide a positive answer. Mineral inclusions and exsolutions are again identified in the microscope by use of the higher magnifications. Determinations of the iron content of the sphalerite and cassiterite and predictions of the optimum grades for the final concentrates are also relatively simple procedures, conducted routinely in mineralogical laboratories.

An important requirement for the study of the base metals, next to the availability of an experienced mineralogist with a good ore microscope, is the use of a competent technician and of an adequate sample preparation laboratory for the preparation of good polished surfaces on the samples under study.

BERYLLIUM SYMBOL: Be

Beryllium was discovered in 1798 as the constituent of gemstones such as aquamarine and emerald. It was isolated in 1828 as beryllium metal by the reduction of its chloride with potassium.

"Beryllium and its compounds have unique properties that promise a growing future despite their relatively high cost. Current consumption patterns indicate that 48 per cent of the

United States production goes into Cu-Be, Ni-Be and Fe-Be alloys containing 2 to 3 per cent Be". (Chamberlin 1985, p. 29-7).

Until 1925 use of beryllium was limited to its gemstones only. Production of beryllium for industrial applications however has increased steadily since 1930. Properties of beryllium such as low atomic number, a relatively high melting point (1278°C), good electrical and thermal conductivity, good mechanical properties at elevated temperatures, oxidation resistance and elasticity, make it well suited for structural and thermal applications and in nuclear reactors. Beryllium is also used in gyroscopes, accelerometers, computers, missiles, heavy duty break drums, springs, and in applications where a good heat sink is important. Aerospace applications, such as for aircrafts and space vehicles, where lightweight is critical are also extensive. Beryllium windows are used for x-rays tubes because of their property of transmitting the x-rays many times faster than other metals such as aluminum.

Most important beryllium deposits are in Brazil, South Africa, Zimbabwe, South West Africa/Namibia, Argentina, and Uganda.

MINERALOGY

Economic beryllium minerals are **beryl**, $Be_3Al_2(SiO_3)_6$, **bertrandite**, $H_2Be_4Si_2O_9$, **phenacite**, Be_2SiO_4, **chrysoberyl**, $BeAl_2O_4$, and **barylite**, $Be_2BaSi_2O_7$. Transparent blue (**aquamarine**), green (**emerald**), red and pink (**morganite**), yellow (**heliodor**), and other colored varieties of the mineral beryl, when clear and free of major inclusions are used as gems.

METALLURGY

Much of the production of industrial beryllium is as byproduct of feldspar and mica.

"Efforts to concentrate the ore have been largely unsuccessful; flotation techniques are being developed, but have had very limited commercial use, and gravity techniques are not used because of the small specific gravity difference between beryl and the associated rock. Hand sorting has long been the main concentrating method, giving a low recovery of about 40 per cent on +1 inch particles." (Chamberlin 1985, p. 29-7).

Beryllium is produced from high grade ores (5 to 10 per cent BeO) by processes of sintering or fusion with the proper fluxes followed by hydrometallurgical processes which lead to the precipitation of beryllium hydroxide. Beryllium oxide for sale is produced by ignition in electric furnace at 800°C. The metal itself is produced from beryllium fluoride by reduction with magnesium and from beryllium chloride by electrolysis.

BORON SYMBOL: B

Boron was first isolated in 1808 by heating boron oxide with potassium.

"The principal consumption of boron minerals is in the glass industry, but significant quantities are used in the manufacture of enamel, agricultural chemicals, and soaps and detergents". (Defoe 1985, p. 29-6).

Boron is also used as a fire retardant in fabrics, for tanning leather, in solutions for electroplating nickel, and as a major constituent in catalysts for many organic chemical reactions. Limited quantities of boron (0.001-0.005%) are used in the steel industry for increased hardness. Boric acid is commonly used as a mild antiseptic for burns and surface wounds and is a major ingredient in eye lotions.

Boron is essential to plant growth.

About 50 per cent of the world's supply of boron comes from Southern California.

MINERALOGY

Some of the more common boron minerals include **borax**, $Na_2B_4O_7\ 10H_2O$, **kernite,** $Na_2B_4O_7\ 4H_2O$ and **tincalconite** [$Na2B4O5(OH)4\ 2H2O$], which are major commercial minerals, common in the arid regions of California. **Ulexite,** $NaCaB_5O_9\ 8H_2O$, **colemanite,** $Ca_2B_6O_{11}\ 5H_2O$, and **tourmaline** (borosilicates of complex-variable compositions) are boron-bearing widely dispersed minerals. **Sassolite,** H_3BO_3, natural boric acid, is common in Italy.

METALLURGY

"The usual treatment for borax and kernite involves crushing to about 3/4" inch size, mixing with a weak borax solution from the plant, and dissolving in heated and agitated tanks. Insoluble rock and clay are then separated from the solution, first by screening and then by a countercurrent thickening system. The solutions from the thickening plant are sent to a vacuum crystallizer where the pentahydrate and decahydrate are produced in separate crystallizers. Very little processing of the ore other than crushing, is practiced, although shale mineralization is removed at some plants. prior to dissolution by high intensity magnetic separation. Colemanite has the property of decrepitation upon heating, and one small plant in the United States is reportedly utilizing this property as a method for separating colemanite from its gangue minerals. The ore is heated in a rotary kiln followed by screening, with the undersize constituting the colemanite rich fraction.

Flotation concentration of boron ores has not yet been practiced in the United States, although flotation is utilized to separate borax and sylvite that crystallize simultaneously in the process. Borax is reportedly floated from sylvite with oleic and naphthenic acids plus xylene, turpentine and kerosene. It has been floated successfully in the laboratory, generally in a carboxylic acid system. Barium salt activation has been shown to be effective in this system.

Boric acid has been shown to be highly floatable without the addition of collector reagents. One system utilizing this property involves first decomposing a colemanite ore with SO_2, thus forming boric acid (H_3BO_3). The ore is first ground and then mixed with a saturated solution of boric acid. Sulfur dioxide gas is then added, and, when the reaction is complete, the pulp is added to flotation cells, where a small quantity of frother is added and the boric acid floated. Flotation cleaning is necessary. The flotation concentrate is dissolved in hot boric acid brine, filtered and relatively pure boric acid recrystallized." (Defoe 1985, pp. 29-6 and 29-7).

Pure crystalline boron is prepared with difficulty by reduction of its bromide or chloride with hydrogen on electrically heated tantalum filament.

CARBON SYMBOL: C

Elemental carbon exists in three forms, **diamond** and **graphite**, crystalline polymorph forms of the element, and **carbon black** (charcoal, lampblack, coal and coke), amorphous, although x-rays sometimes show a low degree of crystallinity. Diamond is colorless or white, sometimes transparent, a poor conductor of electricity, and has the highest known hardness. Graphite is black, opaque, slippery, a good conductor of electricity and is one of the softest minerals. Carbon black is of organic origin and derived from the preservation under certain physical/chemical conditions, of ancient vegetation. Diamond and graphite can be produced synthetically. Carbon is produced by heating any organic substance to very high temperatures in the absence of air.

Uses of diamonds are (i) in jewelry, when forming crystals weighing a tenth of a carat or more and relatively to totally free of inclusions, and (ii) industrial, in the preparation of abrasives for applications such as drills and saws, when forming smaller or badly flawed crystals. The weight of gem quality diamonds is expressed in *carats*, a unit representing 200 mg and in fractions of carats called points, each of which represents 1 hundredth of a carat. The word *carat*, used to indicate the weight of diamonds and other gemstones, as well as the gold content of an object, is conventionally spelled always with a **c** in the United Kingdom, but with a **k** *(karat)* in North America, when expressing gold percentages.

"Graphite is one of the most versatile minerals in industry because of its widely unique and varying physical properties. It has a specific gravity of 2.2 and a hardness of 1-2, the latter property being the basis for many of its uses. " (Zaman 1985, p. 28-2). Graphite is used as a lubricant, in paint, and, mixed with clay, as the lead of pencils. Because of its property of conducting electricity, and its very high melting point (3550°C), applications of graphite are also for the manufacture of electrodes, electric furnaces and dry cells, and of crucibles used to melt other metals.

Carbon is an essential constituent of steel (pages 245-246). Different forms of carbon have each a special character and application. **Coal** and **coke** are used as fuels. **Charcoal** is used as an absorptive and filtering agent, as a fuel and in the manufacturing of gun powder. Charcoal is also used for making carbon paper, typewriter ribbons, paints and inks.

Up to the early 80's, 70 per cent of the world's diamonds came form Africa, major producing countries being South Africa, Botswana, Angola, Zaire, Ghana, and Sierra Leone. Diamonds were then discovered and produced with increasing frequency by the Soviet Union and Australia, while what at first appeared like minor amounts of industrial quality material was found in Arkansas. (For the complete update, see page 277 at the end of the chapter). Graphite is a relatively common mineral mined extensively in many countries some of which are Sri Lanka, Madagascar, North Korea, Mexico (Sonora), Canada (Ontario), the USSR (Western Siberia), and the United States (New York). Coal is produced in the USSR, China, North and South Korea, the United States, South Africa and England.

MINERALOGY

Primary **diamonds** occur in *kimberlites*, or porphyritic brecciated rocks of ultramafic composition, occurring in pipe like bodies called diatremes. Kimberlites are found in *cratons*, or relatively immobile portions of the earth generally of large size, and contain olivine, chrome diopside, phlogopite mica, calcite, serpentine and chromium spinel, as important mineral constituents. "Diamonds are also found in beach sands, alluvial gravels and in ocean muds; their abundance in the deposits from which they are recovered is in the order of one part to 15 to 20 million." (Defoe 1985 p. 28-4).

In South Africa diamonds are mostly found in kimberlite pipes. In the Soviet Union about 450 kimberlite pipes were discovered mostly in Siberia as the result of intensive exploration programs whose first discovery was made in the early 1950's, and are being mined today. Additional diamond concentrations, in the Ural Mountains, are represented largely by deposits of alluvial type. (Sobolev 1981). (For the complete update on diamond exploration and discoveries, see page 277 at the end of the chapter).

"Although it is distributed in igneous, sedimentary and metamorphic rocks, the most common occurrence of **graphite** is in metamorphic rocks, such as schists, gneisses and metamorphosed limestones. Based on geologic origin, there are three categories of natural graphite: flake, amorphous and lump. The flaky form is by far the most valuable, followed by lump and amorphous. The lump graphite is typically massive. The particle size ranges from amorphous to coarse, platy intergrowths of fibrous to crystalline aggregates." (Zaman 1985, p. 28-2).

Coal is a rock-like black to brown product of organic origin and the result of the burial of forest-like vegetation, of its compression in an oxygen-free environment and of a number of dynamochemical processes which have led to the elimination of most of the original elements with consequent carbon enrichment. Coal is therefore of sedimentary origin and found, as the predominant constituent, in deposits of variable proportions.

METALLURGY

"The kimberlite ores, which are soft ultrabasic rocks with inclusions of siltstones, moisture, shales, schists and lava, are first crushed in stages to about 1 and 1/4 inches in size..... In a typical process the ore is scrubbed wet to break up soft conglomerates, and the heavy minerals - including diamond, garnet, chrome diopside and ilmenite - are concentrated on 14 foot diameter washing pans. The washing pan provides both washing and gravity concentration. The heavy minerals, including the diamonds are further concentrated by heavy media separation, the concentrates being passed over grease tables or belts. The greased tables consist of aluminum step-surface tables covered with grease and vibrated electromagnetically at high frequency. Water is added for washing off gangue minerals, and the diamonds adhere to the grease....Many alluvial diamonds have a surface film of mineral salts which renders the diamond water-avid, preventing adherence to the grease tables; but maize oil, saponified with caustic soda, when added to the ore produces a non-wettable surface on the diamond, thus overcoming this difficulty." (Defoe 1985, p. 28-4).

"Both open-pit (United States and Madagascar) and underground mining (Austria, Bavaria, Bohemia, Ceylon, Italy, Korea, Mexico, United States and Russia), are employed for mining graphite ore. ...... Graphite has the distinction of being the first mineral to be concentrated from ore by flotation...with a variety of non-polar hydrocarbon oils (e.g. kerosine) in combination with a frothing agent " (Zaman 1985, p. 28-2), and is separated by such technique from many of its ores.

MAGNESIUM SYMBOL: Mg

Magnesium was known originally through compounds such as Epsom salts (the sulfate), magnesia (the oxide), and magnesia alba (the carbonate, often shortened to magnesia). The metal itself, silvery white in color, does not occur free in nature and was first isolated in 1808 as an amalgam with mercury. Free elemental magnesium was produced in 1828.

In the past, magnesium has been used predominantly in photography, incendiary bombs, and pyrotechnic devices. In recent years, extensive uses for the metal have been found in the aerospace industry, for example, for parts of aircrafts and spacecrafts, because of the light weight and high strength of its alloys (usually with lithium, aluminum, zinc and manganese). Other commercial uses of magnesium alloys include luggage frames, hand trunks, ladders, business machines, camera bodies, power tools, tooling fixtures, race and sports car wheels (mag-wheels), and various household appliances.

Magnesium die castings are extensively used in the automotive industry.

Magnesia (magnesium carbonate), an important end product from magnesium ores and other sources, is used "as a basic refractory in these forms: dead-burned magnesite; double burned magnesite for special refractory applications in basic oxygen furnaces and electrical furnaces; peryclase, a special refractory grade of dead-burned magnesite containing 92 to 98 per cent MgO; and caustic-calcined, light burned magnesite, an active magnesia used for oxychloride and oxysulfate cements, fertilizer, rayon, paper, sugar industries, uranium precipitation, and SO_2 absorption from stack gases. Precipitated magnesium hydroxide is used directly by chemical, paper, and refractory industries and for production of magnesium metal." (Rau 1985, p. 29-14).

An important natural product of magnesium is asbestos (a variety of serpentine) used for the preparation of fireproof cloth, paper, etc.

Largest source for the preparation of magnesium metal and of magnesium compounds, is the ocean. Other sources are carbonate and evaporite rocks, *bitterns* (natural solutions characterized by high magnesium content) and *brines* (water strongly impregnated with salts such as chlorides and sulfates of sodium, magnesium and potassium), both associated with evaporite basins. Canada (Thetford and Asbestos - Quebec), is the major source of asbestos fiber. The United States leads in the manufacture of asbestos products.

MINERALOGY

Some of the commercial magnesium minerals are dolomite, $CaMg(CO_3)_2$, **magnesite**, $MgCO_3$, **brucite**, $Mg(OH)_2$ and **olivine**, $(Mg,Fe)_2SiO_4$. "Magnesite is found in bedded deposits, as in veins, pockets, shear zones in ferro-magnesium rocks, and as replacement bodies in limestone and dolomite. The main impurities in magnesite are lime, silicate and iron, as quartz, calcite, serpentine, chalcedony, dolomite and limonite. The cryptocrystalline magnesite commercial deposits have lesser amounts of the accessory minerals than do the crystalline deposits. The largest deposits are found in Nevada, Washington and California. In addition to magnesite, the minerals brucite, $Mg(OH)_2$, and **hydromagnesite**, $3MgCO_3\ Mg(OH)_2\ 3H_2O$, are important sources. Brucite is chiefly found in Nevada in association with magnesite. Other deposits are located in Arizona, Canada, Ireland and the United Kingdom. Dolomite, a combined magnesium carbonate-calcium carbonate, is principally used without separation of the magnesium and the calcium. However, dolomite is the source of magnesium when using the silicothermic process for producing magnesium metal. Also calcined dolomite, used in the recovery of magnesium from seawater, bitterns and brines contributes magnesium to the final product. These three are increasingly being used to produce synthetic magnesite. Seawater contains 0.2 per cent MgO, bitterns 6 to 9 per cent $MgCl_2$ and 4 to 6 per cent $MgSO_4$, and brines about 10 per cent $MgCl_2$. Olivine is not mined as a source of either magnesium metal or magnesia, but rather is used as a foundry sand." (Rau 1985, p. 29-14).

Other widespread magnesium minerals are **serpentine**, $Mg_3Si_2O_5(OH)_4$, used as a decorative stone, and, when in the asbestos variety, for the preparation of fireproof cloth or paper, **talc**, $Mg_3Si_4O_{10}(OH)_2$, a soft white mineral used as a lubricant and, when appropriately prepared, for personal grooming, **chrysolite**, $(Mg,Fe)_2SiO_4$, a pale yellowish or green variety of olivine, used as a precious stone when transparent and free of flows, and **carnallite**, $MgCl_2$ KCl $6H_2O$, a low temperature chemical deposition mineral exploited mainly as a source of potassium.

METALLURGY

"The processes used for making synthetic magnesite (magnesia) from seawater, bitterns and brines are basically the same. Bicarbonate must be removed from seawater and sulfates from bitterns, while neither of these steps is required with the brines. Seawater is first screened to remove mollusks, fish and seaweed. Then slaked lime from calcined oyster shells, limestone or dolomite is added to precipitate soluble bicarbonates as a carbonate. The precipitated carbonate is removed by thickening and filtration, the magnesium remaining in solution. The magnesium is then precipitated by addition of calculated amounts of calcinated oyster shells, limestone or dolomite. The magnesium forms magnesium hydroxide which is insoluble under these conditions and can be removed by thickening, filtering and drying. One advantage of using dolomite for this step is that the dolomite also adds recoverable magnesium as well as the necessary basic reagent." (Rau 1985, p. 29-14).

"Most of the magnesite and brucite deposits are mined by open pit methods, making it easy to employ selective mining to control lime and silicate contents as much as possible." (Rau 1985, p. 29-14). Heavy media concentration and flotation are the more common techniques practiced for separation of the magnesium minerals from their gangue.

Magnesium metal is commercially produced by electrolysis or electrolytic reduction of molten magnesium chloride, processed mainly from seawater and by direct reduction of the magnesium ores by strong reducing agents such as carbon or silicon. Because less expensive, electrolysis accounts for a large part of the magnesium production in the world.

ALUMINUM SYMBOL: AL

Aluminum is the most abundant metallic element in the Earth's crust (8 per cent), most widely used non-ferrous metal and third most abundant element after silicon and oxygen. Crude aluminum was isolated in 1825, but was not introduced to the public until 1855. "During the 1960s, aluminum moved into first place ahead of copper in world production of non-ferrous metals. Crude salts of aluminum were used as early as the 5th century BC." (Van Horn 1988).

Aluminum is added in small amounts to certain metals to improve their properties for specific uses as in aluminum bronzes and most magnesium base alloys. In aluminum base alloys, moderate amounts of other metals and silicon are added to the aluminum. The metal and its alloys are used extensively for aircraft construction, building materials (including the aluminum siding of many houses), consumer durables (refrigerators, air conditioning units, cooking utensils), electrical conductors, and chemical and food processing equipment.

Bauxite, from which most of the world's aluminum is produced, is found in all the continents except Antarctica. However, because it is the result of tropical weathering (pages 71-72), it predominates, in its geographic distribution, in the tropics and in rocks that were in the past subjected to old tropical climates. Abundant bauxite deposits are in Australia, Guiana, Jamaica, Brazil, Surinam, Ghana, France, Indonesia, Malaysia, Greece, Yugoslavia, the USSR, and China.

MINERALOGY

Bauxite, the most important ore of aluminum, "is a naturally occurring raw material composed principally of a mixture of one or more of the hydrated aluminum oxide minerals **gibbsite,** Al_2O_3 $3H_2O$, **boehmite,** Al_2O_3 H_2O and **diaspore,** Al_2O_3 H_2O, and impurities of silica, iron oxide, titania and other various elements in trace amounts. It is an end or near-end of chemical weathering. Depending on the amount of iron impurities, the color of bauxite varies from dark red and brown to pink and white. Some bauxite is finely divided, free digging earthy material, while other forms are dense and rocky, requiring explosives in mining. Many gradational forms exist. It is one of the most variable of mineral raw materials in chemical composition and physical appearance." (Shaffer 1985, p. 19-2).

Primary minerals of aluminum, of lesser economic importance than the components of bauxite, occur in all types of rocks including acidic to intermediate igneous rocks, sedimentary and metamorphic rocks. Some of the more common aluminum minerals are **feldspars, feldspatoids** and **micas**. These are the minerals which alter to bauxite and iron rich laterites through an intermediate stage represented by the **clays**. Crystalline aluminum oxide, Al_2O_3, (**corundum** and its pulverized form, **emery**), occurring in a few igneous rocks, is mined as a natural abrasive and, when in its finer varieties, as a gemstone (**rubies** and **sapphires**). Aluminum is an essential component of other gemstones, such as **topaz**, $Al_2SiO_4(F,OH)_2$, **garnet** (formula variable) and **chrysoberyl,** $BeAl_2O_4$.

METALLURGY

There are basically two steps involved in the production of aluminum from bauxite. These include (i) the refining of the bauxite to eliminate all the impurities such as iron oxides, silica and titanium oxides, and to obtain pure alumina (usually 99.9 per cent), by the *Bayer process*, and (ii) the reduction of the alumina (Al_2O_3) by an electrolytic process, to obtain aluminum metal. "The principles of the Bayer process are essentially the same today as when first developed by Karl Josef Bayer in 1888. Bauxite is digested in caustic soda to solubilize alumina and precipitate a silica complex. The sodium aluminate solution is collected, separated from the insoluble bauxite constituents, held in seeded agitated tanks to precipitate a hydrated alumina which is calcinated, collected and shipped to reduction plants." (Cundiff 1985, p. 19-4).

Most commonly used approach for the production of aluminum metal from the refined alumina, the *Hall Heroult process*, is by electrolysis. Aluminum is produced by dissolving purified alumina in molten cryolite (Na_3AlF_6) and by electrolyzing it with direct current.

POTASSIUM SYMBOL: K

Potassium was the first element to be isolated by electrolysis, by decomposition of molten potassium hydroxide with a voltaic battery (1807). Being one of the strongest reducing agents, potassium is not isolated by ordinary chemical reduction.

"The term potash originated in *pot ashes* produced by burning wood to ashes in iron pots, and then leaching out the soluble salts, chiefly potassium salts used in agriculture and industry. More specifically, it denotes the oxide of potassium K_2O, and the content of potassium in ores and products is stated as weight percent of K_2O present. Its salts are commonly referred to as muriate of potash, sulfate of potash, etc. The presence of an adequate supply of available potassium in the soil promotes the health and improves the quality of the plant, insures greater efficiency in photosynthesis, increases resistance to certain diseases, offsets the effects of an oversupply of nitrogen, and helps the plant to utilize soil moisture more advantageously. Potassium ranks 7th among the elements in abundance. It is estimated to constitute some 2.6 per cent of the lithosphere and is found in soluble forms in oceans, lakes, rivers and subsurface waters. It is also an important constituent of plant and animal life." (Zandon 1986, p. 22-2).

Potassium metal is used to produce potassium superoxide, KO_2 which refreshes exhaled air by liberating oxygen and by removing carbon dioxide and water vapor. Potassium compounds are used as fertilizers, in the manufacture of high quality glass for the production of television tubes, and in the manufacture of explosives and fireworks.

Important producers of potassium are the USSR, Canada, the United States, Germany.

MINERALOGY

Although most potassium in the Earth's crust is present in minerals such as **muscovite** and **feldspar** which are insoluble in water, common commercial potassium minerals are chlorides and sulfates such as **sylvite**, KCl, **carnallite**, KCl $MgCl_2$ $6H_2O$, **kieserite**, $MgSO_4$ H_2O, and **polyhalite**, $K_2Ca_2Mg(SO_4)_4$ $2H_2O$. Ready solubility of these minerals allows for easy extraction, thus rendering them readily amenable to the chemical processes. Soluble potassium minerals are often found in mixtures with magnesium minerals of similar compositions, both as dry mineral deposits and in brines. "Commercially significant accumulations of potassium salts occur in underground deposits formed in past geologic times by evaporation of ancient bodies of saltwater, inland drainage basins, and shallow sea lagoons. The potash content of such waters was the end result of weathering and leaching of insoluble potash-bearing aluminosilicates such as feldspars, micas, leucites, and others. The concentration of potassium salts in the natural brines of some lakes in the arid and semiarid regions of the continents are also important as sources of potash; the brines of the Dead Sea in Asia Minor, the Great Salt Lake in Utah, and Searles Lake in California typify such concentrations." (Zandon 1985, p. 22-2).

"**Sylvinite** is a term used to indicate a mechanical mixture of KCl and NaCl with minor amounts of sedimentary clays and occasionally other potassium complexes .... Sylvinite ores are noted for their relative abundance, high KCl content, and ease of beneficiation." (Zandon 1985, p. 22-2).

Important occurrences of potassium minerals are in Northeastern Germany, in Spain, the USSR, Tunisia, and in the Southwestern United States. A small percentage of potassium is obtained from plants and animals. Potassium carbonate is one of the final products from the burning of a plant or a tree.

METALLURGY

"Crystallization and flotation are two common processes for beneficiation of potash ores consisting predominantly of sylvinite. Both employ brine as the liquid phase. This brine is fully saturated with respect to potassium and sodium chlorides. Solubility of both vary with temperature and with the concentration of other soluble constituents of the ore, when present...... The crystallization process is normally selected for recovery of KCl from the brines derived by solution mining, from natural brines found in some lakes, and in treatment of potash ores characterized by extremely fine dissemination or abnormal admixture of other saline minerals and clays...Some operations recrystallize their brines by solar evaporation and employ differential flotation for production of their muriates. Employment of a flotation process alone may be justified in the case of sylvite ore noted for good liberation of KCl crystals in coarser sizes". (Zandon 1985, pp. 22-3 and 22-4).

Following the crystallization or flotation processes, the concentrate is debrined (by centrifuges or vacuum filters), dried, sized, and stored for shipment to the customers.

Reduction methods for the production of potassium metal include high vacuum techniques where calcium or sodium vapors are the reducing agents, and electrolysis of the fused hydroxide, cyanide or oxysalt in a fused potassium halide.

TITANIUM SYMBOL: Ti

Titanium oxide was discovered in 1791 in England and again rediscovered in Germany in 1795. It was isolated as a metal in 1910 in the United States through reduction of the titanium tetrachloride with sodium, in an airtight steel cylinder. Titanium changed from a laboratory curiosity to an important structural metal after 1947, with the advent of the space age.

Due to its lightweight, high strength and corrosion resistance, titanium is used in alloy form, for the manufacture of parts of high speed aircraft/spacecraft missiles and ships. Titanium is also used as a deoxidizer and to control the grain size in many steels to which it is added in the form of a ferrotitanium or ferrocarbontitanium alloy. Titanium is added to magnet alloys of iron-nickel-cobalt composition.

Major producers of titanium are Australia, Norway, the United States, the USSR, and Canada.

MINERALOGY

"There are two principal ore minerals of titanium, **rutile**, TiO_2, and **ilmenite**, FeO TiO_2. Rutile is usually found as an accessory mineral in igneous and in certain metamorphic rocks such as gneiss. The weathering of these and subsequent water action has resulted in a concentration of rutile in beach sand deposits. Ilmenite is found in basic igneous rocks such as gabbro and anorthosite. It is associated and sometimes intimately admixed with magnetite and hematite, and, further, ilmenite is also known to take up Fe_2O_3 in solid solution; thus its TiO_2 content may sometimes be lower than theoretical. On the other hand, the ilmenite may be altered by partial oxidation and some removal of iron to produce a mineral of somewhat higher TiO_2 concentration called **leucoxene.** Rutile is mined from placer deposits........ Because of its higher TiO_2 content and greater ease of subsequent processing, rutile is the preferred titanium source. Unfortunately its supply is limited. Ilmenite deposits are the principal source of titanium minerals and the world production of ilmenite is seven or eight times that of rutile. It is produced from both placer and hard-rock deposits." (Aplan 1985, p. 27-14).

METALLURGY

Beneficiation of titanium ores from placer deposits is done by gravity concentration in a *wet* plant and is followed by a number of processes which are directly related to the type of deposit.

Preparation of titanium metal requires an intermediate step in order to transform the oxide, practically impossible to reduce due to the high temperatures involved, into a more amenable phase. Such intermediate step is represented by the chlorination process, during which chlorine in the presence of carbon decomposes the titanium oxide at a relatively moderate temperature, forming titanium tetrachloride from which the metal is then obtained by reduction with magnesium.

CHROMIUM SYMBOL: Cr

Chromium was discovered in 1797 and isolated as a metal a year later. The name (from the Greek meaning color) was chosen because of the distinct tinge which even small quantities of the element give to compounds such as emerald, ruby, chrome-mica and serpentine.

According to Aplan (1985) there are three major uses for chromium, and these are metallurgical, refractory and chemical. The metallurgical applications in the United States, largely related to the manufacturing of stainless steels, are the most extensive, as they amount to approximately 60 per cent. Refractory and chemical applications have amounted to about 20 per cent each in recent years. Because of anticipated increases in stainless steel production, the demand for metallurgical grade chromite is expected to increase.

Stainless steels are alloys of chromium and iron in which the chromium content varies between 10 and 26 per cent. Used in small amounts, chromium hardens steel. Chromium is also mixed with iron and nickel in the form of ferrochromium (approximately 70 per cent Cr) to produce alloys characterized by high resistance to corrosion and oxidation. Chromium alloys are used in the fabrication of oil tubing, automobile trim, and cutlery. Chemical applications and refractory applications involve the use of chromite, rather than of the metal.

The earliest worked deposits of chromite were those in the serpentine of the Bare Hills near Baltimore, Maryland. The principal chromite producing countries today are, in order of importance, South Africa, the USSR, the United States, the Philippines, Zimbabwe, Turkey, Brazil, India, and Finland.

MINERALOGY

"Chromium is derived exclusively from the mineral **chromite**, FeO Cr_2O_3. Unfortunately a mineral of such composition is never found in nature since the Fe^{++} is replaced in major way by Mg^{++}. Furthermore a substantial substitution of Al^{+++} and Fe^{+++} for Cr^{+++} also occur together with a much lesser substitution of Si^{++++} and Ti^{++++}. As might be expected, the properties and uses of chromite are highly dependent upon its composition induced by these substitutions into the crystal lattice at time of formation." (Aplan 1985, p. 27-4).

Chromite (figure 59) is commonly found in brittle masses which may reach large proportions within the host rock, as well in disseminated grains associated with magnetite and secondary iron oxides. Crystalline inclusions in diamonds represent unusual, but reported occurrences.

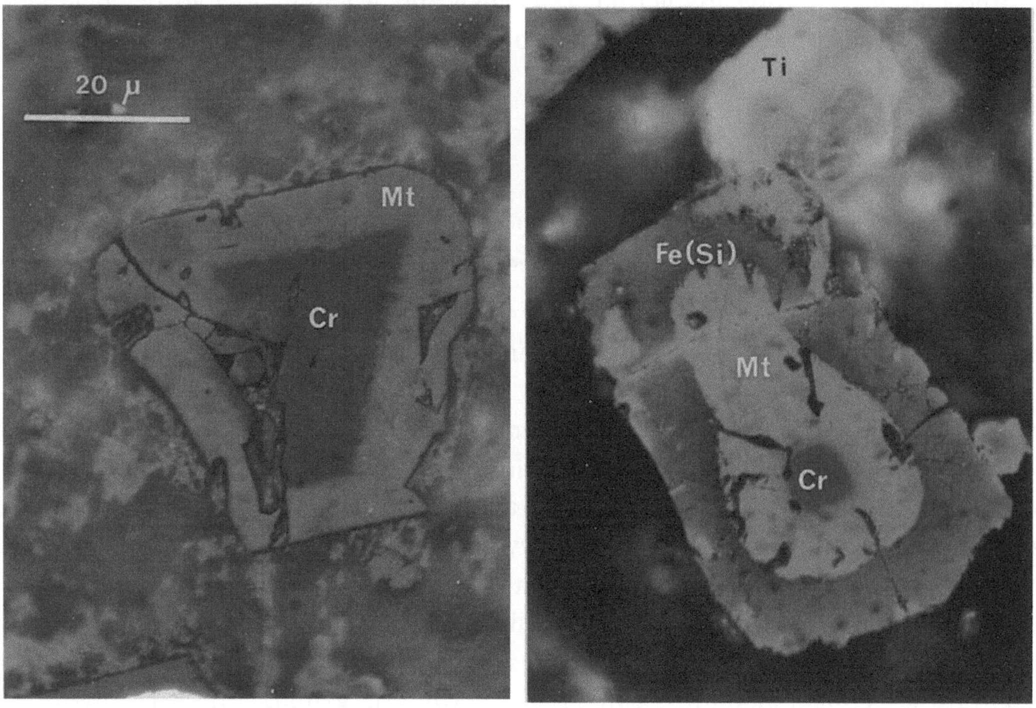

Fig. 59. *Composite grains consisting of chromite and magnetite. The magnetite in the photomicrograph to the right is partly replaced by secondary iron oxides and silicates.*

"Chromite is found primarily in ultramafic or closely related anorthositic rocks either in stratiform (layered) or pod-shaped deposits. The layered deposits, typified by the Bushveld complex in South Africa and the Stillwater complex in Montana, generally have a high iron content, though the Great Dyke in Rhodesia is an exception. The pod deposits are found typically in peridotite-gabbro complexes. A small amount of chromite is found in beach placers as along the coast of Oregon. (Aplan 1985, pp. 27-4 and 27-5). The occurrence of the platinum group elements in mafic and ultramafic formations containing chromium, such as the Bushveld and Stillwater complexes and the Great Dyke, has already been mentioned (page 197 and page 202, table 13).

METALLURGY

"Much of the chromite ore is produced by selective mining: most particularly, the high quality ore produced for metallurgical purposes since lumpy ore is highly desirable for furnacing. For the treatment of low grade and friable ores, or the fines resulting from the mining and handling of the hard, lumpy ores, beneficiation is used. This processing is accomplished by gravity concentration, flotation, or by high intensity magnetic separation, the former being the accepted method." (Aplan 1985, p. 27-5).

Chemical processes may involve reduction with coke and leaching in hydrochloric acid. Chromium metal is also produced by reduction of the oxide in the presence of aluminum or by electrolysis.

MANGANESE SYMBOL: Mn

Manganese was discovered and isolated as an element in 1774.

"The Earth's crust contains, on average, approximately 0.1 per cent Mn, and manganese is the 12th most abundant element. Like many mineral commodities, it is, moreover, unequally distributed over the surface of the Earth and the bulk of its ores are produced by only a few nations. Worldwide, the steel industry consumes over 90 per cent of the manganese produced *(in the form of ferralloys*, sentence in italics added by the author). In addition, certain MnO_2 ores are used as dry cell depolarizers, and manganese ores are used for the manufacture of electrolytic manganese, special alloys, welding rod flux, and for chemical purposes. A small amount is converted to synthetic gamma-MnO_2 for high performance battery use." (Aplan 1985, p. 27-6). Other uses of manganese are in the manufacture of glass and ceramics and in paint pigments.

Manganese producing countries are the USSR, South Africa, Brazil, Gabon, India, Australia, and China. More reserves are in Germany, the United States, Cuba, Morocco, Ghana, and Guinea.

MINERALOGY

"Manganese is commonly found as **rhodochrosite,** $MnCO_3$, or as oxide minerals such as **pyrolusite**, MnO_2. There are a great number of minerals of the dioxide type, and a given deposit will typically contain a variety of these minerals......The manganese dioxide minerals account for nearly all of the world's manganese production, though rhodochrosite ores have been mined in the past and will probably become important again in the future. (*Manganese minerals, as well as the term wad, are also discussed on pages 157-158).*

There are three common types of manganese ore deposits: sedimentary, volcanogene and hypogene. The sedimentary type is by far the greatest in importance and these have usually been deposited in sea water as the carbonate or the oxide." (Aplan 1985, p. 27-6).

METALLURGY

"The majority of the ores being mined are the sedimentary and residual deposits containing the oxide minerals. Since these are weathered, often in lateritic zones, the resistant manganese oxides are often found as nodules and surface rubble interspersed with clay." (Aplan 1985, p. 27-6 and 27-7). Beneficiation of manganese ores include gravity concentration and flotation methods. Flotation is not usually easy, due to the clay slimes.

When preparing steel, manganese ores together with iron ores are first reduced in blast furnaces and electric furnaces with carbon or ferrosilicon to yield ferromanganese, the additive used in steel making.

Leaching in H_2SO_4 is also practiced.

DEEP SEA NODULES. "Manganese sea nodules, also containing nickel, copper and cobalt, *sometimes platinum group elements (page 197*, words in italics added by the author), will probably be mined in the future for these latter, more valuable metals. These nodules often contain 25-30 per cent Mn, and this may well be a very important byproduct. As the valuable metals are intimately associated with the manganese oxide phases, hydrometallurgical techniques will be required to separate the metallic constituents. The only ore-dressing operations likely to be employed will be screening, comminution, drying and those associated with the leaching step." (Aplan 1985, p. 27-9).

IRON SYMBOL: Fe

Iron was discovered in approximately 1200 BC. In the Earth's crust, iron is second in abundance to aluminum among the metals and fourth behind oxygen, silicon and aluminum, among the elements.

Iron, the world's most widely used metal, constitutes about 5 per cent of the Earth's crust. In elemental form or as steel, iron has supplied humankind with most of its tools, many of its products, and the bulk of its structural elements in large scale construction.

Steels are alloys of iron with carbon and a number of other elements. Basically, they can be grouped into three categories: carbon steels, low alloy steels and high alloy steels. The carbon steels contain carbon in amounts between 0.03 to 1.5 per cent and, in a few cases, as much as 2.25 per cent. In addition to carbon, they may contain up to 1.65 per cent of manganese and sometimes smaller quantities of silicon, aluminum copper and other elements. The low alloy steels contain carbon and a total of from 1 to 5 per cent of alloying elements such as cobalt, nickel, chromium, molybdenum, tungsten, titanium, and niobium. The high alloy steels contain carbon and over 5 per cent of at least one alloying element. A steel with more than 10 per cent chromium (and up to 26 per cent) is called a stainless steel. Carbon steels are used for automobile bodies, appliances, machinery, ships, containers and the structure of buildings. The low alloy steels are used most often in applications in which special properties are needed, but cost is of importance. Examples are forged products of highly stressed parts of machines, aircraft landing gear, shafts, hand tools, gears and high strength structural steels. The high alloy steels are used for their unusual properties. Typical are the stainless steels used when resistance to corrosion and oxidation, as well as good appearance are required, for example, parts of jet engines, chemical equipment, tableware, cooking utensils, and cutting tools. Tool steels containing 2 to 12 per cent of cobalt are used on hard materials to produce deep cuts or high speed.

Iron ores are found widely distributed throughout the world. Major producers of iron and steel are the USSR, Canada, Brazil, the United States and Australia.

MINERALOGY

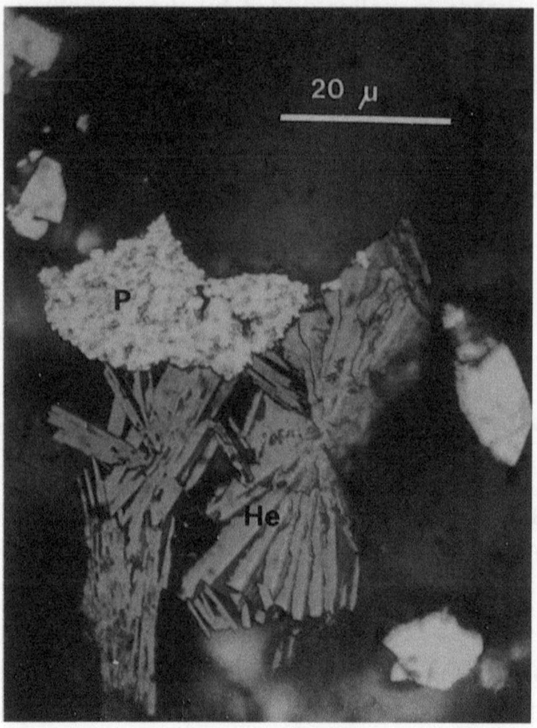

Fig. 60. *Needle-like crystals of hematite associated with pyrite.*

Iron very seldom occurs as a natural metal. An exception is represented by meteorites and occasional terrestrial occurrences such as those in the basalts of Blaafjeld, West Greenland (Dana 1963). Most iron ores occur as oxides (**hematite, Fe_2O_3, magnetite** Fe_3O_4, **limonite**, $2Fe_2O_3\ 3H_2O$, other **secondary iron oxides**, formulae variable), and to a lesser degree, as carbonates (**siderite**, $FeCO_3$). A small amount is recovered as a byproduct from iron sulfides (**pyrite**, FeS_2, and **pyrrhotite**, FeS). (Modified after DeVaney 1985, p. 20-2). Figure 59, on page 243, shows composite grains of magnetite and chromite. Figure 60 shows an occurrence of hematite and pyrite.

Kamacite is a low nickel meteoric iron.

Hematite. Red hematite, when pure, contains 70 per cent of iron. When mined, the content of the ore may be as low as 25 per cent. Hematite deposits of importance are found in Sweden, Belgium, Western Australia, the Alabama and Lake Superior regions of the United States, Cerro Bolivàr - Venezuela, Labrador and Quebec. The most important deposits of hematite are sedimentary, although the ones in Minas Gerais - Brazil are metamorphic.

Limonite, or **brown hematite**, of secondary origin, is a hydrated variety of hematite, containing about 60 per cent of iron, in its pure form. An important ore, it is yellowish to brown in color. It is the chief constituent of bog iron ore. Important deposits are in the United States, the USSR, Germany, France, Spain, and Western Australia.

Magnetite, a magnetic iron ore, is a black oxide occurring in beach sands (also called *black sands*) and rock formations. It is a common mineral the world over. Deposits consisting of magnetite and hematite are found in Western Africa.

Carbonate ores include **siderite** and **spatic iron ore** which contain 48 per cent iron when pure. These ores, also known as *clay ironstones* and *blackband ironstones* and are found in the United Kingdom.

Taconite is an iron bearing chert derived from ferrous silicates. In the United States, the word taconite is used to indicate any grade of extremely hard, lean ore where the iron occurs either in banded or in well disseminated form, and whose main minerals are hematite or magnetite, or a combination of the two. Taconite is the main constituent of the Mesabi ores in Minnesota.

Another source of iron comes from **iron laterites** which consist of soil layer rich in iron oxide and derived from a wide variety of rocks weathered under strongly oxidizing conditions (pages 71-72).

METALLURGY

"Because of its wide occurrence and relatively low value, an iron ore to be merchantable must contain a relatively high percentage of the metal as a source of iron. Few ores containing less than 25 per cent iron would be considered as ore unless they existed in large amounts and could be concentrated very cheaply. The grade of the ore, the ease of concentration, and the transportation cost to market are fundamental determining its worth." (DeVaney 1985, p. 20-1).

"Iron ore must be reduced (oxygen removed) to metal for use. Iron is marketed in many forms and combinations, such as pig iron and a great variety of steels. Iron is practically never pure as marketed, but contains different amounts of carbon, phosphorus, sulfur and alloying elements such as nickel, chromium, etc., all of which affect the physical properties of the end product" (DeVaney 1985, p. 20-2).

The beneficiation stage of the iron metallurgy includes techniques such as gravity concentration, magnetic separation and flotation. Leaching in HCl is also practiced. "Once a concentrate is produced, most iron ores must be agglomerated to produce a physical size acceptable to the steel industry. Originally the concentrate was sintered, but more recently, pelletizing and induration have become the more common approach. Green pellets are produced in drums or discs and fired at high temperature to produce a hardened pellet for charging to the furnaces. The key is to produce a structurally strong particle (hardened pellet) capable of resisting degradation during transport and handling, but still having a high reducibility in the furnace" (J.C. Taylor 1988, personal communication).

The concentrate product is treated by smelting with carbon (coke) and limestone. The molten iron is tapped from the blast furnace at regular intervals and the accumulation of slag is drawn off. The iron is either sent in large ladles directly to steel making furnaces or is pigged (cast into ingots) and stored for later shipment to foundries.

Cast iron or pig iron are generic terms describing a family of iron alloys containing 1.8 to 4.5 per cent of carbon and which are obtained directly from the blast furnace and cast into molds. These alloys are usually made into specified shapes for direct use or for further processing by machining or heat treating.

COBALT SYMBOL: Co

Cobalt was isolated as a metal in 1742, but its compounds have been used for centuries to impart a blue color to glazes and ceramics. Cobalt has been detected in artifacts dating as far back as Egyptian and Persian times (3000 BC).

Most of the cobalt produced is used in special alloys. A quarter of the world production goes into magnetic alloys (magnets) which are cobalt steels containing 2 to 40 per cent of cobalt. Another quarter is used for alloys which retain their properties at high temperatures and superalloys that are used near their melting points (where steel would become too soft). Cobalt is also used in dentistry. The silicates of cobalt yield blue coloring used in ceramics, inks and paints.

Cobalt is an essential element in the nutrition and in the maturation of red blood cells in the form of vitamin B12.

Cobalt is mined in Zaire, Zambia, USSR, Australia, Canada, Finland, Cuba, and Germany.

MINERALOGY

"With very few exceptions, cobalt is rarely found in simple cobalt minerals. More generally, it is found associated with copper, nickel and iron in sulfides, arsenides and oxidized compounds. Typical true cobalt ores nevertheless are found in Morocco where the principal minerals are **smaltite**, $CoAs_2$. in association with gold; **safflorite**, $(Co,Fe)As_2$; and

skutterudite, $CoAs_3$, in association with gold, the gangue being principally carbonate. .... In sulfide ores cobalt occurs: with copper as **carrolite**, CuS Co_2S_3, found in commercial ores in Zambia and the Republic of Zaire; as **linnaeite,** $(Co,Ni,Fe,Cu)_3S_4$, in Zambia and in the Mississippi valley; with nickel and iron as **siegenite,** $(Co,Ni,Fe)_3S_4$, in the cobaltiferous deposits in the Missouri district; as **cattierite** and **vaesite,** $(Co,Ni,Fe)S_2$, in the Shinkolobwe uranium mine (Republic of Zaire); as **pentlandite**, $(Fe,Ni,Co)_9S_8$, in the Sudbury and Lyn Lake districts of Canada; and with iron in **cobaltiferous pyrite** $(Fe,Co)S_2$ of Finland. As arsenide ores, cobalt is mined outside Morocco, in Canada (Cobalt district of Ontario) and in the USA (Blackbird, Idaho); in this latter case it is found as **cobaltite,** CoAsS" (Theis et al 1985, p. 17-5). Figure 61 shows minerals of the linnaeite series associated with pyrite, marcasite and galena.

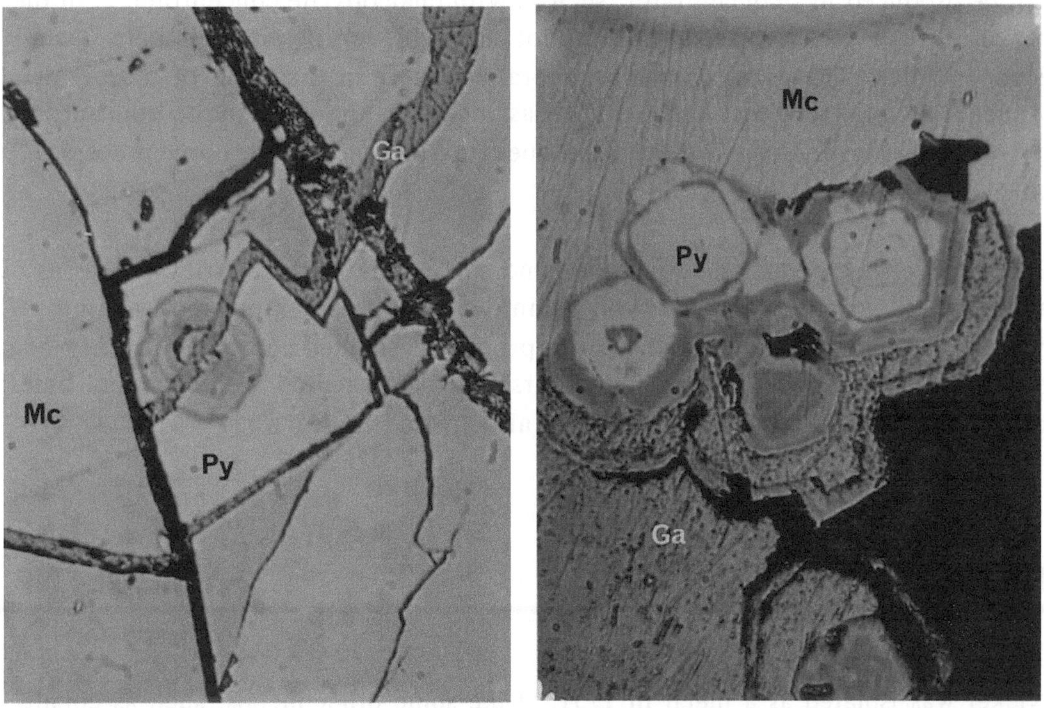

Fig. 61. *Cobalt minerals of the linnaeite series (euhedral crystals with concentric structures), associated with pyrite, marcasite and galena.*

The oxidized cobalt minerals include a number of unusual hydrous compounds, common in lateritic deposits. Some of these are **erythrite,** $Co_3As_2O_8\ 8H_2O$, **bieberite**, $CoSO_4\ 7H_2O$, **roselite,** $(Ca,Co,Mg)_3As_2O_8\ 2H_2O$ and **heterogenite**, CoO $2Co_2O_3\ 6H_2O$.

METALLURGY

A very large portion of the cobalt produced today is derived from the exploitation of ores of copper, copper-nickel or nickel. A small amount comes from pyrite deposits mined for sulfur and from lead-zinc deposits.

"In contrast to the relatively well defined nature of the nickel-bearing ores, cobalt occurs in association with other base metals, notably nickel and copper. Consequently the cobalt industry has grown largely in the wake of the larger base metal operations, and the processing of these cobalt-bearing ores is dictated to a large extent by the metals present in the ores. ....The consequent separation of the cobalt minerals from those containing other base metals becomes complex. Although in some instances a separate cobalt-bearing concentrate can be produced in a concentrator, such as with chalcopyrite concentrate...., in the majority of cases such physical separation is not advisable or possible, neither technically nor economically. The cobalt-bearing minerals are left in the form of minor metallic constituents along with other metals, such as nickel in the pentlandite concentrate, or with pyrite and recovered during the subsequent extraction and refining processes." (Mackiw 1985, p. 17-2).

Beneficiation techniques may include hand sorting and concentrating processes by gravity and magnetic means and a variety of techniques based on heating and leaching. When with nickel or copper concentrates, the entire product is smelted and the cobalt is then separated by electrolysis. Concentrates of cobalt arsenides are smelted with coke and limestone in a blast furnace. Arsenical cobalt ores are also treated by leaching.

NICKEL SYMBOL: Ni

Nickel was isolated as a metal in 1751. Like some other metals, such as cobalt, zinc and antimony, nickel may have been known from an early date (Romans?), but only in the alloy state.

Nickel resembles iron in strength and toughness, but is more like copper in resistance to oxidation and corrosion, a combination accounting for many of its applications.

"Although nickel is one of the more abundant elements in the universe, it ranks only 24th in quantity in the Earth's crust and makes up a scant 0.008 per cent of the whole. Yet in the overall surface-to-core composition of the earth, the only elements occurring more abundantly than nickel are iron, oxygen, silicon and magnesium which make up 90 per cent of the total." (Boldt and Queneau 1985, p. 17-3).

"Typical applications of nickel stainless steel *18/8* (18% Cr, 8% Ni) are in building materials, rapid transit and railway cars, food handling equipment and chemical fertilizer manufacturing and handling equipment....The high-nickel alloys are used for components of the gas turbine or jet engines, nuclear reactors and other applications where resistance to heat and corrosion is essential. Copper-nickel alloys have applications regarding seawater, notably in shipboard piping and desalination plants...Nickel is also used as a catalyst in the production of margarine and in fuel cells." (Mohide, Warden and Mason 1977, pp. 10-11).

Small amounts of nickel were produced in Germany in the mid 19th century. More substantial amounts came from Norway and a little from a mine at Gap, Pennsylvania. A new source, New Caledonia, in the South Pacific, came into production about 1877 and dominated until the development of the copper-nickel ores of the Copper Cliff-Sudbury, Ontario, regions in Canada. These have been the world's largest source of nickel since 1905.

MINERALOGY

Principal nickel minerals, modified after Boldt and Queneau's (1985) list, are: pentlandite, $(Ni,Fe)_9S_8$: **millerite**, NiS, **heazlewoodite**, Ni_3S_2, minerals of the linnaeite series including **polydymite**, Ni_3S_4, **violarite**, Ni_2FeS_4, **siegenite**, $(Co,Ni)_3S_4$, **nickeliferous pyrrhotite**, (Fe,Ni)S, **niccolite**, NiAs, **maucherite**, $Ni_{11}As_8$, **rammelsbergite**, $NiAs_2$, **gersdorffite**, NiAsS, **breithauptite**, NiSb, **annabergite**, $Ni_3As_2O_8\ 8H_2O$, **garnierite**, $(Mg,Ni)_6Si_4O_{10}(OH)_8$, **nickeliferous limonite**, $2(Fe,Ni)_2O_3\ 3H_2O$. Pentlandite is the most common of the nickel minerals, as it accounts for about 65 per cent of the nickel mined in the world. Interesting higher concentrations of niccolite, rammelsbergite and gersdorffite are found associated with the uranium concentrations of the Atabaska Basin in Saskatchewan (figures 50 on page 209 and 62). Most of the other minerals occur in smaller amounts associated with the pentlandite, and only occasionally become predominant. Violarite (a secondary mineral after pentlandite), for example, may be locally enriched. Because of its limited response to the standard techniques, violarite identification is important for the planning of efficient nickel recoveries. Pyrrhotite, a commonly nickel-free iron sulfide, may contain minor amounts of the element (1-2 per cent), when in nickel deposits. Elemental nickel occurs sparingly together with iron, in terrestrial and meteoric deposits. **Kamacite** is a term used to indicate iron meteorites of which nickel is a minor component (5 to 10 per cent).

Pentlandite, pyrrhotite and other nickel minerals, with or without platinum group elements, (pages 197 and 202, table 13) are commonly found in mafic and ultramafic rocks such as gabbros, norites and peridotites. They are predominant in the Sudbury district in Canada, the Bushveld Complex in South Africa, and Bodo in Norway.

Another type of economic ore is the **lateritic nickel ore** resulting from the weathering process of primary nickel sulfide deposits (pages 71-72). **Garnierite**, a hydrated silicate of magnesium and nickel, is a typical mineral of laterites and the one containing the highest nickel. Nickel rich limonite however is the common mineral in many deposits. Most famous nickel laterites are those of New Caledonia in the South Pacific. Other laterite ores are those in Nicaro and Moa Bay - Cuba, Sered - Czechoslovakia, and Hanna - Oregon.

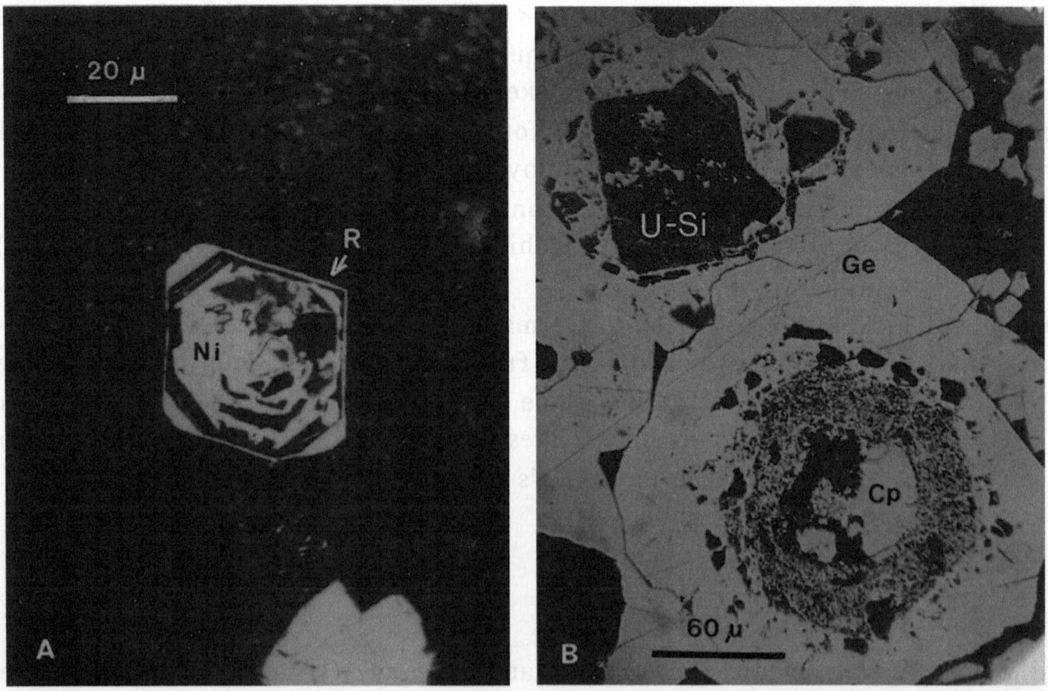

Fig. 62. *Nickel minerals niccolite and rammelsbergite in silicate (A) and gersdorffite associated with chalcopyrite and uranium minerals (B).*

Numerous other laterite deposits are scattered the world over, for example, those of Japan, Indonesia and the Dominican Republic.

METALLURGY

Nickel metallurgy includes pyrometallurgical techniques, such as those employed by Inco and Falconbridge in Ontario, Outokompu in Finland and Noril'sk in the USSR, as well as hydrometallurgical techniques, largely by pressure leaching, such as those employed at Western Mining in Western Australia (Mackiw 1985, p. 17-2 and J.C. Taylor 1988, personal communication).

"The smelting of sulfide nickel concentrates is very similar to the processes used for copper. Indeed some of the smelting furnaces are readily interchangeable, smelting temperatures however are usually higher in the nickel industry." (J.C. Taylor 1988, personal communication).

In **Sudbury** the ore is crushed and ground with water. Sulfide minerals are separated by physical and magnetic means, then by flotation. The nickel containing fraction of pyrrhotite-pentlandite from the flotation process, and the material recovered by magnetic concentration, are roasted to remove a portion of the sulfur, then smelted to remove much of the iron and silicon. The resulting matte, consisting of iron, copper and nickel sulfides, is further oxidized in a Bessemer converter to yield copper and nickel sulfides. These are separated by slowly cooling the matte, a process which is followed by grinding and selective flotation. In subsequent operations, the two sulfides are reduced to metals and converted into anodes for electrolytic refining to yield high purity nickel and copper.

"**Lateritic ores** are also treated by pyrometallurgical (Inco, Indonesia; Falconbridge, Dominican Republic; Societe de Nickel, New Caledonia; and various Japanese smelters), and hydrometallurgical techniques (Nicaro and Moa Bay, Cuba). In laterite processing, it is not possible to concentrate the ore as is the case with sulfide ores. Consequently, large tonnages must be smelted. The majority of laterites contain significant quantities of combined water which must be removed. The fuel required, in addition to that required for reduction, substantially increases the energy requirement for laterite smelters. Ferronickel is the logical end product from a laterite nickel smelter. Some operations produce a nickel matte to provide flexibility in the downstream processing. " (J.C. Taylor 1988, personal communication).

Hydrometallurgical extraction of laterite nickel along with some of the iron and cobalt is obtained by treating the product with sulfuric solution under high temperature and pressure. Extraction is also carried out by using an aqueous solution containing ammonia or ammonia and carbon dioxide.

COPPER SYMBOL: Cu

Copper was first used by neolithic man in the free metallic state in 8000 BC as a substitute for stone. Copper metallurgy dawned in Egypt as copper was annealed, reduced to metal from sulfide ore with fire and charcoal and alloyed with tin to form bronze. The Roman supply of copper came from Cyprus, whence the name.

Copper is part of nearly all coinage metals. Copper has been the metal second in use and importance to iron only, since the bronze age. Only recently, in the 1960's, was it replaced by the less expensive more plentiful aluminum. Considering that copper only accounts for 0.01 per cent of the Earth's crust, against 8 per cent of aluminum, its usefulness and attraction become obvious.

Uses of copper are based on its high conductivity and resistance to corrosion. The major portion of the copper produced in the world is used in the electrical industry, the remainder is combined with other metals to form alloys. Copper alloys include brass (Cu-Zn), bronze (Cu-Sn), nickel silvers (Cu-Zn-Ni, no Ag), aluminum bronzes (Al-Cu) and beryllium copper (2 per cent Be), which can be hardened by heat treatment. Copper is also used for the piping, roofing and siding of many buildings and houses.

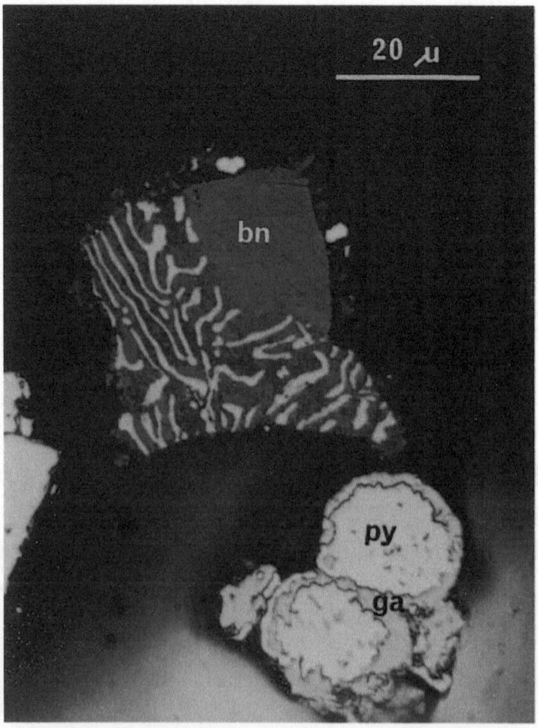

Fig. 63. *Copper minerals bornite (bn), and chalcopyrite/chalcocite (finely intergrown). Also in the photomicrograph are two grains of pyrite coated by galena.*

Copper belongs to the same group IB of the periodic system as gold and silver. Although the interest in copper has always been of a different order of magnitude when compared to the interest in gold and silver, its attractive colors make it probably the only other element *extensively* used in its metallic form for personal and other ornaments.

Copper is present in many sea corals, mollusks, and arthropods and in the human liver. Copper in humans helps catalyze hemoglobin formation.

Although commercial deposits of copper occur in almost every continent, 70 per cent of the world's known reserves are found in seven countries, namely, Chile, the United States, the USSR, Zaire, Peru, Zambia, and Mexico. The greatest known reserve of copper ore in one body is the deposit at El Teniente Mine in Chile. Many lesser deposits are being exploited such as those in Alaska, China, Australia and Europe.

MINERALOGY

"Most metallic elements do not occur in the Earth's crust in native or metallic state. The exceptions are: gold, which commonly occurs so; copper, which has been found primarily as a metal in only one important deposit (in Michigan) and in small amounts in many sulfide ores elsewhere; and silver, bismuth, mercury, *lead, iron, platinum group metals, arsenic, antimony, tellurium and tantalum* (words in italics added by the author). Copper is one of the nine metals found principally in sulfide/sulfosalt minerals; the others are lead, zinc, nickel, *cobalt*, antimony, mercury, bismuth, molybdenum, and cadmium. The copper sulfide ores are of greatest importance; the primary mineral **chalcopyrite,** $CuFeS_2$, and the secondary mineral **chalcocite,** Cu_2S, are responsible for a very high percentage of copper production throughout the world. Following these (not necessarily in order of importance) are: **bornite,** Cu_5FeS_4, **covellite,** CuS, **enargite,** Cu_3AsS_4, and **tetrahedrite,** $(Cu,Fe,Ag)_{12}Sb_4S_{13}$. Copper production from oxidized ores, growing in importance because sulfuric acid *used during the leaching process,* (words in italics added by the author) has been made available at a lower price through programs at copper smelters, to meet government restrictions on SO_2 emission, depends mostly on the minerals **malachite** and **azurite** (basic copper carbonates) **chrysocolla**, and **dioptase** (silicates), **atacamite** (basic chloride), **brochantite** (basic sulfate), and several rarer minerals. **Cuprite**, Cu_2O, and **tenorite** CuO occupy a category of their own. While economically important where they exist or have existed as the principal minerals, such as at Bisbee, Silver Bell and Safford (in Arizona), they are only of minor interest today. These oxide minerals also occur in some sulfide ores in minor proportions and for years have been left in the sulfide tailings because the cost of acid and other economic circumstances did not justify their recovery, either with the flotation of sulfides or from the flotation tailings. Today this picture is changing. The oxide minerals also occur in so called mixed sulfide-oxide ores, where recovery of oxide copper in the processing plant was economical even before there was cheap acid." (Vincent and Weiss 1985, p. 14A-2). Figure 63 shows a composite particle consisting of bornite, chalcopyrite and chalcocite.

Copper deposits can be classified as:

- **Porphyry copper deposits or porphyry coppers.** These deposits, where the copper sulfides and other minerals are more or less uniformly scattered throughout the rock, account for the greater tonnage of the metal in the producing areas of the world. Porphyry coppers are the result of the dissemination of copper minerals in the crushed or fractured periphery of intrusive igneous bodies of granitic rocks, such as quartz monzonites and diorite porphyries. Their grade is low (approximately 1 per cent Cu by average weight), but they are important sources of the metal because they may be worked on a large scale at low cost. Porphyry copper deposits usually produce molybdenum, sometimes gold and silver as by-products. Large porphyry copper deposits occur in the Southern United States, the Solomon Islands, Canada, Peru, Chile and Mexico. The copper minerals are in general oxides, in the upper portions, and sulfides, in the lower levels. The host is a porphyry, schist or other rock.
- **Massive sulfide deposits.** These are of higher metal content, but of more limited extent and may be oxidized in the upper portion with sulfides lower down.
- **Native copper ores.** These consist of the metal, occurring unadulterated in nature, like the one at the Lake Superior area in Michigan.
- **Mixed ores.** In these ores, nickel, zinc and/or lead, with or without the platinum group elements, usually accompany the copper. Where the ores are mined for copper, the other metals are refined and sold as by-products.

METALLURGY

Copper is extracted from its ores by pyrometallurgical and by hydrometallurgical techniques, depending on the nature of the predominant copper bearing minerals. In general, sulfide ores are concentrated by flotation and smelted, while oxide ores are leached. Many oxide ores are not amenable to concentration, so it is necessary to treat and leach the ore as mined, like in the case of the nickel laterites.

Smelting. Smelting of copper ores may be preceded by roasting, "used in some smelters because roasting prior to smelting increases smelting capacity, less energy is required to melt hot roaster calcines than wet sulfide concentrates, roaster off gases are high in SO_2 concentration, 5-15% SO_2, and some volatile impurities are removed from the concentrate prior to smelting." (Moore 1985, p. 14H-3). In roasting, the copper concentrate is heated in air to expel volatile matter, such as arsenic and antimony, and to liberate part of the sulfur. The resulting material known as calcine, is a fine mixture of sulfides and oxides of iron and copper, gangue material and non-volatile impurities. "Many smelters do not use roasters, because the problems associated with handling hot dry calcines outweigh the advantages mentioned." (Moore 1985, p. 14H-3). The ore is generally smelted in large oblong furnaces constructed with heavy, silica brick side walls and a low roof, known as *reverberatory furnaces* because heat is radiated from the roof onto the material treated. The ore may also be smelted in blast furnaces or electric furnaces. The matte resulting from the smelting process consists of a molten sulfide of copper and iron. From this matte, metallic copper is produced by taking advantage of the lower affinity of copper for oxygen than iron and sulfur. The process involves the introduction onto the matte of a stream of air which leads to the formation of iron oxide, sulfur dioxide and copper oxide. The copper oxide reacts with any remaining sulfide to form copper sulfide again, while the sulfur dioxide, being a gas, leaves the pool, and the iron oxide reacts with the silica flux, thus turning into a fusible lighter slag which is skimmed off the top. As air continues to enter the furnace after removal of the slag iron, the copper sulfide is oxidized to copper oxide. This phase reacts with the remaining copper sulfide to form metallic copper or *blister copper.*

Moore (1985 pp. 14H-3 to 14H-7) describes four copper smelting systems. These are the *Mitsubishi Continuous Smelting*, the *Noranda Process*, the *Outokompu Flash smelting* and the *Inco Flash Smelting*.

Fire refining. The blister copper, or metallic copper produced by the smelting process, contains copper oxide and small amounts of other impurities. Some of these impurities are best removed by fire refining which is either done in small reverberatory furnaces or in revolving furnaces equipped with fuel burners which melt the metal and maintain it in a molten condition throughout the refining operation. Air is forced through the molten material to ensure complete oxidation of all impurities, and then the oxides are allowed to rise to the top of the pool from which they are skimmed. This oxidizing treatment is followed by a reducing treatment in which most of the copper oxide is reduced to metal. The reducing treatment is accomplished by highly reducing gases resulting from the distillation of green logs forced into the molten metal, or by reformed natural gas bubbled through the pool.

Electrolytic refining is used for the removal of additional impurities and for the separation of any precious metal values.

Leaching. According to Fletcher (1985), oxidized secondary copper minerals are soluble in sulfuric acid, sulfides are soluble in acidified ferric sulfate. Ammonium carbonate has been used successfully on the native copper ores in Michigan. Leaching today is largely applied to oxidized copper minerals such as cuprite, tenorite, azurite, malachite, and chrysocolla. Sulfuric acid readily dissolves these oxides while leaving most of the gangue minerals unaffected. Some iron may be dissolved at the same time.

Following the leaching process, the pregnant solutions are subjected to purification for the elimination of soluble iron and other impurities. This step often involves neutralization and oxidation, followed by the removal of the precipitated ferric salts.

More common process for the production of metallic copper from the copper sulfate in solution, is electrolysis during which the metal is recovered by passing a current through the solution to break the copper sulfate into positive and negative ions and then by depositing the copper ions onto a copper cathode as metallic copper.

"Proposed advantages of hydrometallurgical processing of sulfide copper concentrate include:

- Amounts of gases and dusts are reduced from those associated with pyrometallurgical processes.
- Final product can be of high enough quality to be used directly without further refining.
- There are lower energy requirements, especially direct use of natural gas and fuel oil.
- Size of the processing plant treating concentrates economically can be smaller that smelters.
- Unit capital and operating costs can be smaller.
- Duplication and back-up processing equipment can be added to lower capital cost.

Disadvantages of hydrometallurgical processes of copper concentrates are:

- Very limited experience to date with large-scale plants designed to treat copper sulfide concentrates.
- Potential pollution sources from the liquids used in leaching. Waste disposal of solutions can be a greater problem than disposing of solids.
- Precious metal and other metal losses can be significant.
- Replacement of existing high cost smelters has not yet been justified by savings possible from hydrometallurgical processing.
- Changing mineralogy in copper concentrates can upset the process.
- Some hydrometallurgical processes cannot recover copper from every copper mineral." (Moore 1985, p. 14H-8).

LEAD AND ZINC SYMBOLS: Pb, ZN

Zinc was known in Roman times only in combination with copper as the alloy brass. The metallurgists of India seem to have isolated the metal as early as the 13th century and those of China had achieved large scale production of zinc by the 16th century. In the West, commercial zinc production got under way by the middle of the 18th century in England.

Zinc is a metallic element of moderate hardness that can be made ductile and easily worked with moderate heating. Its most important use, as a protective coating for iron, derives from two if its outstanding characteristics: high resistance to corrosion and sacrificial protection, corroding in place of the iron. Its second most important use is in die-casting alloys. Zinc alloys melt easily, quickly assume the shape of even the most complicated die, and do not attack the steel dies into which they are injected.

Zinc is an essential trace element in the human body and believed to control the growth of body hair.

Lead is a soft silvery white or grayish metal very malleable, ductile and dense and poor conductor of electricity. Known in antiquity and believed by the alchemists to be the oldest of metals, lead is highly durable and resistant to corrosion, as is indicated by the continuing use of lead water pipes installed by the ancient Romans.

Because of its general chemical resistance, considerable amounts of lead are used in roofing, as coverings for electrical cables placed in the ground or under water, and as linings for water pipes and conduits and structures for the transportation and processing of corrosive substances.

Uses of zinc and lead are summarized by Brown (1985, p. 15-1): "Zinc ranks third and lead fourth in the world nonferrous metal consumption... In 1972 zinc uses were for galvanizing (39%), die casting (26%), brass (15%), rolled zinc (8%), oxide excluding American Process zinc made directly from concentrate (6%) and other uses (6%). The corresponding distribution of lead uses in 1972 was for storage batteries (42%), lead alloys (7%), pigments and chemicals (11%), cable sheathing (9%), gasoline additives (13%), pipes and sheets (8%), and miscellaneous uses (10%)."

Major zinc producing countries are Canada, the USSR, Australia, Peru, the United States, Japan and Mexico. The United States is the world's major lead producing country. More than 95 per cent of the mined lead is produced from primary ores in which galena is the predominant lead mineral.

MINERALOGY

"In most primary ore deposits, the sulfides galena, PbS, and **sphalerite**, ZnS, are the only minerals of lead and zinc respectively. They commonly occur together with sphalerite predominating, but there are all gradations in relative proportions, and each may occur to the exclusion of the other, depending on the district. In a few deposits **jamesonite**, $Pb_4FeSb_6S_{14}$, is an important constituent of the ore, generally in addition to galena." (Mc-Knight 1985, p. 15-2). **Anglesite**, $PbSO_4$, and less commonly **cerussite**, $PbCO_3$, **pyromorphite**, $Pb_4(Pb,Cl)(PO_4)_3$, **mimetite**, $Pb_4(Pb,Cl)(AsO_4)_3$ and **phosgenite**, $PbCO_3$ $PbCl_2$, are found in oxidized environments replacing the galena to various extent. **Smithsonite**, $ZnCO_3$, is a common secondary mineral after sphalerite. "Commonly associated sulfides in lead-zinc deposits are pyrite, marcasite and chalcopyrite; less common are arsenopyrite, tetrahedrite, tennantite, enargite, bornite, and pyrrhotite. Stibnite is an occasional constituent. Copper is often a minor constituent in lead-zinc ores. Silver may occur as argentite, as native silver or in other minerals. The common gangue minerals include several carbonates such as calcite, dolomite, ankerite and siderite; rhodochrosite is less common. Quartz and jasperoid, fluorite and barite are also common gangues; rhodonite is less common." (McKnight 1985, p.15-2). Most of the world's zinc is extracted from geological replacement deposits, or deposits where a mineral in solution gradually replaced an earlier formed mineral, in limestones and dolomites. Examples of ore bodies of this type include the large Sullivan Mine in British Columbia, Canada, and Broken Hill in New South Wales, Australia. The latter deposits contain lead, silver and minor but valuable amounts of copper and cadmium, besides zinc. Other zinc deposits occur in Belgium, Poland and North Africa.

Besides those associated with zinc, some galena deposits occur as replacements of limestone or dolomite (Santa Eulalia - Mexico), some are of contact metamorphic origin (Darwin - California). Galena is found in cavities and brecciated zones in limestone and chert, like the one in the Mississippi Valley deposits where 90 per cent of the United States lead is mined. Galena deposits also occur at Greiberg - East Germany, Broken Hill - Australia, Coeur d'Alene - Idaho, Clausthall Zellerfeld Niedersachsen - West Germany, and Cornwall - England.

METALLURGY

Recovery of zinc is obtained by flotation, roasting and sintering and separation by electrolytic process or zinc blast furnace. "Major zinc extraction processes operating on a commercial scale in the free world, are:

- Electrolytic (roast-leach-electrowinning from sulfate solution).
- Imperial Smelting Process (ISP, ideally a zinc-lead process).

Horizontal retorting, vertical retorting (New Jersey process) and electrothermic smelting (St. Joseph Lead Co. process), used in relatively recent past, are presently obsolete, due to energy and environmental restraints. (J.C.Taylor 1988, personal communication, paragraph added by the author).

The electrolytic process currently produces about two thirds of the free world's primary zinc and seems to be increasing this proportion against a background of rising costs of electric power and fossil fuels.......The Imperial Smelting Process, which has shown rapid expansion in the last ten years, is encountering higher fuel costs, which it is able in part to offset by its flexibility with regard to raw material requirement. The trend in the zinc industry to higher specification zinc encourages electrolytic processing, while the expected decrease in raw material quality will favor ISP smelting." (Milner 1985, p.15-9).

In general, the electrolytic process is used in areas of cheap electrical power, while the zinc blast furnace is more indicated for complex and mixed ores. Metallic zinc is produced by thermal reduction of the oxide with carbon, followed by condensation of the zinc vapor, and electrolysis of the purified zinc sulfate solutions.

Recovery of lead includes flotation, roasting, sintering, and smelting in a blast furnace. Extractions are also obtained by roasting the ore in an open hearth furnace.

"Blast furnace smelting, both the more common lead blast furnace and the ISP zinc-lead furnace, is by far the more important extraction process for lead, accounting for well over 85 per cent of the primary lead being produced." (Milner 1985, p. 15-12). "However, new processes such as the Kivet and the Queneau-Schuman-Lurgi (QSL) systems have been developed in recent years and adopted for new projects and for the modification of older smelters such as the Cominco's Trail lead smelter in Canada". (J.C. Taylor 1988, personal communication).

Refining is accomplished by electrolysis. Kettle refining is a more complex process suited for ores where gold and silver are by-products. Large capacity steel pots or reverberatory furnaces are used to hold the bullion at red heat. Blown oxygen air and other suitable reagents oxidize the impurities. Gold and silver are removed by adding zinc and by raising the temperature sufficiently to reach their melting point. The molten zinc reacts upon stirring and forms compounds with gold and silver and with any traces of copper and nickel. These zinc compounds are lighter than the metallic lead: upon cooling, they form a crust which is removed. The gold and silver are then recovered in a parting plant and the lead is treated for removal of the remaining zinc by a vacuum process.

The refining process yields pig lead of good marketable uniformity. Large quantities of pig lead are available which are 99.999 per cent pure.

ARSENIC SYMBOL: As

Although compounds of arsenic were known as early as the 4th century BC, the element was not identified as such until 1648.

Elemental arsenic is used in certain alloys to increase strength at elevated temperatures, in bronzing and in pyrotechnics, and to impart nearly spherical shapes in the manufacture of lead shots. Arsenic is also used in some chemotherapeutic agents, in certain paints, wallpapers and ceramics, in the electronic industry and, to some extent, in the chemical industry. Arsenic and some of its compounds are lethal and are used as insect, rodent and weed killers.

MINERALOGY

More common arsenic minerals are arsenopyrite, FeAsS, **realgar**, AsS, and **orpiment**, As_2S_3. Nickel and cobalt arsenides and several sulfoarsenides occur, but perhaps not as frequently. Elemental arsenic has been identified, but is rare.

Arsenopyrite is the most common mineral containing arsenic (although not commonly an ore) and is very widespread worldwide in its occurrence. Realgar and orpiment are important ores of arsenic and are found at the Getchel gold mine in Nevada, at Boron - California, Bigadic - Turkey, Carrara - Italy, and Binnatal - Switzerland. Other localities rich in realgar/orpiment are in the USSR, Macedonia - Rumania, Luceram - France and Mercur - Utah.

METALLURGY

Despite the wide array of applications, arsenic minerals are not always recovered for their arsenic value, they must however frequently be separated as contaminants from ores of other minerals.

Separation of contaminant arsenic is commonly conducted by flotation.

To the author's knowledge, there are no major deposits mined exclusively for arsenic, the element being usually a byproduct of other metals. "In some non-ferrous smelters, such as Boliden's Ronnskor smelter in Sweden, arsenic has been recovered for a number of years by selective cleaning of the smelter gas streams. Lepanto in the Philippines roast their copper concentrate prior to smelting, to remove the arsenic, and recently in Japan several new processes have been developed to recover arsenic as a byproduct from smelter operations." (J.C. Taylor 1988, personal communication).

NIOBIUM AND TANTALUM SYMBOLS: Nb, Ta

Niobium was first discovered in 1801 and named columbium. It was rediscovered and named again in 1844. International agreement has established the name niobium, although the United States mining industry continues to refer to it as columbium.

Tantalum was discovered in 1802. Its name refers to the mythological Greek character and was chosen because of the tantalizing problem of dissolving the oxide in acids.

"Niobium and tantalum are relatively rare, heavy and refractory metals. They are chemically similar. Niobium in ferro-alloy additions to steel accounts for 90 per cent of the metal consumed. The niobium-bearing steel and superalloys are used as structural members of buildings and bridges, aircrafts, rockets and ships; as furnace parts; heat exchangers; and in electronic components. The main use of tantalum is as capacitors and tubes of electronic circuits. It is also used in heat exchangers, as cutting tools (tantalum carbide), and in aerospace structures and jet engines as superalloy. Pure tantalum metal has been used in humans as a substitute for bones in surgery." (Zaman 1985, p. 27-3).

Because of its compatibility with uranium, resistance to corrosion by molten alkali metal coolants and low thermal-neutron cross section, another application of niobium is in nuclear reactor cores.

"At present Brazil is the major producer of niobium concentrates in the world with Canada, Nigeria, and the Congo also as major producers. Brazil is also the major producer of tantalum concentrates closely followed by Australia, Nigeria, Mozambique and the Congo." (Zaman 1985, p. 27-3).

MINERALOGY

The most important commercial minerals of niobium and tantalum are **pyrochlore**, $(Na,Ca)_2(Nb,Ti)_2(O,F)_7$ and members of the **columbite-tantalite** series, $(Fe,Mn)(Nb,Ta)_2O_5$. **Microlite**, $Ca_2Ta_2O_7$, is a minor commercial source of tantalum. **Native tantalum**, which usually contains minor niobium and traces of manganese, is rare and occurs sparingly in placer deposits in the Ural Mountains and possibly in the Altai Mountains in central Asia.

METALLURGY

"Tantalum concentrates are obtained from the byproducts of tin ore concentration as well as from placer and alluvial mining. Although some niobium concentrates are derived as byproducts of tin mining, major ore bodies are mined specifically for its recovery in Brazil and Canada. The beneficiation of these ores varies from handpicking and simple gravity concentration to highly sophisticated process systems." (Zaman 1985, p. 27-3). Separation of niobium from tantalum is effected by solvent extraction. Other methods such as fractional crystallization and distillation, have also been employed. Among the methods for reducing the purified niobium compounds to metal are reduction of the oxide by carbon or niobium carbide in vacuum, hydrogen or magnesium reduction of the chloride, sodium reduction of double fluorides and fusion electrolysis.

MOLYBDENUM SYMBOL: Mo

Molybdenum was isolated in 1782. The element was known before, but thought to be graphite or a lead ore, whence the name which in Greek means lead.

"Molybdenum, the 42nd element in the periodic table, is now a very important metal in the industry of the United States and the world. Chemically it is a very complex element having valences of 0, +2, +4, +5 and +6 and can exist as a mixture of these valences in many compounds." (Vincent 1985, p. 16-2).

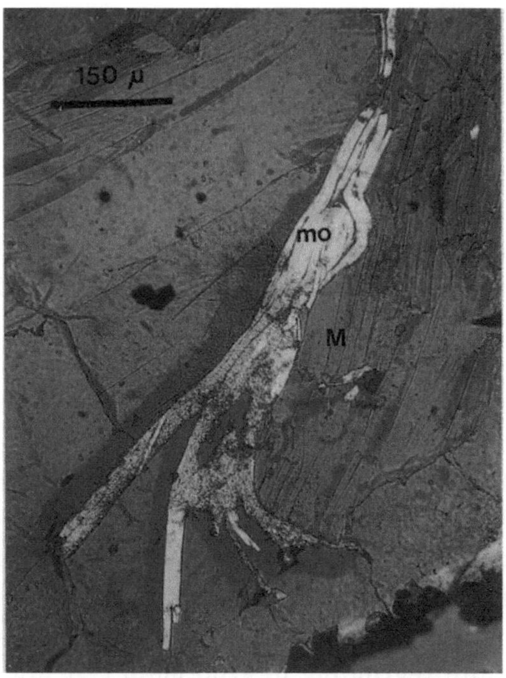

Fig. 64. *Molybdenite plates associated with muscovite mica.*

" There are four major applications in which molybdenum is used: in alloy addition, in the metallic form as pure molybdenum, molybdenum chemicals and catalysts, and as molybdenum disulfide lubricants." (Briggs and Vertes 1985, p. 16-32).

Molybdenum base alloys and the metal itself have useful strength at temperatures above which most other metals and alloys are molten. The major use of the molybdenum, however, is as an alloying agent in the production of ferrous and non-ferrous alloys, to which it uniquely contributes hot strength and corrosion resistance. Molybdenum is one of the most effective elements for increasing hardenability of iron and steel, and it also contributes to the toughness of quenched and tempered steels. Metallic molybdenum has been used for electric and electronic parts such as filament supports, anodes and grids. Rod or wire is used for heating elements, in electric furnaces operating up to 1700°C.

Major molybdenum producing countries are Canada and the United States.

MINERALOGY

"At the present time **molybdenite**, MoS_2, (figure 64), is the only important source of molybdenum, *some minor amounts being extracted from powellite, $Ca(Mo,W)O_4$, the Mo equivalent of scheelite, $CaWO_4$* (sentence in italics added by the author). Molybdenum occurs in some 12 to 14 recognized minerals, but practically all of the production is from molybdenite......Molybdenum has also been produced in the US from **wulfenite**, $PbMoO_4$, from Arizona and Nevada and a small amount of molybdenite was produced up to the early 1970's as a coproduct from two molybdenum-bismuth mines in Quebec, Canada." (Vincent 1985, p. 16-2).

Important molybdenite mines are those at Sherritt Gordon - Manitoba, Climax - Colorado and Tyrnyauz - USSR.

METALLURGY

"Molybdenite production from mines in which it is the sole or preponderantly important mineral is confined to ore bodies in granites of igneous-type rocks or pegmatite dikes. More than one state of mineralization is sometimes present as at Climax (Colorado), York (New Mexico), Hardy, and Questa (New Mexico). In the case of Climax, important byproduct recovery of tungsten, tin, pyrite and monazite is also attained. The molybdenite is commonly disseminated very finely in the siliceous granitic gangue but does occur in stringers and blebs in silicified mineral solution channels. Molybdenite is also extracted from porphyry coppers *(page 255)*.

All primary molybdenum ores are low in molybdenite, the content ranging from less than 0.20% MoS_2 to a high of about 0.40% MoS_2 .... Essentially, the treatment consists of intensive rougher flotation at a relatively coarse grind with production of a low grade rougher froth containing a substantial percentage of the molybdenite in the form of low grade middling particles, followed with multistage regrinding to liberate the locked molybdenite and several stages of flotation cleaning to produce a finished grade concentrate. Rougher flotation on pulp ground to approximately 35 mesh is common practice, even if complete liberation of molybdenite from siliceous gangue may require regrinding to -200 mesh." (Cuthbertson and Vincent 1985, p. 16-2).

The concentrated molybdenite is usually roasted in an excess of air to yield molybdenum trioxide, MoO_3. Subsequent treatment depends on the ultimate use of molybdenum. Ferromolybdenum (50 to 75 per cent of molybdenum) is produced from the technical oxide by either thermite or electric furnace process. Molybdenum metal is produced in the form of a powder by hydrogen reduction of purified molybdenum trioxide or ammonium molybdate, $(NH_4)_2MoO_4$. The powder is converted to massive metal by the powder metallurgy process or by the arc casting process.

"One of the most important practical applications of activated carbon, was in the molybdenum extraction plant operated by AMAX. In the process, the oxide-molybdenum minerals are leached with sulfurous acid followed by sorption by activated charcoal, separation of coarse carbon by screening, desorption of leaded char with ammonia and eventual recovery of molybdenum as MO_2." (Bhappu 1985, pp. 13-50 and 13-51).

TIN SYMBOL: Sn

Tin has been known and used since antiquity. Mining and smelting of tin ores marked the beginning of the bronze age.

Because of the low melting point and firm adhesion to clean surfaces of iron, steel, copper and copper alloys, an important application of tin is as an oxidation resistant coating material. Other applications are in the plating of steel cans used as food containers, in metal used for bearings, and in solder. Tin is also used for the construction of pipes, wire, tubes, sheet, foil and anodes for electroplating. Alloys of tin are numerous and include soft solder (tin-lead), pewter (91% Sn, 7.5% Sb, 1.5% Cu, in modern alloys; only Pb and Sn in the Roman pewters), bronze (copper-tin), bell metal, babbitt metal (antimony-tin-lead-copper, used in components of machine engine bearings), white metal, die casting alloy, and phosphorus bronze. A crystalline alloy with niobium is a superconductor at temperatures as low as 18°K (-427°F). Such alloy retains its properties in low magnetic fields.

"China, Indonesia, Malaysia, Thailand and Burma are among the world's leading tin producers." (Aplan 1985, p. 27-10). Other tin producing countries are Bolivia, Zaire, Brazil, Nigeria, Australia, England, the USSR and Spain.

MINERALOGY

"The only tin mineral of major importance is **cassiterite**, SnO_2, (figure 65) though a host of complex sulfide minerals such as **stannite**, Cu_2FeSnS_4, may have local importance as, for example, in certain Bolivian ores. Cassiterite is found associated with granitic rocks as an accessory mineral in trace amounts, as segregation in pegmatites, or as localized concentrations in veins and fissures. Some of these relatively high concentrations of cassiterite exist as lode deposits, but the world chief source of tin today is from alluvial and eluvial deposits resulting from the weathering of primary deposits. Cassiterite is not only heavy, but it also strongly resists weathering and, as a consequence, it is often found in placer deposits." (Aplan 1985, p. 27-10).

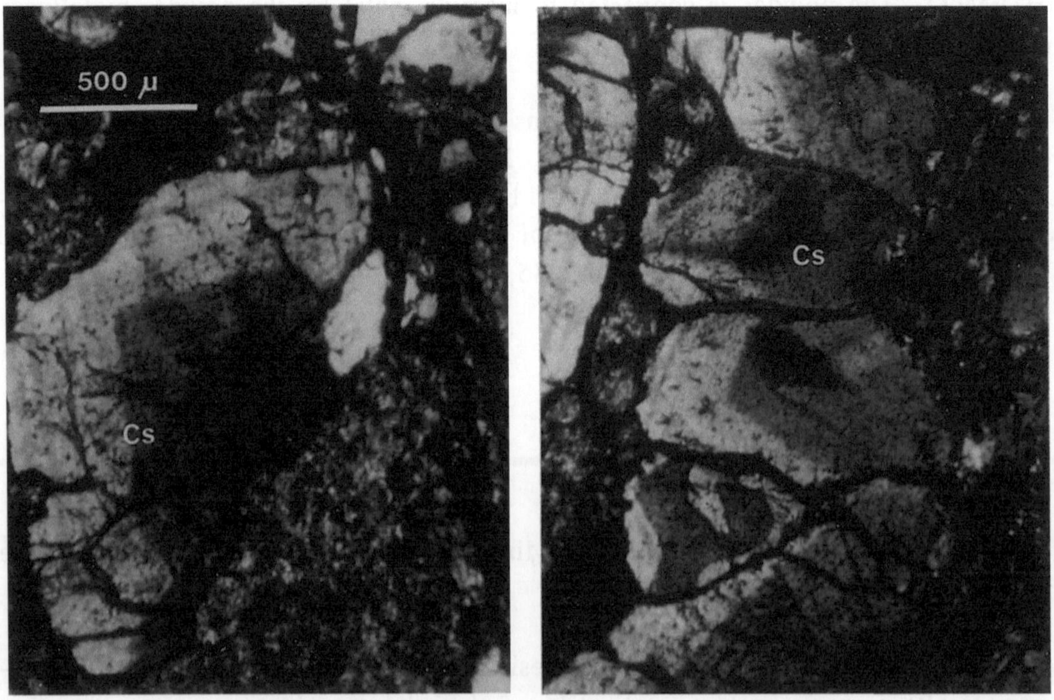

Fig. 65. *Cassiterite showing an obvious zoning due to the presence of variable amounts of iron. The dark red areas are rich in iron.*

The majority of the alluvial deposits of tin in Southeast Asia form on top of a bedrock source of cassiterite, as chemical decay removes the other less stable minerals. In Zaire, residual alluvial material grades downward into weathered lodes and ore minerals, either as alluvial deposits or as open pit lodes.

Although the oldest tin mines were in England and Spain, today "the world's major tin deposit is an elongated zone of over 2000 Km extending from Indonesia to Malaysia, Thailand and Burma into China. Commercial production here is almost exclusively from placer deposits. Other placer deposits are known throughout the world and are of major importance in Brazil, Nigeria, and Zaire. Bolivia is the major producer of tin from lode ores, though mining of these deposits is also done in Australia, Canada, England (Cornwall), the USSR, Spain, etc. The United States has negligible tin resources, and in recent years, it has been mined in Alaska, Colorado and New Mexico." (Aplan 1985, p. 27-10).

METALLURGY

"Alluvial deposits lend themselves to mining by water washing methods and gravity separation of the cassiterite. Lode deposits require more complex operations, including rock blasting and crushing prior to gravity separation. Placer tin is mined either by marine methods using bucketline dredges and gravel pumps or through land mining techniques, especially open-pit methods....The tin-bearing mud is excavated at depths down to 43 meters below the water level. Final concentration and removal of the associated minerals is carried out on shore using a variety of techniques. Final concentrates, ready for direct smelting, contain almost pure cassiterite (between 70 and 77 per cent tin)." (MacIntosh 1988, p. 425).

Tin is obtained from cassiterite by reduction with coal in reverberatory furnaces. Tin concentrates from alluvial deposits are usually smelted without further beneficiation. Lode deposits are roasted and acid leached in hydrochloric acid to remove sulfur and metallic impurities such as iron, copper, lead, zinc, and bismuth. The smelting process is then carried out in reverberatory tin smelting furnaces and involves treatment of the concentrate and re-treatment of the slag at higher temperatures to remove the unmelted tin silicate. Refining is accomplished by further heat or by electrolysis.

Tin is sometimes recovered as a byproduct from the tail products of other ores.

ANTIMONY SYMBOL: Sb

Antimony is one of the oldest metals. The natural sulfide was known to ancient peoples. Women of biblical times used it as medicine and as a cosmetic for eye brow painting. A vase found in Chaldea, cast in metallic antimony, suggests that about 4000 BC the Chaldeans knew the art of reducing the metal. Later in the 15th century, German monk Basil Valentine and then in the 16th century German physician and scholar Georgius Agricola (the father of mineralogy) described the refining of antimony metal. (**US Bureau of Mines Material Survey** 1951).

The presence of antimony in type metal, which also includes lead and small amounts of tin, increases the hardness of the type and gives it a sharp definition. Even when added in small quantities, antimony imparts strength and hardness to other metals, particularly lead with which it forms alloys used in plates of automobile storage batteries, in bullets and in coverings for telephone cables. Combined with tin, lead and copper, antimony forms antifriction alloys called babbitt metals that are used as components of machine bearings. Antimony compounds are used in ammunition and friction matches, fire works, in flameproofing fabrics, in paints and plastics, both as pigment and as a fire retarding agent.

Antimony is also used in medicine, glass making, pottery glaze, printing processes, and stained glass.

A property typical of antimony (and bismuth and water) is of expanding upon solidifying.

Over 80 per cent of the world production of antimony in 1968 was contributed by the Republic of South Africa, China, Mexico, Bolivia and the USSR.

MINERALOGY

"Among over 100 known minerals of antimony, only **stibnite, Sb_2S_3, stibiconite, $H_2Sb_2O_5$, tetrahedrite**, $(Cu,Fe)_{12}Sb_4S_{13}$, and lead ores containing antimony, yield substantial commercial quantities of the metal. (Zaman 1985, p. 28-6).

Stibnite is found in hydrothermal veins or hot spring deposits in Hunan province - China, Kremnica - Czechoslovakia, Calston - England, Djebel Hamminate - Algeria.

METALLURGY

"In antimony mining, hand methods (China, Mexico and Bolivia), underground stoping (United States, Mexico and South Africa) on a moderate scale, and open pit (United States) mining on a large scale are employed......The largest underground antimony mine, near Gravellote, Republic of South Africa, uses shrinkage stoping methods....Both oxide and sulfide antimony ores are concentrated by hand jigging in Bolivia and Mexico. Tabling has been used for additional recovery of fines. Flotation for concentrating stibnite has been successful in producing high grade concentrate containing over 60 per cent antimony with recoveries as high as 95 per cent." (Zaman 1985, pp. 28-6 and 28-7).

Methods of obtaining antimony metal from stibnite include ore treatment by (i) roasting, to form the oxide Sb_2O_4 which is then reduced to the element by heating with carbon; or by (ii) melting with scrap iron in a furnace. As the iron combines with the sulfur to form a liquid layer of iron sulfide, the heavier liquid antimony settles to the bottom and is drawn off.

BARIUM SYMBOL: Ba

Although barium compounds have been known for many centuries, barium was only isolated in 1808.

Barium is used as a getter in electron tubes to perfect the vacuum by combining with final traces of gases, and as a deoxidizer in copper refining. Some of the barium compounds are used in pyrotechnics, petroleum mining and radiology. Barium sulfate is used as the opaque medium in the x-ray examination of the gastro-intestinal tract, and as a body or filler for paper and cloth.

The major use for barite is as a weighting agent in drilling muds. It is also used in the manufacture of paint, glass and rubber.

The major source of production of barium in the United States is Arkansas where the barite is found in large bedded deposits. A similar type deposition is found in Nevada and California. Nevada is an important producer of barium today.

MINERALOGY

The principal mineral of barium is **barite**, $BaSO_4$.

Significant deposits, besides those in Arkansas, Nevada and California, occur in Tennessee, Georgia and Missouri. Barite is also abundant in Castile and Andalusia - Spain, Nordrhein-Westfalen - Germany, and various localities in the Southern Appalachians.

METALLURGY

"In the United States, the processing of barite varies from district to district and is somewhat dependent upon the character of the deposit involved. In Missouri, the separation is mostly done by trommels which reject the quartz, dolomite and other gangue minerals. In Arkansas, flotation is the predominant process, while log washing is the predominant processing method used in Georgia and Tennessee. "(Davis 1985, p. 29-6).

Most barium compounds are produced from the sulfate via reduction to the sulfide. Barium is produced by electrolysis of fused barium chloride, or by reduction by aluminum of a mixture of barium monoxide and peroxide in an electrically treated vacuum furnace.

RARE EARTHS

The first rare earth elements to be discovered were yttrium, in 1794 and cerium, in 1803. Investigations were carried out through the 19th century and first half of the 20th century, when the modern classification was finalized.

"The rare earth metals comprise 15 elements that have atomic numbers from 57 through 71 and include lanthanum, cerium, praseodymium, neodymium, promethium, samarium, europium, gadolinium, terbium, dysprosium, holmium, erbium, thulium, ytterbium, and lutetium, plus scandium and yttrium. The rare earths having atomic numbers from 57 to 61 are often classified as the cerium group (or *light rare earths*), and those having atomic numbers from 62 through 71 plus scandium and yttrium are classified as the yttrium group (or *heavy rare earths*, words in italics added by the author)......

The principal use of rare earths is as a catalyst in the petroleum refining industry, with significant use in the manufacture of optical instruments, glass manufacture, electronic ware fabrication and alloying and TV picture tubes." (Davis 1985, p. 27-3 and 27-4).

Applications of the individual rare earth elements have increased considerably since the 1970's, following the advent and popularity of color television, where yttrium, gadolinium and europium are being used for the *red phosphors* in the picture tubes.

Rare earths are mined in Brazil, Norway, Madagascar, Sri Lanka, India and North Carolina.

MINERALOGY

No minerals representing the end members of the different rare earth solid solutions have been identified. All of the recognized rare earth minerals consist of mixtures of several elements.

"The rare earths may also be referred to as lanthanides. They occur in the Earth's crust in appreciable quantities and in more than 150 minerals. The two minerals of greatest importance are monazite and bastnaesite (*both containing rare earths of the cerium group or light rare earths*, words in italics added by the author). **Monazite**, $(Ce,La,Di)PO_4$, is essentially a phosphate of the cerium group of lanthanides. The composition of monazite is reasonably constant throughout the world with 50 per cent of its *rare earth* content as cerium and the remaining 50 per cent as the other members of the cerium group. Commercial concentrates also contain a few percent of both yttrium oxide and thorium oxide. Most commercial monazite occurs in beach sand and placer deposits......**Bastnaesite** is a rare earth fluocarbonate having the basic formula $(RE)FCO_3$...Unlike monazite, its composition varies from locality to locality ....The most frequent deposits result from hydrothermal replacement of sedimentary rocks in contact metamorphic zones.....Major ore minerals of the yttrium group (or *heavy rare earths*) are **xenotime,** YPO_4, **gadolinite,** $Be_2FeY_2Si_2O_{10}$, **euxenite** *(see uranium minerals for formula)*, **samarskite** *(see uranium minerals for formula)*, and **yttro-fluocarbonate** *(formula not available)*." (Davis 1985, p. 27-3 and 27-4).

Monazite is one of the main sources used commercially to obtain the light rare earths. Sandbars containing monazite are found along the coast of Brazil and Southwest India. Concentrated deposits are also found in certain uplands thought to be the beaches of ancient seas or oceans which were later uplifted. Such deposits are found in Australia, the USSR, South Africa, South Carolina, Florida and Idaho.

METALLURGY

In placer mining, monazite is first recovered in a bulk concentrate along with ilmenite, leucoxene, rutile, staurolite and zircon. It is then separated from these minerals through other processes such as high tension, magnetic and gravity separation. Leaching in hydrochloric acid may increase the concentrate grade.

Extraction of the rare earths from their minerals is accomplished through a variety of essentially hydrometallurgical techniques (Hampbel 1954). Ion exchange is used to isolate the individual elements. Some rare earth elements are obtained as by-products of uranium operations.

TUNGSTEN SYMBOL: W

"Tungsten is a metal discovered in 1783 but its main commercial applications did not come until the 20th century.....Tungsten is used chiefly for cutting and wear-resistant alloys such as tungsten carbide and this accounts for over half of its use. Other uses are for the production of alloy steel, metal powder, chemical ceramics, and for electrical use." (Aplan 1985, pp. 27-17 and 27-18).

First commercial applications of tungsten were in the fabrication of filament material for lamps, and thereafter in many other electrical and electronic applications. Use of cemented tungsten carbide is for very hard and tough dies, tools, gauges and bits. Some of the tungsten steels have been used in the aerospace industry to fabricate rocket-engine nozzle throats and leading edge reentry surfaces.

Tungsten carbide, with a hardness of 9.5 on the Mohs scale, is used in combination with the carbides of other metals to impart hardness to cast iron and for the cutting edges of saws and drills.

"The People's Republic of China has historically been the world's largest producer of tungsten concentrates, though the USSR, the Koreas, United States, Thailand, Bolivia, Portugal and Australia are also important producers, roughly in that decreasing order......The largest producing mine in the United States is the Pine Creek mine of Union Carbide Corp. near Bishop, California.....In Canada, the mine of Canada Tungsten Mining Corp. in the

Northwest Territories is an important scheelite producer." (Aplan 1985, pp. 27-17 and 27-18).

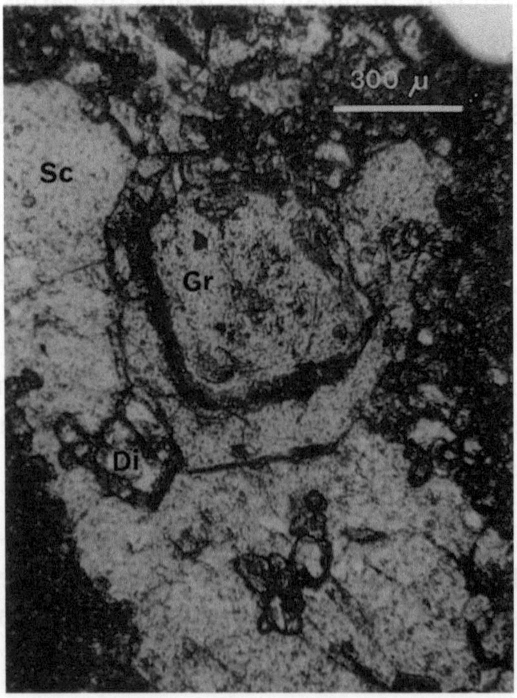

Fig. 66. *Scheelite enclosing a large grain of garnet and a small grain of diopside.*

MINERALOGY

"The principal minerals of tungsten are **scheelite,** $CaWO_4$ (figure 66), and **wolframite,** $(Fe,Mn)WO_4$.... Powellite, the Mo equivalent of scheelite, is sometimes found associated with the scheelite. The deposits containing wolframite are typically of igneous origin and wolframite may occur in veins or segregated in pegmatites. It is not infrequently associated with cassiterite in these pegmatites and this tin/tungsten association is especially important in Bolivia and Thailand. Scheelite is normally found in skarn deposits formed through contact metamorphism by replacement of a limestone bed. The scheelite-bearing tactite at the granite-limestone contact will contain calcite, garnet, pyroxene, fluorite, sulfides and quartz. Virtually all mining is of lode deposits, though in years gone by, placer mining, especially of wolframite, was an important means of production." (Aplan 1985, p. 27-17).

METALLURGY

In tungsten extractions, "the beneficiation used is dependent upon the type of ore; gravity concentration is generally used for wolframite ores whereas flotation is used for scheelite ores." (Aplan 1985, p. 27-18).

Recovery of tungsten from wolframite involves the fusion with sodium carbonate under oxidizing conditions, or by digestion with a concentrated sodium hydroxide solution. These processes are followed by leaching with H_2O to produce sodium tungstate liquor and then treatment with $CaCl_2$ solution to produce calcium tungstate insoluble in hydrochloric acid. Recovery from scheelite includes decomposition by digestion with concentrated hydrochloric acid to produce a crude tungstic acid and further purification leading to an ammonium paratungstate. Reduction of both the calcium tungstate (derived from the wolframite) and the ammonium paratungstate (derived from the scheelite) is done by carbon or hydrogen. The result is tungsten powder.

MERCURY SYMBOL: Hg

Also known as quicksilver, mercury has been known and used for more than 3500 years. Mercury was known to the ancient Chinese and Hindus and has been found in an Egyptian tomb dating back to approximately 1500 BC. In early times, mercury was used in religious ceremonies, in medicinal ointments and cosmetic preparations, and in the recovery of gold and silver by amalgamation. Since the 17th century, the importance of mercury has steadily increased with advancing science and industry.

Mercury does not wet glass or cling to it and this property, coupled with its uniform volume expansion throughout its liquid range, makes it useful in thermometers. Barometers and manometers utilize its high density and vapor pressure. The good electrical conductivity of mercury makes it exceptionally useful in sealed electrical switches and relays. An electrical discharge through mercury vapor produces a bluish glow rich in ultraviolet light, a phenomenon exploited in ultraviolet, fluorescent and high-pressure mercury-vapor lamps. Mercury is also used in dental fillings. A very important application of mercury is in the production of chlorine and caustic soda by the electrolysis of brine in a mercury cell.

Mercury is still widely used for the recovery of coarse gold and silver by amalgamation (pages 120 and 165). Such is possibly the only major use that any metal has retained unchanged for about three millennia (page 12), with the exception perhaps of the use of gold and silver for jewelry and other ornaments.

Air samples taken in the vicinity of some sulfide and precious metal ores have shown a mercury content several times greater than that found in samples taken over non-mineralized areas. Use of air sampling for mercury is a recently discovered exploration tool for many metals including gold, and could become very widespread in the future.

Ease of reduction to the elemental form during mine production and then release into the air and water, make mercury a dangerous contaminant (pages 131-132), because it may be absorbed by the living environment, with consequences that may be lethal. Close control of the amount of contaminant mercury is required by the law in many countries.

Most important deposits and sources of mercury are at the Almanden Entredicho Mine in Spain where the metal has been mined for 2000 years. Other deposits are in Huancavelica, Santa Barbara Mine - Peru, Iudrio and Monte Amianta - Italy, Idria Mine - Yugoslavia, and the New Almanden Mine of California. Mercury is also found in Algeria, Japan, China, Czechoslovakia, Turkey and the USSR.

MINERALOGY

"The most important mercury mineral is **cinnabar**, HgS, red mercury sulfide, and this accounts for 90 per cent or more of the mercury produced in the world. **Metacinnabar**, HgS,

livingstonite, HgS $2Sb_2S_3$, and **native mercury** are also mined. Cinnabar occurs as impregnations and as vein fillings, usually in sedimentary rocks, near recent volcanic rocks and hot springs, having been deposited near the source from alkaline solutions. It is associated with pyrite, marcasite, stibnite and other sulfides in gangue minerals which include opal, chalcedony, quartz, barite, calcite and fluorite. Cinnabar is often associated with clays, such as montmorillonite." (Butler 1985, p. 28-5). Cinnabar is also found in placer deposits, although probably not usually in economic amounts. Native mercury is scarce and may be found in isolated drops and occasionally in larger fluid masses, usually with cinnabar, near volcanos and hot springs. Compounds with gold **(amalgam, goldamalgan and weischanite)**, silver **(moschellendsbergite)** and palladium **(polarite)**, although reported, are extremely scarce.

METALLURGY

"In modern practice, where the ore is too low in grade to stand the cost of furnacing, it may first be upgraded......Because cinnabar is a heavy mineral, gravity concentration would seem to be a logical method of concentration, but cinnabar slimes during grinding; because of this tendency, and the fact that the mineral floats away on the wash water, the fine cinnabar may be lost in gravity concentration......Because of the ready floatability of cinnabar, it would seem that flotation would be the natural method to use for processing these ores. Difficulties arise from three characteristics of cinnabar:

- It slimes readily during grinding.
- It is frequently associated with clayey (montmorillonite, etc.) material.
- It frequently occurs in highly opalized deposits, and the mineral is not liberated at normal grinds of -65 to -100 mesh.

When the cinnabar occurs as a relatively coarse crystalline mineral and is readily liberated without overgrinding, and the gangue does not contain excessive clay, a high recovery and a high grade of concentrate can be produced by flotation." (Butler 1985, p. 28-5). After the beneficiation stage, mercury is extracted from cinnabar, by roasting in air, at temperatures near or above 580°C, followed by condensation of the vapor, a practice dating as far back as before the Romans. This procedure is one of the simplest of all the metallurgical extraction processes, due to the mercury's very low temperature of vaporization (375°C). The mercury sulfide vapors react with the oxygen in the air to form mercury and sulfur dioxide and are then passed through a dust collector and into a series of condenser pipes. The gaseous mercury condenses and collects in the hoppers under a water seal. Mercury recovered in this manner is pure enough for most commercial uses.

Due to the difficulties associated with the concentration of the cinnabar, in some deposits, roasting is applied directly to the crushed ore without beneficiation.

The presence of contaminant mercury in the air, produced during the mining and metallurgical processes applied to many base and precious metals (pages 138-139), is controlled by retort systems.

BISMUTH SYMBOL: Bi

Bismuth was discovered and described in 1450 by German monk Basil Valentine.

Like antimony and water, bismuth expands slightly upon solidifying, a property that renders its alloys particularly suited for the manufacturing of sharply detailed metal castings. Because the alloys have low melting points, they are used in special solders as well as in automatic sprinkler heads, fire door release lines, fuses, safety plugs for compressed gas cylinders, and many other fire detection devices. Salts of bismuth are used in medicine and for medical treatment of injuries, skin infections, and ailments of the digestive tract. They are also used to outline the gastro-intestinal tract during x-ray examinations. Bismuth is an important constituent of *Pepto Bismol*, widely used in the United States for the treatment of upset stomach.

Bismuth may be a contaminant in precious and base metal ores, along with mercury and arsenic.

An important producer of bismuth is Bolivia.

MINERALOGY

"In addition to the **native element**, four common mineral types are as follows: **bismuthinite**, Bi_2S_3, **bismite**, Bi_2O_3...,**bismutite,** Bi_2O_3 CO_2 H_2O...and **tetradymite**, $Bi_2(Te,S)_3$. Good primary bismuth ores are restricted to a few areas of the world. Native bismuth is a common constituent of hydrothermal veins in some mining localities, such as Cordova Province, Spain." (Zaman 1985, p. 28-8).

Appreciable concentrations of bismuthinite are found in Cornwall - England, Schneeberg - Germany, the Tasna district in Bolivia, associated with the tin-tungsten veins of Bedford, Westchester County - New York and Boulder County - Colorado.

METALLURGY

"With the exception of the Tasna mine in Bolivia, where the ore is worked primarily for bismuth, there are no deposits commercially mined for bismuth content alone. It is mostly produced as a byproduct from processing cobalt, copper, gold, lead, molybdenum, silver, tin, tungsten and zinc ores....The ore beneficiation methods employed for concentration of lead and copper minerals also concentrate the associated bismuth minerals. The high specific gravity of bismuth minerals makes gravity an attractive concentration method. Tabling is practiced in conjunction with tin and tungsten ores in Bolivia. Because of the small quantities of ore mined, such as in the Tasna mine in Bolivia and Cordova in Spain, hand sorting is the usual practice to obtain concentrate from the ore. The flotation properties of bismuth sulfides are similar to base metal sulfides (lead particularly) and float well in the presence of the usual sulfide collectors. The oxide minerals, such as bismutite, however, are not amenable to flotation unless sulfidized with sodium sulfide or similar reagents.... The technology of bismuth extraction and refining is well established and is mainly associated with lead refining plants. In copper and lead smelting operations, bismuth collects with the lead. Part of the bismuth, during smelting operations, is volatilized and collected as dust by *Cottrell precipitators.* These products are then sent to lead-refining plants for treatment. A number of processes are employed for the production of impure bismuth bullion, such as crystallization, the electrolytic (*Betts*) process and the *Betterton-Kroll* process. The bismuth bullion thus obtained is further refined to 99.99 per cent by *liquating* in large kettles or crucibles or melting with fluxes. Depending upon the impurities, fluxes commonly employed are caustic soda, sodium or potassium chloride, sodium carbonate, sulfur and borax." (Zaman 1985, pp. 28-8 and 28-9).

Other technique for the production of metallic bismuth from its compounds are by reduction of the oxide with carbon or by roasting of the sulfide in the presence of charcoal and metallic iron to remove the sulfur.

Case study 23. Bismuth contamination in zinc concentrate. Microanalytical studies may be necessary to locate and identify contaminant bismuth minerals during the beneficiation stages of the process of extraction. The final concentrates may contain bismuth values which are above those of the sales specifications, and a later separation during the refining stages may not be acceptable. In some such situations, bismuth may not be eliminated by the standard techniques because it either occurs in very fine particles or in minerals which do not respond to the reagent used. Their proper identification becomes essential for decisions concerning the correct approach for their separation.

In this particular case study, the bismuth content of a zinc concentrate varied markedly, despite the small changes in the composition of the mill feed. The microanalytical study of samples containing bismuth in amounts below 0.013 and above 0.019 per cent (the sales specifications) determined that in both samples native bismuth was present as grains 1 to 2 microns in size, always enclosed in or attached to galena. A complex sulfosalt of bismuth with iron, lead and antimony, identified as **kobellite**, $Pb_2(Bi,Sb,Fe)_2S_5$, accounted for the bismuth content of the high bismuth sample. The kobellite was in the form of grains approximately 10 to 15 microns in diameter, and liberated.

It was concluded that although the larger grains of native bismuth were liberated by grinding and consequently were eliminated, the finer ones enclosed in galena accounted for the bismuth in the zinc concentrates of lower bismuth content (below the sales specifications). The increase in the element in the high-bismuth concentrate was attributed to the presence of liberated grains of kobellite which evidently had flotation characteristics different from those of native bismuth, thus giving rise to a differential flotation problem in the existing mill circuit.

The variation in bismuth content of the zinc concentrate was probably a reflection of a variation in the chemistry and conditions of crystallization within the deposit, comparable to that of the silver case study 5 (pages 30-31).

* * * * * *

RECENT DIAMOND DISCOVERIES (updated information to be read together with the data on page 234). Starting during the late 70's, diamond exploration projects were conducted in Australia, Canada and the United States. Diamonds of largely industrial and "near-gem" quality were first found in Australia in the early 80's in kimberlite pipes and associated with rocks of unusual, predominantly alkaline compositions, known as *lamaroites*.

Later, during the second half of the 80's, many new diamond concentrations were found in Australia, Canada and the United States, as a consequence of new and systematic exploration techniques (D.A. Lowrie personal communication 1992, Pasteris 1984).

While production from South Africa declined, the Argyle mine in Australia quickly became the largest diamond producer in the world. Most of the material extracted was of industrial grade, although the mine also yielded a few excellent quality gemstones some of which were colored and valued at US$ 85,000 and more.

In North America, diamond concentrations were found in Ontario, Saskatchewan, the North West Territories, Southeastern British Columbia, Idaho, Montana and Wisconsin, while commercial operations were planned at the old *tourist* pipe in Arkansas.

Findings were particularly successful in Canada. Diamonds largely of industrial quality with a few gemstones (1-2 mm in size) were found in the Kirkland Lake, Ontario, area, associated with 8 to 10 kimberlite pipes. Exploration continued and between 1990 and 1992 about 100 pipes were discovered in Saskatchewan, although the large majority of the associated diamonds were of industrial quality, with a few gem quality microdiamonds (less than 0.5 mm in size). Between 1991 and 1992, 6 to 7 pipes were found in the North West Territories, and no less that 50 more pipes were expected to be in existence in the same general area. Excitement about this new discovery was derived from the fact that many of the gem quality stones were of larger size. While work continued in the North West Territories, additional exploration shifted to the Rocky Mountains of Northwestern Alberta where gem quality diamonds of alluvial origin were discovered in the drainage of the foothills.

So many new finds, the extensive exploration programs and the presence in North America of the largest *unexplored* craton areas in the world, suggest that Canada's diamond production may be competing with the one of South Africa, the Soviet Union and Australia in the not too distant future. (Information kindly provided by D. A. Lowrie 1992).

TABLE 17 - BASE METALS

METAL	ISOLATED	AT.#	MARKET VALUE	ABUNDANCE	MAJOR DEPOSITS	MINERALS	EXTRACTION
Aluminum	1825	13	$1.01-1.07/lb	3x10 5	Worldwide	Gibbsite, boehmite, diaspore (components of bauxite)	Bayer process (leaching in NaOH)
Antimony	4000BC	51	$1.05-1.10/lb	0.16	In South Africa, China, Mexico, Bolivia, the USSR	Stibnite, stibiconite, tetrahedrite, various sulfo-salts	Gravity concentration/ flotation. Roasting or smelting
Arsenic	1648	33		2.4	Getchel mine, Nevada; Boron, California; in sulfosalt localitites	Arsenopyrite, realgar, orpiment	Flotation. Roasting/selective cleaning of smelter
Barium	1808	56		310	In Arkansas, Nevada, California, Tennessee, Georgia, Missouri. Worldwide	Barite	Gravity concentration/ Log washing. Reduction with Al or electrolysis
Beryllium	1828	4		31	In Brazil, South Africa, Argentina, Uganda, Zimbabwe, Namibia,	Beryl, bertrandite, phenacite crysoberyl, barylite	Hand sorting. Reduction with Mg or electrolysis
Bismuth	1450	83	$6.10-6.15/lb	0.08	Cordoba Province, Spain; Tasna Mine, Bolivia. Also in	Bismuthinite, bismite, bismutite, tetradymite, various	Usually byproduct of Co, Cu, Au, Pb, Mo, Ag, W and Zn;

TABLE 17 - CONT.

METAL	ISOLATED	AT.#	MARKET VALUE	ABUNDANCE	MAJOR DEPOSITS	MINERALS	EXTRACTION
					Cornwall, England, New York and Colorado	sülfosalts	also hand sorting/ gravity separation/ tabling/flotation. Reduction with C
Boron	1808	5		93	In S. California	Borax, kernite, tincalconite	Crushing/ screening. Reduction with H2
Carbon	NA	6		1,700	Diamonds: South Africa, the USSR Graphite: world-wide. Coal: USSR, China, United States, South South Africa, Korea	Diamond, carbon, coal	Diamonds: heavy media separation, grease tables. Graphite: flotation
Chromium	1798	24		190	Stillwater, Montana; Bushveld Complex, South Africa; Great Dyke, Rhodesia	Chromite	Selective mining or gravity concentration/ flotation/magnetic separation. Roasting/leaching. Reduction with Al or electrolysis
Cobalt	1742	27	$7.05-7.20/lb	42	Sherritt Gordon, and Lyn Lake, Manitoba; Mississippi Valley, Missouri District; Shinko-	Cobaltite, pyrite, safflorite, smaltite, linnaeite, pentlandite	Byproduct of copper and nickel. Smelting or leaching in H2SO4

TABLE 17 - CONT.

METAL	ISOLATED	AT.#	MARKET VALUE	ABUNDANCE	MAJOR DEPOSITS	MINERALS	EXTRACTION
					lobwe, Zaire; Cobalt, Ontario; Also in Zambia, The USSR, Australia, Finland, Cuba, West Germany		
Copper	8000BC	29	$1,71/lb	87	In Chile, the United States, the USSR, Zaire, Peru, Zambia, Mexico	Chalcopyrite, chalcocite, bornite, covellite, enargite, tetrahedrite, cuprite azurite, malachite	Gravity concentration/ flotation. Smelting for sulfides. Leaching in H2SO4 for oxides
Iron	1200BC	26	$0.87/long ton	1x10 5	In the USSR, Canada, the United States, Australia	Magnetite, hematite, other oxides, siderite	Gravity concentration/ magnetic separation/ flotation/leaching in HCl. Smelting
Lead		82	0.47/lb	6.0	Cerro de Pasco, Peru; Cominco, BC; St.Joe Minerals, Missouri. Also in Canada and Australia	Galena, anglesite	Flotation. Smelting
Magnesium	1828	12	$1.51/lb	9.6x10 4	In Nevada, Washington, California, Arizona, Canada,	Dolomite, magnesite, brucite, olivine,	Gravity concentration, flotation, removal from seawater

TABLE 17 - CONT.

METAL	ISOLATED	AT.#	MARKET VALUE	ABUNDANCE	MAJOR DEPOSITS	MINERALS	EXTRACTION
					Ireland, United kingdon, the ocean, bitterns, brines		
Manganese	1774	25		17,000	In the USSR, South Africa, Brazil, Gabon, India, Australia, China	Pyrolusite and other Mn oxides, rhodochrosite	Gravity concentration/ flotation. Leaching in H2SO4
Mercury	1500BC	80		0.04	Almanden Entredicho, Spain; Huencavelica, Peru; Monte Amianta, Italy. Also in the United States, Algeria, Japan, China, Czecho-slovakia, Turkey and the USSR.	Cinnabar, metacinnabar, livingstonite	Flotation. Roasting and condensation
Molybdenum	1782	42	$3.49-3.53/lb	1.6	Sherritt Gordon, Manitoba; Climax, Colorado; Tyrnyauz, the USSR	Molybdenite,	Flotation. Smelting or reduction
Nickel	1751	28	$4.98-5.18/lb	130	Sudbury, Ontario; Sherritt Gordon, Manitoba; laterites of Nicaro,Cuba; Sered, Czechoslovakia and New Caledonia in the South Pacific	Pentlandite, pyrrhotite, garnierite	Gravity concentration/ flotation/smelting (for sulfides). Leaching in H2SO4 or NH3 (for some laterites)

TABLE 17 - CONT.

METAL	ISOLATED	AT.#	MARKET VALUE	ABUNDANCE	MAJOR DEPOSITS	MINERALS	EXTRACTION
Niobium	1801	41		22	In Brazil, Canada, Nigeria, Congo	Pyrochlore	Hand sorting/ gravity concentration. Reduction with various elements. Ion exchange
Potassium	1807	19		5.3x10 4	Brines of Dead Sea, Great Salt Lake,Utah Searles Lake, Calif. Also deposits in Spain, the USSR, Germany, Tunisia, SW Unites States	Sylvite, carnallite, kieserite, polyhalite	Leaching in water. Crystallization or flotation
Rare earths	1794	57-71		0.28 to 43	Mountain Pass, California; In Sweden, Russia, New Mexico, Colorado	Monazite, bastnaesite, xenotime, gadolinite	Gravity concentration/ magnetic separation. Leaching in H2SO4. Ion exchange. Byproducts of uranium
Tantalum	1802	73		1.1	In Brazil, Australia, Nigeria, Mozambic, Congo	Members of the columbite- tantalite series, microlite	Hand sorting/ gravity concentration. Solvent extraction. Byproduct of tin
Tin	2000BC	50	$3.36/lb	1.7	In Malaysia, Thailand, Burma, China,	Cassiterite, stannite	Gravity concentration. Roasting/leaching

TABLE 17 - CONT.

METAL	ISOLATED	AT.#	MARKET VALUE	ABUNDANCE	MAJOR DEPOSITS	MINERALS	EXTRACTION
					Brazil, Nigeria, Zaire, Bolivia		in HCl. Smelting
Titanium	1910	22		1.2x10 4	In Australia, Norway, the United States the USSR, Canada	Ilmenite, rutile, leucoxene	Gravity concentration. Chlorination/ reduction with Mg
Tungsten	1783	74	$52-62/mtu	0.82	Pine Creek, California; Canada Tungsten, NWT; also in China, Korea, Thailand, Bolivia, the USSR, Portugal, Australia	Wolframite, scheelite, powellite	Gravity concentration/ flotation. Fusion (for wolframite), leaching in HCl (for scheelite). Reduction with C or H2
Zinc	1200?	30	$0.86/lb	110	In Canada, the United States, Australia	Sphalerite	Flotation. Electrolytic process. Imperial smelting process

PART III

CHAPTER 12

MINERAL EXPLORATION

Having discussed the importance of the knowledge of the microcomposition of the ore in the extraction of the different metals, the book would not be complete without a few words on the importance of the same knowledge in mineral exploration. This chapter, far less comprehensive than the ones addressing the metal recoveries, is only included to give an indication of the potential of the microanalytical techniques. It does not provide actual leads in exploration.

Limited information is available on the applications of the microanalytical techniques in mineral exploration. The data is in fact not sufficient to establish whether microanalytical information affects the precious and base metals in a differential manner. Based on the high market value and low concentrations, common sense does however suggest that studies and understanding of ore microcompositions are more critical in the exploration for precious metals.

DETERMINATION OF THE ORIGIN OF A MINERALIZATION AND LOCATION OF THE HIGHER METAL CONCENTRATIONS. The mineral in which an element is concentrated, the particle size and shape, the associated minerals, and the manner of distribution, are all indicative of the relation of the mineralization to its host rock. These variables, determined by microanalytical studies, may provide clues as to the origin of the mineralization, which can be primary or secondary (for example, figure 54 on page 213). They may give an indication of the size to be expected for a newly discovered deposit. They may lead to the discovery of the higher metal concentrations within the deposit (for example, case study 4 on pages 29-30).

Silver in minor amounts in chalcopyrite, chalcocite, etc., may be indicative of the presence of more abundant concentrations in locations nearby, where the metal may occur in major minerals such as native silver, acanthite, and other silver sulfides and sulfosalts.

New **uranium** deposits can be discovered, and their size, can be predicted. Although the secondary uranium minerals sometimes occur in concentrations of economic value, their significance lies more often in the clues which they provide to the possible existence of primary uraninite or pitchblende deposits within the same geological environment. Their radioactivity, together with their bright colors in the form of stains, films and patches, are important properties which may lead to the discovery of higher concentrations of the metal. Several other elements can be indicators of the presence of uranium. For example, titanium, if in the proper environment, may lead to the discovery of brannerite because alteration of the mineral releases uranium and disseminates fine titanium oxide. Coal, petroleum, and their progenitors and derivatives, with which uranium is often associated, may also suggest the presence of the metal nearby. (Steacy and Kaiman 1978).

The identification of uraninite and pitchblende in ore samples is significant because both these minerals are often associated with deposits of large dimension. On the other hand, except for brannerite, the complex multiple oxides usually occur in smaller deposits than those containing uraninite and pitchblende.

Nickel in olivines, and the chemistry/mineralogy of the general environment, may provide leads as to the presence or absence of nickel sulfides.

An example of the use of microanalytical techniques applied to mineral exploration, is provided by the following case study conducted by Carson and Jambor (1974), and reported by Harris (1975, p. 157).

Case study 24. Microanalytical techniques applied to the mineral exploration. "In a study of porphyry copper deposits in the Babine Lake area, British Columbia, Carson and Jambor (1974) showed that all deposits have similar zonal patterns of sulfides and hydrothermal alteration minerals. The normal outward zoning is from biotite-chalcopyrite to chlorite-carbonate-pyrite. On the basis of field work and extensive thin section studies, a general zonal alteration pattern was outlined in which each copper deposit is contained within a zone of hydrothermal biotite, which is in turn surrounded by a zone of chloritic alteration. It was found that the areal extent and character of hydrothermal biotite was related to the economic potential of the enclosed copper mineralization. That is deep brown, coarse grained sugary-textured biotites are associated with relatively strong copper mineralization, whereas greenish and/or fine-grained biotites are indicative of weak copper mineralization. In porphyry copper deposits, biotite zones are larger and better exposed than the copper occurrences, and since the color of the biotite is not too reliable as an indicator of copper mineralization, being dependent on too many unknown factors, a need exists to quantify the relationship of color to grade." This has been done by Jambor (1975) by using the microprobe to determine ratios of Fe to Fe+Mg+Mn in biotites, and by using these ratios rather than the color of the mineral to predict copper grades.

Another application of the microanalytical techniques in mineral exploration was attempted at the Colorado School of Mines Research Institute. The instrument was used to trace variations in metal content (molybdenum, gold and selenium) across roll-front uranium concentrations. (A.P. Beavan 1979, personal communication).

EVALUATION OF A DEPOSIT. Once the mineralization has been located and defined, it needs to be evaluated. The microanalytical techniques may provide additional assistance. We have seen in chapter 3 that difficulty of recovery varies, depending on the mineral bearing the valuable metal, on its grain size and on the associated minerals and general ore microcomposition. Microanalytical studies of representative sections of the ore will provide information permitting estimates of recovery costs and will help determine the economic viability of the deposit. For example,

Gold is less expensive to recover when:

- Present in native form and in coarser grains (more than 10 microns).
- Associated with the sulfides in a manner which will promote liberation or exposure during the grinding process.
- In oxidized hosts/environments.

Gold occurring in particles 1 or 2 microns in size in a compact fresh pyrite and in concentrations of 0.1 ounce per ton or 3.4 ppm, is often too costly to recover. But, when occurring in an oxidized environment, that same gold is usually economically viable, because it is amenable to heap leaching, the least expensive of the recovery techniques (case studies 6 and 7, on pages 32-35).

Uranium is less expensive to recover from uraninite, coffinite and secondary minerals than from most of the complex multiple oxides.

Nickel, is more easily concentrated when in the minerals pentlandite and millerite, than when in violarite.

Uses of mineralogy and petrology in mineral exploration are discussed by Amstutz (1983).

Finally we should not forget that, as **fractal geometry** (Mandelbrot 1982, Gleick 1987, Middleton 1991) gains recognition, the importance of the knowledge of the precious metal distributions on a microscopic scale may become essential in the understanding of the ore body as a whole. The principles on which fractal geometry is based postulate that, when magnifying objects which are *scale symmetric*, their parts are seen to bear an exact resemblance to the whole. Scale symmetry has been proven to exist within nature, for example, in coast lines and mountain ranges. If scale symmetry can be recognized within ore deposits as well, then the precious metal distributions as observed in the polished mounts, may very well reflect their distributions within the ore body (Gasparrini 1991, Gasparrini 1992-B). Consequently, information obtained from the microanalytical studies could be used for the determination of the size and shape of the entire body.

CONCLUSIONS

TO WHAT EXTENT DOES MICROANALYSIS HELP. During the course of the book we have seen that there are variables which affect the metal extractions, or at least the decisions concerning the planning of the metal extractions. One of these variables is the **ore microcomposition**, the correct determination of which gives an insight into the distribution of the metal within the ore. Knowledge of the microcomposition leads to the understanding of the processes which take place during extraction, and of the reasons why some recoveries are successful and some are not. We have also seen that knowledge of the ore microcomposition is much more critical when dealing with precious metal recoveries than when dealing with base metal recoveries, because their much lower concentrations make the extraction process more difficult, and because precious metals are so valuable that we want to recover as much of them as possible.

There are two common arguments against the validity/usefulness of the microanalytical techniques, for the determination of ore microcompositions:

- **Precious metal extractions work just as well without a detailed knowledge of the ore microcomposition, they always have, why change.** Perhaps precious metal extractions are **not** as satisfactory as it is generally assumed. In fact this argument may not have very strong grounds, if we consider that essentially one locality (the Witwatersrand Reef) within a relatively small country such as South Africa, produces annually a quantity of gold which is comparable to the quantity produced by the rest of the free world (page 89). In the light of such an observation, the possibility that perhaps South Africa's high gold output is not entirely the result of natural resources and that the extraction techniques **do** play some role,

- should definitely be considered. And even if we were to ignore the thoughts about South Africa and assume that precious metal extraction techniques do work in a satisfactory manner just the way they are, reluctance to change still makes a very poor argument, when viewed against the natural tendency of humankind towards progress and improvement. Candle light and horse-drawn carriages worked well too, during the past century. Yet their satisfactory performance did not stop the development of electricity and fast racing cars.

- **The microanalytical techniques may give reasons for poor recoveries, they do not improve them, or if they do, they do so only to a very limited extent (2 to 3 per cent at the most).** This argument is again missing the point when estimating the potential of microanalysis. The microanalytical techniques are not and should not be regarded **only** as an aid for the understanding of why recoveries are poor. Their real value is in the fact that they represent the first step towards the introduction of revisions in the actual approach to the process of extraction.

Revisions are made possible by the insight into the ore provided by the microanalytical studies. For example, let us assume that a deposit yields 40 per cent of gold or silver by heap leaching of an entire area of which very little is known about the occurrence and distribution of either metal. Once the occurrence and distribution are correctly determined, it may be possible to focus the attention on the higher grade levels which leach the best and to develop more accurate techniques other than heap leaching, for those levels only (see case studies 4 on pages 29-30 and 19 on pages 173-174). In both case studies, the extractions could very possibly be doubled, as well as expedited.

In those not uncommon situations where answers to problems are not immediately available, with or without the microanalytical studies, knowledge of the distribution of the metal of interest, will provide leads as to the appropriate research to be conducted. For example, if dealing with a gold ore rich in clay which is only yielding a recovery of 20 to 30 per cent gold, we may, without any microanalytical studies, come to the conclusion that the presence of clay is causing the poor recoveries. We may try a few of the routine laboratory techniques and conditions, in the hope that one of them will give satisfactory results. If none of them is successful, we may decide to abandon the project and move on to the exploration for more amenable ores. On the other hand, we may conduct microanalytical studies and find that we are dealing with (i) coarse gold showing impermeable coatings of clay which have developed after the grinding process, or with (ii) invisible emulsion-like gold, finely mixed with the clay, or with (iii) a mixture of the two occurrences. Once aware of the correct gold occurrence, we will be capable of choosing intelligently the technique for recovery: if dealing with coarse gold, we may decide to separate a rough concentrate free of clay, then grind it down to finer sizes for liberation of a native gold free of coatings; if dealing with fine emulsion-like gold, we may find that an easy technique for its extraction is not readily available. Yet, we are now in a position to experiment with techniques which apply to this type of occurrence only, with better chances of success. Fine emulsion-like gold and coarse gold which becomes impermeable following the grinding process, if occurring together, may require two different approaches, one to separate the coarse gold, one to separate the fine gold. Again, such an approach can be planned, if the distribution of the gold is known beforehand.

Unfortunately a large amount of the skepticism towards the potential of the microanalytical techniques, results possibly from the expectations on the part of the mining industry. When suggestions are made as for a revised approach to extraction, a common response is that very little can be changed because the project has already been planned and can only be conducted in a certain manner. If the ore is being heap leached, the microanalytical study **must** suggest improvements without changing the basic technique for recovery. Such an approach leads to the conclusion that the knowledge of the microcomposition of the ore may give the reasons for poor recoveries, yet it provides little assistance as to improvements in recovery rates. Perhaps this following sentence is unnecessary, but, as long as little is changed in the basic approach, expectations for major improvements within such approach, are unreasonable and unrealistic.

A good metaphor for the attitude of the mining industry towards acceptance of microanalysis, could be represented by a man living in a forest or other open space in which there is no light whatsoever, not even from the stars. The man survives for several months on food scattered around him, which he finds exclusively by feeling the ground. After a period of time during which he has become totally comfortable with his technique for maintaining himself alive, it starts dawning and he can finally look at the place around him. The food on which he has been living is represented largely by over-ripe, sometimes rotting fruit fallen from the trees. In addition to what is on the ground, there is much more, fresher fruit on the trees, of which he was unaware while in the dark. Instead of reaching for the better quality food, he goes on eating the rotten fruit on the ground and concludes that, yes, perhaps he **can now** see where the food is, but, with the exclusion of a few extra pieces which he might have missed when he was in the dark, he can still gather just as much and of similar quality.

Or another metaphor could be the one involving a number of people trapped in a dark enclosed space, who finally locate a tunnel leading to the outside. The tunnel is very long and very narrow and only accessible to the smaller thinner individuals. The lights come on and, as well as the tunnel, an unlocked wide door becomes visible. Yet, our people ignore the door, their only comment being that, yes they can now see the tunnel, but it is still narrow and long and difficult to cross.

SUGGESTIONS FOR FUTURE RESEARCH. Despite the author's attempts at making the book as comprehensive as possible, understanding an ore is not always as straight forward and easy as it may have appeared in many of the descriptions, even when microanalytical studies are conducted systematically. Organization and planning are essential, but they may not be sufficient without the proper experimenting, because in many situations the answers are just not available. We have seen that ores rich in clay are not always readily treatable. Silver ores are not well understood, even if extensive microanalytical data are available. If ten different silver minerals are identified in an ore from which poor recoveries are experienced, it is very possible that little is known about the leaching and flotation properties of five of these minerals. Characterization of silver ores is important whether the deposit is being exploited primarily for silver or silver is a byproduct, because some of the metal may be lost during the beneficiation stages. As for the platinum group elements, one of the reviewers of an earlier draft of the book stated that the chapter covering their description was far too short and that he would have liked to know more about these elements. Well, so would the author, and probably so would anybody else involved in platinum group metal extractions. Unfortunately available microanalytical information on platinum group elements, which can be used for recovery planning, is not much more extensive than what is summarized in chapter 9.

Systematic studies which are comprehensive and conducted independently of and without the pressure from the impending recovery process, are necessary. For example, when searching for answers to poor recoveries, it would be helpful to investigate those sections of the ore giving acceptable recoveries in the same detail as those giving the poor recoveries. Comparisons of microcompositions of ores giving good and poor recoveries can be invaluable in the understanding of the poor recoveries. Yet these types of investigations, conducted for comparison purposes, are not usually considered, upon the assumption that, if no problems are encountered, expenditures of time and money are not justified.

Listed below, some topics are suggested where the acquisition of chemical/microanalytical information may assist in the development of plans for improved and efficient recoveries. Some such topics could be the subject of university research projects.

Fire assay. Precious metal values for an ore represent the very first information which is acquired and which, together with the market value, assist in the decision whether the deposit is exploitable or not. Fire assay values for gold and silver are usually accurate, and yet there are situations where such values do not reflect the actual gold content of the sample (for example, case studies 8 and 9 on pages 93-94). Experiments and accurate chemical determinations similar to and including the ones suggested on pages 94-95, may help pin down the faults within the fire assay technique, and assist in all decision regarding their most efficient handling.

Fire assay techniques for the determination of all six of the platinum group elements are presently not sufficiently understood, for accurate routine analyses (page 196). There are no published data of platinum group element contents in ores obtained **routinely** by other analytical techniques. Research on the subject is long overdue.

Permeability of rocks during heap leaching. Precious metals leach easily from some rocks, but not from others. This is sometimes due to variable precious metal grain sizes, but the permeability of the rock and host mineral to cyanide and other leaching agents plays probably the most important role. Classification of rocks and minerals on the basis of their permeability to leaching solutions will expedite decisions concerning the use of heap leaching against other techniques.

Gold and silver mineral solubility in cyanide and other leaching agents. This should be determined systematically for **all** of the gold and silver minerals. The solubility of some gold tellurides is known, but there is still disagreement regarding the behavior of most of them during the leaching process. In some laboratories, many gold recovery problems are blamed on the presence of tellurides, even before average chemical analyses for tellurium are obtained. Perhaps it is time that some positive answers regarding the actual solubility of these minerals became available.

The common occurrence of many of the silver minerals, much more so than that of the gold tellurides, renders the knowledge of their solubility far more critical, for efficient silver recoveries.

Alternate leaching agents capable of dissolving gold and silver minerals which are insoluble in cyanide.

Specific gravity and other physical properties of all the gold and silver minerals. Knowledge of these values will greatly accelerate the separation of coarse gold and silver when using physical methods such as gravity concentration. Some of these values are already available in the literature, however, their collection in one comprehensive manual may eliminate time wastes searching through papers and outdated books which may not be readily available.

Flotation properties of the silver minerals, for their concentration by physical means, without lengthy experimentations.

Silver sulfide coatings developed during gold and silver recoveries. Silver ores are often troublesome during flotation and cyanidation due to the development of impermeable coatings over the surface of the particles to be separated. The process is not completely understood. Techniques to prevent the development of these coatings, such as the use of high pH during leaching (page 123), are empirical and may not take care of the problem entirely.

Platinum group elements. These are the least known, metallurgically, of the precious metals, at least based on the available literature. The field is wide open for research and data organization.

Chemical properties of the host mineral. Knowledge of these properties is necessary where dealing with very fine grain sizes which do not become liberated or exposed after the grinding process. In such a situation, if the host is impermeable to the leaching solution, its chemical breakdown is necessary prior to the cyanidation step.

Cyanide solubility of any mineral, in a rock to be treated for gold or silver extractions, for a final positive classification of the cyanicides and non-cyanicides. Microanalytical studies indicate that some of the minerals which are traditionally assumed to be cyanide soluble, for example marcasite, are not, at least not under normal conditions.

Chemical and physical properties of the cyanicides, for their elimination prior to cyanidation.

Interfering minerals during cyanidation, other than the cyanicides, such as carbon, graphite, clay, sericite, which cause early gold and silver precipitation or insolubility during recoveries. Although processes have been developed for the control of the carbonaceous material, the information on the clay minerals and on means of controlling their negative effects is limited. Two limitations contributing to the poor dissolution of the gold and silver are fine impermeable coatings and emulsion like particles. There may be more. These could be clarified through the systematic examination of a number of gold and silver-bearing clay rocks, for an indication of the various types of precious metal/clay associations and for a classification of their response to the common recovery agents.

Environment contamination. The systematic development of methods for the control of environment contaminations, such as those caused by arsenic, mercury and sulfur dioxide released into the air during the roasting of pyrite and arsenopyrite, may become essential, if we consider that the two minerals are common hosts to gold and fine gold grain sizes often require host breakdown by chemical methods. There are other types of environment contaminations, such as those related to the presence of mercury in the ore or in the gold-bearing minerals (pages 138-139), for example. Awareness of contamination problems and government restrictions have already raised considerable interest within the world of precious metals, and papers are being published and discussed at various seminars and conventions. One organization promoting such research and exchange of information is the **International Precious Metals Institute** (page 14). Papers addressing the problems related to the environment are published in the **IPMI** proceedings volumes once a year and environment news are provided in the newsletter **Precious Metals News and Review** on a regular basis.

Comparisons of precious metal distributions on a micro and on macroscopic scale within the same ore. These studies would tie in with the principles of fractal geometry briefly discussed at the end of the previous chapter (page 289), and with the similarities assumed to be present between the whole and its parts.

The above type of research would take place independently of the recovery process. The pressure for answers would be limited and the data would be available for practical use, only after their validity has been confirmed by accurate and extensive studies.

GLOSSARY OF MINERALS, ROCKS AND GEOLOGICAL TERMS

Much of the information in the glossary was taken selectively from Amstutz (1971), Dana (1963), Pough (1983), and from the **Dictionary of Geological terms** (1962). If the name of a mineral, rock, or geological term is not found in the glossary, refer to the subject index.

Adularia. A variety of orthoclase (see orthoclase) forming colorless prismatic crystals.

Alabandite, MnS. An uncommon sulfide of manganese. Occurs in ore veins associated with rhodochrosite, sphalerite and other sulfides.

Alaskite. An intrusive rock consisting of feldspar and quartz with very minor or no mafic constituents.

Albite, $NaAlSi_3O_8$. See also plagioclase.

Allanite, H_2O $4R^{+2}O$ $3R^{+3}{}_2O_3$ $6SiO_2$, where R^{+2} = Ca and Fe and R^{+3} = Al, Fe and rare earths. An accessory mineral found in igneous and metamorphic rocks. It may contain minor amounts of uranium.

Alunite, $K_2Al_6(OH)_{12}(SO_4)_4$. A basic hydrous sulfate of aluminum and potassium. Alunite is most commonly associated with acid volcanic rocks which have undergone extensive alteration. The conditions of crystallization usually indicate high pressure and temperature.

Amphibole. A group of mafic rock forming minerals found in many rock types. The general formula, representative of a large variety of different species could be expressed as $A_{2-3}B_5C_8(O)_{22}(H,F)_2$, where A = Ca, Na, K, or Li; B = Mg, Fe^{+2}, Fe^{+3}, Al, Ti; C = Fe^{+3}, Al, Si.

Amphibolite. A metamorphic rock consisting largely of amphibole and plagioclase, with minor or no quartz.

Anglesite, $PbSO_4$. A mineral which usually results from the alteration/weathering of galena. It can be an important ore of lead. It may occur as coatings over the surface of the galena, when the mineral is in a less advanced stage of alteration.

Ankerite, $CaCO_3$ $(Mg,Fe,Mn)CO_3$. Carbonate mineral often found associated with dolomite and iron ores.

Annabergite, $Ni_3As_2O_8$ $8H_2O$. An uncommon nickel mineral of secondary origin occurring in the alteration zone near the surface of nickel deposits.

Anorthite, $CaAl_2Si_2O_8$. See also plagioclase.

Anorthosite. An intrusive rock consisting almost exclusively of plagioclase.

Apatite, $(Ca,F)Ca_4(PO_4)_3$. or $(Ca,Cl)Ca_4(PO4)_3$ A widely distributed accessory mineral occurring in many rock types including some metalliferous veins such as those of tin. When in pegmatites, it may form large attractive crystals.

Arsenopyrite, FeAsS. A sulfoarsenide of iron occurring in high temperature veins, pegmatites and sometimes in igneous rocks. An ore of arsenic, usually byproduct of other operations.

Augite, $(Ca,Na)(Mg,Fe,Al,Ti)(Si,Al)_2O_6$. Aluminous pyroxene common in igneous rocks.

Azurite, $Cu_3(CO_3)_2(OH)_2$. Hydrous copper carbonate. See also secondary copper minerals.

Basalt. An intrusive rock consisting of calcium rich plagioclase, pyroxene, and sometimes olivine.

Bindheimite. An amorphous hydrous antimonate of lead, the result of decomposition of other antimonial ores.

Biotite, $K(Mg,Fe)_3AlSi_3O_{10}(OH)_2$. A rock forming mica mineral common in igneous and metamorphic rocks of mafic composition. Also found in minor amounts in felsic rocks.

Bornite, Cu_5FeS_4. See also secondary copper minerals.

Calamine, $Zn_4Si_2O_7(OH)_2$ $2H_2O$. See also hemimorphite.

Calcite, $CaCO_3$. Calcium carbonate. One of the most common minerals and main constituent of limestones.

Chalcocite, Cu_2S. See also secondary copper minerals.

Chalcopyrite, $FeCuS_2$. An important copper ore. Common in sulfide veins.

Cerargyrite, AgCl. A secondary silver mineral commonly found in the upper portions of silver deposits. Associated with lead, copper and zinc ores and their alteration products, and with other silver minerals.

Chert. Cryptocrystalline variety of silica, composed mainly of chalcedony and fine quartz particles.

Chlorite, $(Mg,Fe,Al)_6(Si,Al)_4O_{10}(OH)_8$. Secondary phyllosilicate minerals of variable composition common in rocks of low to medium metamorphism.

Chrysocolla, $Cu_2H_2Si_2O_5(OH)_4$. Hydrous copper silicate. See also secondary copper minerals.

Clay. Complex group of minerals of Si-Al composition with or without other elements, characterized by submicroscopic sizes for the individual crystal grains. They are the product of secondary alteration of rock forming silicates, particularly feldspars.

Conglomerate. A sedimentary rock consisting of rounded smooth fragments of rock (pebbles) cemented together by a matrix of different mineralogical composition. The size of the pebbles is larger than 2 mm.

Covellite, CuS. See also secondary copper minerals.

Cuprite, Cu_2O. Red copper ore. An important secondary ore of copper. Found in the oxidation zone of copper sulfide ore deposits.

Dacite. An extrusive rock consisting of plagioclase, quartz, pyroxene/hornblende and minor feldspar and biotite.

Digenite, Cu_9S_5. See also secondary copper minerals.

Dike. A flat body of intrusive material cutting through the structure of adjacent rock formations, or through fractures within a massive rock formation.

Diopside, $CaMg(SiO_3)_2$. A calcium-magnesium clinopyroxene.

Diorite. An intrusive rock consisting of plagioclase of andesite composition, hornblende, biotite, with minor pyroxene, quartz and orthoclase.

Dolomite, $CaMg(CO_3)_2$. The main mineral constituent of the rocks bearing the same name. Found in sedimentary rocks and in metamorphic rocks occurring in high temperature surroundings.

Domeyckite, Cu_3As. A copper arsenide found in reniform, botroidal, massive and disseminated forms in copper deposits of complex mineralogy.

Emplectite, $Cu_2S\ Bi_2S_3$. A copper-bismuth sulfide found in copper deposits of complex mineralogy.

Enargite, Cu_3AsS_4. An important ore of copper. Found in medium temperature ore veins.

Epidote, $HCa_2(Al,Fe)_3Si_3O_{13}$. A rock forming mineral formed by the metamorphism of impure calcareous sedimentary rocks, or of igneous rocks containing high proportions of lime. Relatively common.

Erythrite, $Co_3As_2O_8\ 8H_2O$. A mineral of secondary origin, commonly found in the upper portions of cobalt mineral deposits.

Feldspar. General name for a number of rock forming minerals the more common of which are orthoclase, microcline and the members of the plagioclase series.

Feldspatoid. General name for a number of alumino-silicates of potassium, sodium and calcium, which are similar in composition to the feldspars, but lower in silica. More common feldspatoids are leucite and nefeline.

Felsic. Term used to indicate light colored rocks which are rich in feldspar-feldspatoid (fel) silica (si), and muscovite. The term may be applied to the minerals themselves.

Fluorite, CaF_2. Fluoride of calcium found in formations of widely varying conditions. It occurs most commonly as a vein mineral either in deposits in which it is the chief constituent or as a gangue mineral with various metallic ores, especially those of lead, silver and zinc. It may form attractive crystals of large sizes and a variety of bright colors.

Franklinite, $(Fe,Zn,Mn)O\ (Fe,Mn)_2O_3$. An oxide of iron, zinc and manganese, varying widely in their relative amounts and conforming to the general formula of the spinel group. It can be an ore of zinc.

Gabbro. An intrusive rock consisting of plagioclase of labradorite composition, clinopyroxene and minor orthopyroxene and olivine. The term may also indicate a coarse grained dark igneous rock.

Galena, PbS. Most important ore of lead. Found in ore veins, in igneous and sedimentary rocks.

Garnets. A series of orthosilicates the general formula of which is $R^{+2}{}_3R^{+3}{}_2(SiO_4)_3$, where R^{+2} may be Ca,Mg, Fe^{+2} and Mn and R^{+3} Al, Fe^{+3}, Cr and Ti. Garnets are commonly found in accessory amounts in a variety of rock types more common of which are mica, hornblende and chlorite schists, serpentinites, gneisses, crystalline limestones, metamorphosed rocks in contact zones, granites, syenites and other volcanic rocks. It may also be found in massive form as the predominant constituent of some metamorphic rocks. The larger transparent crystals are used as gems.

Gelamine, $H_2Zn_2SiO_4$. An uncommon silicate of zinc.

Gneiss. Metamorphic rock derived from sedimentary or granitic facies. Component minerals are quartz, feldspar and mica. Metamorphic minerals such as garnet and staurolite may be abundant. Augen gneiss is characterized by the presence of eye shaped nodular aggregates or porphyroblasts, commonly of potassium feldspar composition.

Granite. The most abundant intrusive rock consisting of quartz and orthoclase feldspar with usually very small amounts of plagioclase and mafic minerals. Mineral grain sizes can be coarse, up to 1/2 inch in diameter or more.

Granulite. A metamorphic rock consisting of interlocking granular components of similar sizes. Common minerals are quartz, orthoclase, plagioclase, pyroxene and garnet.

Graphite, C. Hexagonal crystal form of carbon. Found predominantly in granites, limestones and metamorphic rocks.

Gratonite, $Pb_9As_4S_{15}$. A sulfosalt of As and Pb found in complex arsenic-lead mineralizations.

Hematite, Fe_2O_3. See also magnetite.

Hemimorphite, $Zn_4Si_2O_7(OH)_2\ 2H_2O$. Secondary silicate after sphalerite, found in oxidized zones of zinc ores. Found usually associated with smithsonite in veins and beds cutting through calcareous rocks which may contain the unaltered sulfides of zinc, lead and iron. Also known as calamine.

Hornblende, $(Ca,Na)_{2-3}(Mg,Fe^{+2},Fe^{+3},Al)_5(Al,Si)_8O_{22}(OH)_2$. A rock forming mineral of the amphibole group, found in metamorphic rocks and in mafic and ultramafic igneous rocks.

Hydrozincite, $Zn_5(CO_3)_2(OH)_6$. Secondary hydrous carbonate found in the weathered zone of zinc deposits.

Jarosite, $K_2Fe_6(OH)_{12}(SO_4)_4$. A secondary mineral formed probably under solfataric conditions with high temperature and pressure. It is sometimes associated with clay and may contain silver in its lattice.

Leucite, $KAlSi_2O_6$. A mineral of the feldspatoid group, common in high potassium-low silica volcanic rocks.

Limestone. A sedimentary rock consisting predominantly of calcite (calcium carbonate). The most common of the carbonate rocks.

Lindstromite, $2PbS\ Cu_2S\ 3Bi_2S_3$. An unusual sulfosalt of lead, copper and bismuth. Found associated with quartz and copper minerals.

Lollingite, $FeAs_2$. An iron diarsenide which may contain variable amounts of cobalt. Found in medium temperature veins, associated with various carbonates, sulfides and with cobalt, gold and silver ores.

Mafic. A term used to indicate dark colored rocks or minerals rich in silicates of magnesium (ma) and iron (fe).

Magnetite, Fe_3O_4 or FeO Fe_2O_3, **hematite**, Fe_2O_3, more common secondary oxides **goethite**, FeO(OH) or Fe_2O_3 (H_2O), and **limonite**, $2Fe_2O_3$ $3H_2O$. All important ores of iron. Magnetite is found in igneous metamorphic and pegmatitic rocks and in beach sands. Hematite is widespread. Secondary iron oxides occur in oxidized environments. Their formulae are variable.

Malachite, $Cu_2(CO_3)(OH)_2$. Hydrous copper carbonate. See also secondary copper minerals.

Marcasite, FeS_2. Polymorph orthorhombic equivalent of pyrite (cubic). Found in sedimentary rocks and low temperature veins.

Mica or **micas**. General term for a number of rock forming minerals characterized by sheet-like structure and belonging to the phyllosilicate group. All characterized by the presence of potassium in their lattice. More common of these are muscovite/sericite, (aluminum-silicon predominant) biotite (iron-magnesium variety), lepidolite (lithium variety) and phlogopite (magnesium variety). Found in all types of rocks.

Microcline, $KAlSi_3O_8$. Polymorph triclinic equivalent of orthoclase (monoclinic). Occurs in similar environments to those of orthoclase.

Monzonite. An intrusive granular rock consisting of similar amounts of orthoclase and plagioclase, with minor hornblende, diopside and biotite. The rock has an intermediate composition between diorite and syenite.

Muscovite/sericite, $KAl_3Si_3O_{10}(OH)_2$. Common rock forming mica minerals, particularly abundant in granites and schists. Muscovite is also common in pegmatites. Sericite is a fine grained variety of muscovite.

Native copper, Cu. See also secondary copper minerals.

Norite. A variety of gabbro in which the orthopyroxene predominates over the clinopyroxene. Common in the Sudbury complex.

Olivine. A group of minerals consisting of solid solutions of end members forsterite, Mg_2SiO_4, and fayalite, Fe_2SiO_4. An important constituent of mafic and ultramafic rocks.

Opal, SiO_2 nH_2O. Amorphous silica. May be used as a gem.

Orpiment, As_2S_3, often with realgar. See also realgar and orpiment.

Orthoclase, $KAlSi_3O_8$. A common feldspar mineral found in igneous and metamorphic rocks of felsic composition and in high temperature veins. Used in the ceramic and glass industry.

Pegmatite. Coarse grained igneous rock forming usually dikes associated with intrusive formations characterized by finer grain sizes. Most pegmatites have granitic composition. Some pegmatites contain rare minerals rich in elements such as lithium, fluorine, boron, rare earths, uranium, niobium and tantalum.

Peridotite. An intrusive rock the main components of which are olivine and other mafic minerals. It is characterized by the absence of quartz and feldspar.

Phyllosilicates. Rock forming silicate minerals characterized by flat platy (or sheet-like) crystal structures. They include the micas, the chlorites and the serpentines.

Plagioclase. Minerals belonging to the feldspar group and consisting of a complete solid solution, the end members of which are albite and anorthite. The other members of the series are oligoclase, andesite, labradorite and bytownite. They occur in igneous and metamorphic rocks, the sodium rich end members being abundant in the more acidic types.

Pod. A body of cylindrical shape, the diameter of which decreases at the ends, like a cigar. A flat elongated ore body.

Porphyry. A term used to indicate any rock consisting of phenocrysts (crystal grains of larger sizes) in a fine grained matrix.

Pyrargyrite, Ag_3SbS_3. A common silver sulfosalt found in the upper portions of silver veins. An ore of silver, also known as ruby silver.

Pyrite, FeS_2. Sulfide of iron crystallizing in the cubic system, and most common sulfide mineral. Found in all rock types in major or minor amounts. Most common host to gold. Once an important source of sulfur.

Pyromorphite, $(Pb,Cl)Pb_4(PO_4)_3$. Green lead ore. A secondary lead product often found in lead ores, but never in large amounts.

Pyroxenes. A group of mafic rock forming minerals common and abundant in mafic and ultramafic rocks. Also found in metamorphic rocks. The general formula, representative of a variety of different species could be expressed as ABC_2O_6, where A = Mg, Fe^{+2}, Ca, Na, Li; B = Fe^{+3}, Al; C = Si, Al.

Pyrrhotite, $Fe_{1-x}S$ (sometimes expressed as FeS, but containing some dissolved sulfur). Widespread mineral, common in magmatic segregations and high temperature veins. Also in pegmatites and contact metamorphic deposits. It may be magnetic.

Quartz, SiO_2. Silicon dioxide and most common of all minerals. Found in most rock types. Important industrial mineral. Quartz from sands is used for the manufacture of glass. The clear crystals are important in the electronics industry. Crystals of different color, such as amethyst, citrine, smokey quartz, rose quartz, etc., if free of major flaws, are used as gems.

Realgar, AsS, and **orpiment**, As_2S_3. Arsenic sulfides found usually associated in low temperature veins. Important ores of arsenic.

Rhodonite, $MnSiO_3$. Silicate of manganese, frequently found in Mn ore bodies, and usually associated with the carbonate equivalent rhodochrosite.

Rhyolite. The extrusive equivalent of a granite.

Rutile, TiO_2. An ore of titanium. Common in beach sands. Also in accessory amounts in intrusive and metamorphic rocks.

Sandstone. A compacted sedimentary rock consisting predominantly of quartz and minor feldspar as clastic components. Sandstones are classified as argillaceous, siliceous, calcareous, ferruginous, etc., depending on the nature of the cementing medium.

Scapolite group, $(Na,Ca,K)_4Al_3(Al,Si)_3Si_6O_{24}(Cl,SO_4,CO_3)$. Complex silicates found in a variety of metamorphic environments.

Schist. Metamorphic rock characterized by the presence of abundant mica the orientation of whose plates, all along the same direction, is responsible for the typical appearance from which it derives its name. Besides mica, feldspar and metamorphic minerals such as garnet, staurolite, andalusite and kyanite may be abundant. Chlorite is present in the less metamorphosed forms.

Scorodite, $FeAsO_4\ 2H_2O$. An alteration product from arsenic minerals, especially arsenopyrite.

Secondary copper minerals, bornite, Cu_5FeS_4, **covellite,** CuS, **chalcocite,** Cu_2S, **digenite,** Cu_9S_5, **djurleite,** $Cu_{1.96}S$, **native copper,** Cu, **azurite,** $Cu_3(CO_3)_2(OH)_2$, **malachite,** $Cu_2(CO_3)(OH)_2$, **chrysocolla,** $Cu_2H_2Si_2O_5(OH)_4$. Important secondary copper ores. All occurring in oxidized zones of copper deposits.

Secondary iron oxides. See magnetite, hematite and secondary iron oxides, goethite and limonite.

Sericite. Fine grained variety of muscovite.

Serpentine, $Mg_3Si_2O_5(OH)_4$. A common rock forming mineral of secondary origin, resulting from the alteration of non-aluminous silicates containing high magnesium. Antigorite and chrysotile are two common lamellar and fibrous varieties. Most asbestos, the economic form of serpentine, consist of chrysotile.

Shale. A sediment which has been compacted to show a certain amount of lamination and the constituents of which are predominantly clay minerals.

Siderite, $FeCO_3$. Sometimes an ore of iron. Found in sedimentary rocks and in pegmatites.

Sill. An flat intrusive rock formation characterized by uniform thickness, which has been emplaced in a parallel direction to the bedding or schistosity of the rock.

Skarn. Contact metamorphic rock derived from the reaction and recrystallization of lime-stone and/or dolomite with rocks rich in silicon, aluminum, iron and magnesium. Com-mon minerals are calcium silicates such as garnet and pyroxene to form a tactite.

Smithsonite, $ZnCO_3$. Sometimes an important ore of zinc. Found in secondary weathered zones of zinc deposits.

Sphalerite, ZnS. The most important ore of zinc. Common in sulfide ore veins.

Sphene, $CaTiSiO_5$. An accessory mineral common in igneous rocks of intermediate com-position. Also known as titanite.

Spinels. A mineral group, the general formula of which is $R^{+2}R^{+3}{}_2O_4$, where R^{+2} = Mg, Fe^{+2}, Mn^{+2}, Zn and Ni and R^{+3} = Al, Fe^{+3}, Mn^{+4} and Cr. Important members are mag-netite, chromite and spinel itself, MgO Al_2O_3

Staurolite, $HFeAl_5Si_2O_{13}$. Metamorphic mineral most commonly found in schists and gneis-ses associated with garnets and tourmaline. Also found in placers deposits. Forms often typically cruciform twins which allow for easy identification.

Stibnite, Sb_2S_3. An ore of antimony. Found in low temperature veins and rock impreg-nations.

Stockwork. A mass of rock irregularly fractured in various directions by short fissures along which mineralization is spread. The resulting ore deposit has a form which allows the working in floors or stories. Stockworks are distinguished from vein deposits because of their thickness, the criss-crossed structure and the presence of veins and veinlets.

Sulfosalts. Salts of arsenic, antimony and bismuth in which the sulfur takes the place of the oxygen. More common cations are copper, silver, lead, zinc mercury and iron.

Syenite. A volcanic rock in which more than 60 per cent consists of microcline, orthoclase or albite and in which mafic minerals rarely exceed 20 per cent. The remainder is quartz, plagioclase and accessory minerals.

Tactite. A rock of complex mineralogical composition formed by contact metamorphism and metasomatism of carbonate rocks.

Tetrahedrite, $(Cu,Fe)_{12}Sb_4S_{13}$. An important ore of copper. Can carry silver (freibergite, up to 34.8 per cent Ag), mercury (schwazite, up to 17 per cent Hg) and lead (malinowskite, up to 16 per cent Pb). Found in medium to low temperature ore veins.

Tourmaline. A complex silicate of B and Al with Mg, Fe and alkali metals. Found in granites and gneisses as a result of fumarolic action or of mineralized gases from the fluid magma. Also frequent in the pegmatite veins associated with these rocks.

Tremolite, $CaMg_3Si_4O_{12}$. An amphibole, product of metamorphism. Common in limestones, particularly the magnesitic and dolomitic varieties.

Troilite, FeS. Ferrous sulfide similar in composition to pyrrhotite. More common natural forms are found in meteorites, but terrestrial occurrences are reported.

Ultramafic. Term used to indicate igneous rocks characterized by the presence of less than 45 per cent silica and consisting essentially of mafic minerals.

Vein. Any ore or non-ore material exhibiting a tabular or sheet-like structure.

Vesuvianite, $Ca_6[Al(OH,F)]Al_2(SiO_4)_5$. Complex hydrous silicate commonly found as a contact mineral formed from the alteration of impure limestones. Usually associated with grossularite garnet, phlogopite mica, diopside and wollastonite. Also found in some metamorphic rocks.

Villamaninite, $(Cu,Ni,Co,Fe)(S,Se)_2$. A copper reach sulfide belonging to the pyrite group. May contain gold. The most important occurrence is in Villamanin, Spain.

Violarite, $(Ni,Fe)_3S_4$. A secondary nickel mineral which may be host to the platinum group minerals. It does not respond to the standard flotation techniques for nickel concentration.

Wad. A mixture of Mn oxides with other secondary products, sometimes found in laterite deposits.

Willemite, Zn_2SiO_4.. Silicate of zinc secondary after sphalerite, occurring in oxidized portions of zinc veins.

Wollastonite, $CaSiO_3$. Silicate of calcium found mostly as a contact mineral in crystalline limestones, formed by the action of silicic acid at relatively low temperatures.

Zincite, ZnO. An ore of zinc, but rather unusual.

Zircon, $ZrSiO_4$. A common accessory mineral of igneous rocks. The chief ore of zirconium. Used as a refractory, and as a gem when free of impurities.

REFERENCES

A Generic Study of the Capital and Operating Costs for the Recovery of Precious Metals from Refractory Ores, (1988), report published by Kilborn Limited, Toronto, Canada.

AMSTUTZ, G.C. (1960), The preparation and use of polished-thin sections: The American Mineralogist, vol. 45, pp. 1114-1116

AMSTUTZ, G.C., (1961), Microscopy applied to Mineral Dressing: in Proceedings - 50th Anniversary of Froth Flotation in the U.S.A: Colorado School of Mines Quarterly, vol. 56, pp. 444-481.

AMSTUTZ, G.C., (1971), **Glossary of Mining Geology,** Ferdinand Enke, Stuttgart, (in four languages).

AMSTUTZ, G.C., (1972), Stereological methods applied to mineralogy, petrology, mineral deposits and ceramics: Journal of Microscopy, vol. 95, pp. 145-164.

AMSTUTZ, G.C. (1982), Recollections of industrial applications of mineralogy: pp. 3-9 in Process **Mineralogy II: Applications in Metallurgy, Ceramics and Geology,** symposium volume published by the Society of Mining Engineers of AIME, New York.

AMSTUTZ, G.C., (1983), Mineralogy and petrology applied to exploration - Plenary lecture on exploration: Special Publication of the Geological Society of South Africa, no. 7, pp. 1-4.

AMSTUTZ, G.C., ZIMMERNINK, W.G., and ZIMMERMANN, R.A., (1987), Ore textures and locking types in concentration products of certain skarn deposits as compared to those in non-metamorphic stratabound ore deposits: paper presented at the 7th International Congress for Stereology, Caen, France.

ANTHONY, M., (1988), Technical change - The impending revolution: paper presented by P.M.J. Gray at the Australian IMM conference held in Sydney.

APLAN, F.F., (1985), Gravity concentration - Introduction: in **SME Mineral Processing Handbook,** published by the Society of Mining Engineers, New York, N.L. Weiss, Editor.

APLAN, F.F., (1985), Oxides of heavy metals - Titanium: in **SME Mineral Processing Handbook,** published by the Society of Mining Engineers, N.L. Weiss, Editor.

APLAN, F.F., (1985), Oxides of heavy metals - Chromium: in **SME Mineral Processing Handbook,** published by the Society of Mining Engineers, N.L. Weiss, Editor.

APLAN, F.F., (1985), Oxides of heavy metals - Manganese: in **SME Mineral Processing Handbook**, published by the Society of Mining Engineers, N.L. Weiss, Editor.

APLAN, F.F., (1985), Oxides of heavy metals - Tin: in **SME Mineral Processing Handbook**, published by the Society of Mining Engineers, N.L. Weiss, Editor.

APLAN, F.F., (1985), Oxides of heavy metals - Tungsten: in **SME Mineral Processing Handbook**, published by the Society of Mining Engineers, N.L. Weiss, Editor.

ARBITER, N., (Section editor), (1985), Flotation - Chemistry of flotation: in **SME Mineral Processing Handbook**, published by the Society of Mining Engineers, N.L. Weiss, Editor

BACON, W.G., HAWTHORN, G.W., and POLING, G.W., (1989), Gold analyses - myths, frauds and truths: The CIM Bulletin, vol. 82, (Nov. issue), pp. 29-36.

BAJON, P.H., GATEAU, C., JACQUIN, J.P., LAFORET, C., MORIZOT, G., and POLGAIRE, J.L., (1982), The contribution of quantitative mineralogy in the beneficiation of ores containing precious metals (silver in particular): in proceedings of the XIV International Mineral Processing Congress held in Toronto.

BAN, T.E., (1985), Agglomeration - Partial fusion methods-sintering and nodulizing: in **SME Mineral Processing Handbook**, published by the Society of Mining Engineers, N.L. Weiss, Editor.

BAUM, Wolfgang, and GILBERT, S. Rick, (1989), The routine use of process mineralogy in a cost-efficient gold-silver operation: in **Productivity and Technology in the Metallurgical Industries**, proceedings of the International Symposium on Productivity and Technology in the Metallurgical Industries, co-sponsored by The Minerals, Metals & Materials Society and Gesellschaft Deutscher Metallhutten-und Bergleute, held in Cologne, West Germany, Michael Koch and John C.Taylor, Editors.

BEAMISH, F.E., and VAN LOON, J.C. (1972), **Recent Advances in the Analytical Chemistry of the Noble Metals**, Pergamon Press, Oxford/New York/Toronto/Sydney/Braunschweig.

BEAMISH, F.E, and VAN LOON, J.C., (1977), **Analysis of Noble Metals: Overview and Selected Methods**, Academic Press, New York.

BEREZOWSKY, R.M.G.S, and WEIR, D.R., (1983), Pressure oxidation for treating refractory uranium and gold ores: paper presented at the 22nd anniversary CIM conference of metallurgists, held in Edmonton.

BHAPPU, R.B., (1985), Hydrometallurgy - Solution purification and concentration-carbon adsorption: in **SME Mineral Processing Handbook**, published by the Society of Mining Engineers, N.L. Weiss, Editor.

BIRKS, L.S., (1963), **Electron Probe Microanalysis,** Interscience Publishers, a division of John Wiley and Sons, New York/London.

BOLDT, J.R., Jr., and QUENEAU, P., (1985), Nickel and cobalt - Nickel and cobalt-bearing ores - Nickel-bearing ores: in **SME Mineral Processing Handbook**, published by the Society of Mining Engineers, N.L. Weiss Editor.

BOLLES, J.L., (1985), Properties of minerals: in **SME Mineral Processing Handbook,** published by the Society of Mining Engineers, N.L. Weiss, Editor.

BOND, F.C., (1985), General aspects of comminution - History: in **SME Mineral Processing Handbook,** published by the Society of Mining Engineers, N.L. Weiss Editor.

BOORMAN, R.S., PETRUK, W., and GILDERS, R., (1982), **Silver Carriers in Concentrate and Tailings from Brunswick Mining and Smelting Corporation Limited,** CANMET division report MRP/MSL 82-83.

BOWIE, S.H.U, SIMPSON, P.R., and RICE, C.M., (1973), Application of fissation track and neutron activation methods to geochemical exploration: pp. 359-372 in **Geochemical Exploration 1972,** published by the Institution of Mining and Metallurgy, London.

BOYDELL, D.W., LAXEN, P.A., BOSCH, D.W., and CRAIG, W.M., (1979), The new uranium recovery circuit at Blyvooruitzicht: The CIM Bulletin, vol. 72, (May issue), pp. 127-134.

BOYLE, R.W., (1968), **The Geochemistry of Silver and its Deposits,** Geological Survey of Canada Bulletin no. 160, published by the Ministry of Supply and Services, Canada.

BOYLE, R.W., (1979), **The Geochemistry of Gold and its Deposits,** Geological Survey of Canada Bulletin no. 280, published by the Ministry of Supply and Services, Canada.

BOYLE, R.W., (1987), **GOLD - History and Genesis of Deposits,** Van Nostrand Reinhold Company, New York.

BRIGGS, J. and VERTES, M., (1985), Molybdenum - Molybdenum products and uses: in **SME Mineral Processing Handbook,** published by the Society of Mining Engineers, N.L. Weiss, Editor.

BROWN, J.A., (1985), Lead and zinc - Uses of lead and zinc: in **SME Mineral Processing Handbook,** published by the Society of Mining Engineers, N.L. Weiss, Editor.

BROWNLOW, A.H., (1979), **Geochemistry,** Prentice-Hall, Inc., Engelwood, Cliffs, N.J.

BUGBEE, E.E., (1981), **A Textbook of Fire Assaying,** The Colorado School of Mines Press, Golden, Colorado.

BUTLER, J., (1985), Carbon, sulfur, mercury, miscellaneous sulfides - Mercury: **in SME Mineral Processing Handbook**, published by the Society of Mining Engineers, N.L. Weiss, Editor.

CABRI, L.J. (1972), The mineralogy of the platinum group elements: Minerals, Science and Engineering, vol. 4, pp. 3-29.

CABRI, L.J., (1981-A), Relationship of mineralogy to the recovery of platinum-group elements from ores: pp. 233-250 in **Platinum Group Elements: Mineralogy, Geology, Recovery**, CIM Special Volume no. 23.

CABRI, L.J., (1981-B), The Platinum Group Minerals: pp 83-150 in **Platinum Group Elements: Mineralogy, Geology, Recovery**, CIM Special Volume no. 23.

CABRI, L.J. (1991), AND GENKIN, A.D. (1991), Re-examination of Pt alloys from lode and placer deposits, Urals : The Canadian Mineralogist, vol. 29, pp. 419-425.

CARLISLE, H.P., (1985), Properties of ores and recovery methods in South Africa: in **SME Mineral Processing Handbook**, published by the Society of Mining Engineers, N.L. Weiss Editor.

CARSON, D.J.T., and JAMBOR, J.L., (1974), Mineralogy, zonal relationships and economic significance of hydrothermal alteration at porphyry copper deposits, Babine Lake area, British Columbia: The CIM Bulletin, vol. 67, pp. 110-133.

CHAMBERLIN, P., (1985), Non metallic industrial minerals - Beryllium: in **SME Mineral Processing Handbook**, published by the Society of Mining Engineers, N.L. Weiss, Editor.

CHAPMAN, F.H., (1985), Hydrometallurgy - Electrolysis: in **SME Mineral Processing Handbook**, published by the Society of Mining Engineers, N.L. Weiss, Editor.

CHEN, T.T., DUTRIZAC, J.E., OWENS, D.R., and LAFLAMME, J.H.G., (1980), Accelerated tarnishing of same chalcopyrite and tetrahedrite specimens: The Canadian Mineralogist, vol. 18, pp. 173-180.

CHEN, T.T., and PETRUK, W. (1980), Mineralogy and characteristics that affect recoveries of metals and trace elements from the ore at Heath Steele Mines, New Brunswick: The CIM Bulletin, vol. 73, (823), pp. 167-179.

CLARK, G.B., (1988), Industries - Extraction and processing - Mining: pp. 339-349 in **Encyclopaedia Britannica-Macropaedia**, Vol. 21, Chicago.

CLARK, L.A., (1960), The Fe-As-S system, phase relations and applications: Economic Geology, vol. 55, pp. 1345-1381.

CLARKE, F. W., (1924), **The Data of Geochemistry - Gold,** US Geological Survey Bulletin no. 770, pp. 656-663.

CLEVENGER, G.H., and CARON, M.H., (1917), **The Caron Process,** US Bureau of Mines Bulletin no. 226.

CORNWELL, W.G., and HISSHION, R.J., (1976), Leaching of telluride concentrates for gold, silver and tellurium. Emperor Process: Transactions of the Society of Mining Engineers, vol. 260, pp. 108-112.

CORRANS, I.J., and LEVIN, J., (1979), Wet high intensity magnetic separation for the concentration of Witwatersrand gold-uranium ores and residues: Journal of the South African Institute of Mining and Metallurgy, vol 79, no. 8. pp 210-228.

CORRIVAUX, L., and LAFLAMME, J.H.G., (1990), Mineralogie des elements des groupe de platine dans le chromitites de l'ophiolite de Thetford Mines, Quebec: The Canadian Mineralogist, vol. 28, pp. 579-595.

COTE', Bruno, BENGUEREL, Elyse, and DEMOPOULOS, George, P. (1992), Solvent extraction separation of platinum group metals: p. 351 in **Precious Metals 1992,** proceedings of the 16th International Precious Metals Institute Conference held in Scottsdale, Arizona, A.K. Mehta and R.M. Nadkarni, Editors.

CUNDIFF, W.H., (1985), Alumina - Introduction and the Bayer process: in **SME Mineral Processing Handbook**, published by the Society of Mining Engineers, N.L. Weiss, Editor.

CUTHBERTSON, R.E., and VINCENT, J.D., (1985), Molybdenum - Primary molybdenite ores: in **SME Mineral Processing Handbook,** published by the Society of Mining Engineers, N.L. Weiss, Editor.

DAHLSTROM, D.A. and EMMETT, Jr., (1985), Hydrometallurgy - Solid-liquid separations: in **SME Mineral Processing Handbook,** published by the Society of Mining Engineers, N.L. Weiss, Editor.

DANA, E.S., (1963), **A Textbook of Mineralogy**, 4th edition revised and extended by W.E. Ford, John Wiley and Sons, New York and London.

DAVIS, F.T., (1985), Nonmetallic industrial minerals - Barite: in **SME Mineral Processing Handbook**, published by the Society of Mining Engineers, N.L. Weiss, Editor.

DAVIS, F.T., (1985), Oxides of heavy metals - Rare earth minerals: in **SME Mineral Processing Handbook**, published by the Society of Mining Engineers, N.L. Weiss, Editor.

DAVEY, T.R.A, (1985), Pyrometallurgy - Introduction: in **SME Mineral Processing Handbook**, published by the Society of Mining Engineers, N.L. Weiss, Editor.

DEFOE, M.E., (1985), Nonmetallic industrial minerals - Boron: in **SME Mineral Processing Handbook**, published by the Society of Mining Engineers, N.L. Weiss, Editor.

DEFOE, M.E., (1985), Carbon, sulfur, mercury, miscellaneous sulfides - Diamonds: in **SME Mineral Processing Handbook**, published by the Society of Mining Engineers, N.L. Weiss, Editor.

DEMOPOULOS, G.P., (1989), Refining of platinum-group metals: The CIM Bulletin, vol. 82, (923), pp. 165-170.

DESBOROUGH, G.A., (1970), Silver depletion indicated by microanalytical studies of gold from placer occurrences, Western United States: Economic Geology, vol. 65, pp. 304-311.

DESBOROUGH, G.A., and, CRIDDLE, A.J., (1984), Bowieite: a new rhodium-iridium-platinum sulfide in platinum alloy nuggets, Goodnews Bay, Alaska: The Canadian Mineralogist, vol. 22, pp. 543-552.

DESCHENES, G., (1986), Literature survey of the recovery of gold from thiourea solutions and comparison with cyanidation: The CIM Bulletin, vol. 79 (895), pp. 76-83.

DeVANEY, F.D., (1985), Iron ore - Introduction: in **SME Mineral Processing Handbook**, published by the Society of Mining Engineers, N.L. Weiss, Editor.

DE WAAL, S.A., and HERZBERG, W., (1969), Uranium and gold mineralization in the Black Reef series of the Kaapse Hoop area, Nelspruit District, Eastern Transvaal: Annals of the Geological Society of South Africa, vol. 7, pp. 111-124.

Dictionary of Geological Terms, (1962), prepared under the direction of the American Geological Institute, Dolphin books, Doubleday & Company, Inc., Garden City, New York.

DOOLEY, J.R., Jr., (1958), The radioluxograph - a fast simple type of autoradiograph: paper no. 1762, presented at the United Nations International Conference on Peaceful Uses of Atomic Energy.

DOOLEY, J.R., Jr., CRONWELL, C.N., BEREDSEN, P., OTTON, J.K., PIERSON, C.T., HOISINGTON, W.D., LINDSEY, D.A., and ROSHOLT, J.N., (1977), Examples of uranium distribution graphics in geologic rock specimens illustrated with the radioluxograph, induced fission tracks and other tracks methods: U.S.G.S. Circular 753, p. 32.

DORR, J.V.N., and BOSQUI, F.L., (1950), **Cyanidation and Concentration of Gold and Silver Ores**, McGraw-Hill Book Company, New York.

DYCK, W., (1978), The mobility and concentration of uranium and its decay products in temperate surficial environments: pp. 57-100 in **MAC Short Course in Uranium Deposits: Their Mineralogy and Origin**, published by the Mineralogical Association of Canada, Toronto, M.M. Kimberley, Editor.

ELLIOTT, J.F., (Editor), (1988), Industries - Extraction and processing - Iron production: pp. 360-388 in **Encyclopaedia Britannica-Macropaedia,** vol. 21, Encyclopaedia Britannica, Inc., Chicago.

ELLSWORTH, V.H., (1928), Thucholite, a remarkable primary carbon mineral from the vicinity of Parry Sound, Ontario: The American Mineralogist, vol. 13, pp. 419-441.

Encyclopaedia Britannica-Micropaedia, (1988), volumes describing the metals addressed in this book, Encyclopaedia Britannica, Inc., Chicago.

ENGELBRECHT, C.J., (1984), **Operating Gold Mines and Gold Recovery Plants in the Republic of South Africa,** report published by the Minerals Bureau, Department of Minerals and Energy Affairs, Republic of South Africa.

EVANS, A.M., (1987), **An Introduction to Ore Geology,** Blackwell Publishing Company, Oxford.

FLETCHER, A.W., FINKELSTEIN, N.P., and DERRY, R., (1985) Hydrometallurgy - Leaching systems: in **SME Mineral Processing Handbook,** published by the Society of Mining Engineers, N.L. Weiss, Editor.

FLETCHER, J.B., (1985), Copper - Typical copper oxide leaching operations - In-situ leaching at Miami: in **SME Mineral Processing Handbook,** published by the Society of Mining Engineers, N.L. Weiss, Editor.

FORWARD, F.A. and PETERS, E., (1985), Hydrometallurgy - Leaching principles: in SME **Mineral Processing Handbook,** published by the Society of Mining Engineers, N.L. Weiss, Editor.

FORWARD, F.A. and WARREN, I.H., (1985) Hydrometallurgy - Leaching technology: in **SME Mineral Processing Handbook,** published by the Society of Mining Engineers, N.L. Weiss, Editor.

FREEMAN, H.D., HARTLEY, J.N., and WICK, O.J., (1981), Scanning electron microscopy and microprobe analysis of refractory gold ores and process implications of the analysis: pp. 305-323 in **Process Mineralogy in Metal Extraction, Mineral Exploration and Energy Materials,** symposium volume published by the Metallurgical Society of AIME, Warrendale, Pennsylvania.

GASPARRINI, Claudia, (1976), The energy-dispersive spectrometer attached to the microprobe: a new tool of investigation for the earth scientist: Minerals, Science and Engineering, vol. 8, pp. 3-22.

GASPARRINI, Claudia, (1980), The role of the ore microscope and electron microprobe in the mining industry: The CIM Bulletin, vol. 73, (817), pp. 73-85.

GASPARRINI, Claudia, (1983-A), The mineralogy of gold and its significance in metal extraction: The CIM Bulletin, vol. 76, (851), pp. 144-153.

GASPARRINI, Claudia, (1983-B), **Study of the Mineralogy of a Gold-Bearing Sample of Barite from the El Indio Deposit in Chile**, report prepared by Minmet Scientific Limited for St. Joe Minerals Corporation, no. 383.

GASPARRINI, Claudia, (1983-C), **Study of a Second Gold-Bearing Sample of Barite from the El Indio Area (Reina Vein),** report prepared by Minmet Scientific Limited for St. Joe Minerals Corporation, no. 385.

GASPARRINI, Claudia, (1984-A), The mineralogy of silver and its significance in metal extraction: The CIM Bulletin, vol. 77, (866), pp. 99-110.

GASPARRINI, Claudia, (1984-B), The metallurgy of the precious metals as affected by their mineralogy and manner of occurrence in their ores: pp. 101-109 **in Precious Metals, Mining, Extraction and Processing,** proceedings of The Metallurgical Society of AIME and IPMI International Symposium on Precious Metals held in Los Angeles, Kudryk-Corrigan-Liang, Editors.

GASPARRINI, Claudia, (1985), Microanalytical techniques for the determination of the occurrence of the precious metals in their ores: pp. 11-19 in **Precious Metals 1984,** proceedings of the 8th International Precious Metals Institute Conference held in Toronto, T.P. Mohide, Editor.

GASPARRINI, Claudia, (1986), Microanalytical techniques for the study of precious metal ores. An assessment of the present state of the art: pp. 189-201 in **Precious Metals 1986,** proceedings of the 10th International Precious Metals Institute Conference held in Lake Tahoe, U.V. Rao, Editor.

GASPARRINI, Claudia, (1988), Microanalytical techniques for the study of gold and silver ores: pp. 85-90 in **Precious Metals Sampling and Analysis,** proceedings of a seminar of the International Precious Metals Institute held in Chicago, Silve Kallmann, Editor.

GASPARRINI, Claudia, (1991), The paragenetic relations in gold and silver ores, as observed through the microscope: invited lecture presented at the Institute of Mineralogy and Petrography of the University of Heidelberg, January 8.

GASPARRINI, Claudia, .(1992-A), Could metallurgists learn from mineralogy and microanalysis?: Mining Magazine, vol. 166, no. 3, p. 169.

GASPARRINI, Claudia, (1992-B), Fractal Geometry applied to the study of precious metal ores: how the macro relates to the micro: pp. 353-36 in **Precious Metals 1992,** proceedings of the 16th International Precious Metals Institute Conference held in Scottsdale, Arizona, A.K. Mehta and R.M. Nadkarni, Editors.

GASPARRINI, Claudia, and LOWELL, Gary. R., (1983), Argentiferous galena from the Silver Mine discredited: paper presented at the GAC-MAC meeting held in Victoria, British Columbia.

GASPARRINI, Claudia and LOWELL, Gary, R., (1985), Silver-bearing inclusions in "argentiferous" galena from the Silvermine district in Southeastern Missouri: The Canadian Mineralogist, vol. 23, pp. 99-102.

GASPARRINI, Claudia, and WILLIAMSON, R.G., (1981), The mineralogy of the uranium ores with some considerations on their significance in metal extraction and mineral exploration: pp. 325-337 in **Process Mineralogy in Metal Extraction, Mineral Exploration and Energy Materials**, symposium volume published by The Metallurgical Society of AIME, Warrendale, Pennsylvania.

GAUDIN, A.M., (1939), **Principles of Mineral Dressing**, McGraw-Hill Book Company, New York.

GILCHRIST, J.D., (1980), **Extraction Metallurgy**, Pergamon Press, Oxford, New York, Toronto, Sydney, Paris, Frankfurt.

GIRARDI, S., ALRUIZ, O., and ANFRUNS, J., (1989), Pressure cyanidation: an option to treat refractory gold resources: paper presented at the International Symposium on Productivity and Technology in the Metallurgical Industries, held in Cologne, West Germany.

GLEICK, James, (1987), **CHAOS, Making A New Science**, Penguin Books, New York, New York.

Glossary of Precious Metal Industry Terms, (1987), published in cooperation by the International Precious Metals Institute and Met-Chem Research Inc., Boulder, Colorado.

Gold Extraction for the Small Operator, (1936), Imperial Chemical Industries Ltd., Liverpool, England.

GOLDSTEIN, Joseph, I., NEWBURY, Dale, E., ECHLIN, Patrick, JOY, David, C., FIORI, Charles, and LIFSHIN, Eric, (1984), **Scanning Electron Microscopy and X-ray Microanalysis, A Text for Biologists, Materials Scientists, and Geologists**, Plenum Press, New York and London.

GUAY, W.J., (1980), **Recovery of Gold from Carbonaceous Gold-Bearing Ores**, U.S. Patent 4188208.

HABASHI, F., (1966), The theory of cyanidation: transactions of the Society of Mining Engineers, Sept. issue, pp. 236-239.

HABASHI, F., (1967), **Kinetics and Mechanisms of Gold and Silver Dissolution in Cyanide Solution,** Montana Bureau of Mines, Geological Bulletin no. 59.

HABASHI, F., (1986), **Principles of Extractive Metallurgy**, Gordon and Breach Science Publishers, New York, London, Paris, Montreaux, Tokyo.

HAGEL, Enrique, (1991) **Mineria del Oro**, Progama de Geologia Economica Aplicada, Serie Monografias y Textos - Ano 1, no. 1.

HALLBAUER, D.K., and JOUGHIN, N.C., (1972), **An Investigation into the Distribution, Size and Shape of Gold Particles in Some Witwatersrand Reefs and their Effects on Sampling Procedures,** research report no. 43/72, Chamber of Mines of South Africa Research Organization, Johannesburg.

HAMILTON, E.M., (1920), **Manual of Cyanidation**, McGraw-Hill Book Company, New York.

HAMPBEL, C.A., (1954), (Editor), **Rare Earth Metals Handbook,** Reinhold Publishing Corporation, New York.

HAQUE, K.E., (1992), The role of oxygen in cyanide leaching of gold ore: The CIM Bulletin, Vol. 85, (963), pp. 31-37.

HARRIS, D.C., (1976), Applications of the electron microprobe in exploration, evaluation, development and genetic aspects of ore deposits: pp. 157-172 in **Short Course in Microbeam Techniques**, published by the Mineralogical Association of Canada.

HARRIS, D.C., and CABRI, L.J., (1991), Nomenclature of platinum group element alloys: review and revision: The Canadian Mineralogist, vol. 29, pp. 231-237.

HAUSEN, D.M., (1981), Process Mineralogy of auriferous pyritic ores at Carlin, Nevada: pp. 271-289 in **Process Mineralogy in Metal Extraction, Mineral Exploration and Energy Materials,** symposium volume published by the Metallurgical Society of AIME, Warrendale, Pennsylvania.

HAYES, Peter, (1985), **Process Selection in Extractive Metallurgy**, Hayes Publishing Company, Brisbane, Australia.

HAZEN, W.C., (1985), Hydrometallurgy - Solution purification and concentration-solvent extraction: in **SME Mineral Processing Handbook**, published by the Society of Mining Engineers, N.L. Weiss, Editor.

HEAD, R.E., (1936), Physical characteristics of gold lost in tailings: pp. 256-262 in proceedings of the AIME Conference held in New York City.

HEDLEY, R.E., and TABACHNICK, H., (1968), **The Chemistry of Cyanidation**, the American Cyanamid Company.

HENLEY, K.J., (1975), Gold ore mineralogy and its relation to metallurgical treatment: Mineral, Science and Engineering, vol. 7, no. 4, pp. 289-312.

HENLEY. K.J., (1981), Ore-dressing mineralogy - A review of techniques, applications and recent developments: in proceedings of the 1st ICAM Conference, held in Johannesburg, South Africa.

HISKEY, J.B., and PEIHAO QI, (1991), Dissolution and electrochemical behavior of gold in iodide solutions: pp. 69-81 in **Precious Metals - Recovery and Refining**, proceedings of a seminar of the International Precious Metals Institute held in Tempe, Arizona, Martin T. Durney, Editor.

HORST, W.E., (1985), General aspects of comminution - Purpose and terminology: in **SME Mineral Processing Handbook**, published by the Society of Mining Engineers, N.L. Weiss, Editor.

JAMBOR, J.L. (1975), **Geology and Hydrothermal Alteration of the Maggy Porphyry Copper Deposit, North Central British Columbia,** Geological Survey of Canada special paper.

JOCHENS, P.R., and LAXEN, P.A., (1975), **Some Developments in the Extraction of Gold from Witwatersrand Ores,** NIM (presently Mintek) unrestricted report no. 1768.

JOHNSTON, W.E., (1933), Gold tellurides are soluble in cyanide: Mining and Metallurgical Bulletin, vol. 26, (254), pp. 224-225.

KAESER, Robert, C., (1992), Managing rhodium in a volatile market: pp. 27-30 in **Rhodium - Sampling and Analysis,** proceedings of a seminar of the International Precious Metals Institute held in Orlando, Florida, Richard C. Kaltenbach and Larry Manzick, Editors.

KALLMANN, Silve, (1986), Analysis of sweeps. The cuprous sulphide collecting system: Talanta, vol. 33, no. 1, pp. 75-83.

KING, J.W., and ADLER, H.H., (1985), Uranium - Geology and mineralogy of uranium: in **SME Mineral Processing Handbook**, published by the Society of Mining Engineers, N.L. Weiss, Editor.

KNIGHTS, J.G., (1982), **Hilton Mine Silver Mineralogy - Its Distribution, Textural Associations and its Mineralogical Significance**, unpublished typed manuscript.

KRUPP, R.E., and WEISER, T., (1992), On the stability of gold-silver alloys in the weathering environment: Mineralium Deposita, vol. 27, pp. 268-275.

KUHNEL, R.A., PRINS, J.J., and ROORDA, H.J., (1980), **The "Delft" System for Mineral Identification, I. Opaque Minerals**, Delft University Press, Delft, The Netherlands.

LAKSHMANAN, V.I., and CHILDS, A.M., (1984), Treatment of refractory gold ores: paper presented at the 14th CIM annual hydrometallurgy meeting, held in Timmins, Ontario.

LAN XINZHE, ZHANG JIAN, XIU JINGFU, LI WENGSING, and LAI JUEWEN, (1992), On mechanism of agents consumption in extraction of gold and silver with thiourea: pp. 235-242 in **Precious Metals 1992**, proceedings of the 16th International Precious Metals Institute Conference held in Scottsdale, Arizona, A.K. Mehta and R.M. Nadkarni, Editors.

LANGHANS, J.W., LEI, K.P.V., and CARNAHAN, T.G., (1992), Copper-catalyzed thiosulfate leaching lo low grade gold ores: Bureau of Mines publication - Hydrometallurgy, vol. 29, pp. 191-203.

LAW-WEST, D., (1986), Gold: pp. 29.1-29.12, in **Canadian Minerals Yearbook, Review and Outlook - Centennial Edition**, published by the Ministry of Energy, Mines and Resources.

LAXEN, P.A., (1964), **Factors in the Dissolution of Uranium from South African Ores and Observations on the Nature of the Undissolved Uranium in Unit Processes in Hydrometallurgy**, Gordon and Breach Science Publishers, New York.

LIMA DE FARIA, J., (1964), Identification of metamict minerals by x-ray diffraction photographs: Lisbon, Junta Invest. Ultramar. Estudos Ensaidos, doc. 112.

LORIG, C.H., (Editor), (1988), Industries - Extraction and processing-metallurgy: pp. 349-360 in **Encyclopaedia Britannica-Macropaedia**, vol. 21, Encyclopaedia Britannica, Inc., Chicago.

LUCAS, J.M., (1985), Gold: pp. 445-480 in **Minerals Yearbook 1985**, Volume 1, Metals and minerals, prepared by the staff of the US Bureau of Mines.

MACINTOSH, R.M., (Editor), (1988), Industries - Extraction and processing-tin: pp. 424-430 in **Encyclopaedia Britannica-Macropaedia**, vol. 21, Encyclopaedia Britannica, Inc., Chicago.

MACKIW, V.N., (1985), Nickel and cobalt - Preface: in **SME Mineral Processing Handbook**, published by the Society of Mining Engineers, N.L. Weiss, Editor.

MANDELBROT, Benoit, (1982), **The Fractal Geometry of Nature**, W.H. Freeman, New York.

MARLEY, F.L., and HAGNI, R.D., (1983), Mexican silver ores: their mineralogical occurrence and treatment: in **Process Mineralogy II: Applications to Metallurgy, Ceramics and Geology**, symposium volume published by the Society of Mining Engineers.

MASON, P.G., and O'KANE, P.T., (1985), Refractory gold - process options: paper presented at the 24th CIM annual conference of metallurgists, held in August in Vancouver, British Columbia.

MAULE, Almorie, A., (1992), Will South Africa produce rhodium at the expense of platinum?: pp. 1-9 in **Rhodium - Sampling and Analysis**, proceedings of a seminar of the International Precious Metals Institute held in Orlando, Florida, Richard C. Kaltenbach and Larry Manzick, Editors.

McLEAN, DANIEL C. (1982), Practical considerations in the sampling and estimation of reserves in low grade ore deposits: paper presented at the International Precious Metals Institute International Seminar held in London. Available from Davy McKee, San Francisco Operations.

McMILLAN, R.H., (1978), Genetic aspects and classification of important Canadian uranium deposits: pp. 187-204 in **MAC Short Course in Uranium Deposits: Their Mineralogy and Origin**, published by the Mineralogical Association of Canada, Toronto, M.M. Kimberly, Editor.

McNIGHT, E.T., (1985), Lead and zinc - Mineralogy: in **SME Mineral Processing Handbook**, published by the Society of Mining Engineers, N.L. Weiss, Editor.

McQUISTON, F.W., Jr., (1985), Properties of gold ores and recovery methods: in **SME Mineral Processing Handbook**, published by the Society of Mining Engineers, N.L. Weiss, Editor.

McQUISTON, F.W., Jr., (1985), Silver: in **SME Mineral Processing Handbook**, published by the Society of Mining Engineers, N.L. Weiss. Editor.

McQUISTON, F.W., Jr., (1985), Platinum group metals: in **SME Mineral Processing Handbook**, published by the Society of Mining Engineers, N.L. Weiss, Editor.

MERRIT, R.C., (1971), **The Extractive Metallurgy of Uranium**, published by the Colorado School of Mines Research Institute, Denver, Colorado.

MERRIT, R.C., SEIDEL, D.C., BURNHAM, D.A., and BUSH, P.D., (1985), Hydrometallurgy - Recovery from solution-precipitation: in **SME Mineral Processing Handbook**, published by the Society of Mining Engineers, N.L. Weiss, Editor.

MIDDLETON, G.V., (Editor/Organizer), (1991), **Nonlinear Dynamics, Chaos and Fractals with Applications to Geological Systems**, Geological Association of Canada, Short Course Notes, vol. 9, Toronto.

MILLER, G.L., (1988), Industries - Extraction and processing-tungsten: pp. 432-433 in **Encyclopaedia Britannica-Macropaedia**, vol. 21, Encyclopaedia Britannica, Inc., Chicago.

MILNER, E.F., (1985), Lead and zinc - Lead and zinc concentrates: extraction processes and schedules: in **SME Mineral Processing Handbook**, published by the Society of Mining Engineers, N.L. Weiss, Editor.

MINARCIK, E.J., (Editor), (1988), Industries - Extraction and Processing-lead production: pp. 403-406 in **Encyclopaedia Britannica-Macropaedia**, vol. 21, Encyclopaedia Britannica, Inc., Chicago.

MITCHEL, R.S. (1973), Metamict minerals: a review: The Mineralogical Record, Jul.-Aug. issue, pp. 177-182 and Sept.-Oct. issue, pp. 214-223.

MOHIDE, T.P., (1979), **Platinum Group Metals - Ontario and the World,** Ontario mineral policy background paper no. 7, published by the Ministry of Natural Resources, Toronto.

MOHIDE, T.P., (1981), **GOLD,** Ontario mineral policy background paper no. 12, published by the Ministry of Natural Resources, Toronto.

MOHIDE, T.P., (1985), **SILVER,** Ontario mineral policy background paper no. 20, published by the Ministry of Natural Resources, Toronto.

MOHIDE, T.P., WARDEN, C.L., and MASON, J.D., (1977), **Towards a Nickel Policy for the Province of Ontario,** Ontario mineral policy background paper no. 4, published by the Ministry of Natural Resources, Toronto.

MOORE, D.C., (1985), Copper - Processing of concentrates and development trends - Hydrometallurgical processes: in **SME Mineral Processing Handbook,** published by the Society of Mining Engineers, N.L. Weiss, Editor.

MORTON, R.D., (1978), The identification of uranium minerals: pp. 141-183 **in MAC Short Course in Uranium Deposits: Their Mineralogy and Origin**, published by the Mineralogical Association of Canada, Toronto, M.M. Kimberley, Editor.

NADKARNI, Ravi, and SAVOLAINEN, Arnold, (1992), Statistical process control and fire assay: pp 39-60 in **Precious Metals 1992,** proceedings of the 16th International Precious Metals Institute Conference held in Scottsdale, Arizona, A.K. Mehta and R.M. Nadkarni, Editors.

NALDRETT, A.J., (1981), Platinum group element deposits: pp. 197-231 in **Platinum Group Elements: Mineralogy, Geology, Recovery,** CIM Special Volume no. 23.

NEKRASOV, I., (in press), **Mineralogy, Geochemistry and Genesis of Gold Deposits,** to be translated from Russian into English. In the process of selling the copyright to a Western publisher, as the present manuscript goes to the presses.

NIXON, G.T. and CABRI, L.J., (1990), Platinum group element mineralization in lode and placer deposits associated with the Tulameen Alaskan-type complex, British Columbia: The Canadian Mineralogist, vol. 28, 503-535.

OGDEN, H.R., (Editor), (1988), Industries - Extraction and Processing-titanium: pp. 430-432 in **Encyclopaedia Britannica-Macropaedia**, vol. 21, Encyclopaedia Britannica, Inc., Chicago.

PASTERIS, Jill Dill, (1984), Use of indigenous kimberlite minerals, particularly spinels, in the evaluation of diamond potential: pp. 157-179 in **Process Mineralogy III**, symposium volume published by the Society of Mining Engineers of AIME, New York.

PETERSON, E.C., (1985), Uranium - Introduction: in **SME Mineral Processing Handbook**, published by the Society of Mining Engineers, N.L. Weiss, Editor.

PICKET, D.E., (1985), Hydrometallurgy - Feed preparation: in SME **Mineral Processing Handbook**, published by the Society of Mining Engineers, N.L. Weiss, Editor.

PICOT, P. and JOHAN, Z., (1977), **Atlas des Mineraux Metalliques**, Memoires du Bureau de Recherches Geologiques et Minieres.

PLASKIN I.N., and KOZHUKHOVA, M., (1941), The solubility of gold and silver in thiourea: Dok. Ak. Nauk., SSR 31, pp. 671-674.

PLASKIN, I.N., and KOZHUKHOVA, M, (1960), Sb. Nauch. Tr. Ins. Tsvt. Met., 33, p. 107.

Platinum Group Elements: Mineralogy, Geology, Recovery, (1981), CIM Special Volume no. 23.

PORTER, R.R., (1985), Hydrometallurgy - Solution purification and concentration-ion exchange: in **SME Mineral Processing Handbook**, published by the Society of Mining Engineers, N.L. Weiss, Editor.

POUGH, F.H., (1983), **Rocks and Minerals**, Peterson Field Guides, Houghton Mifflin Company, Boston.

Precious Metals Science and Technology, (1991), published in cooperation by the International Precious Metals Institute and Ilse V. Nielsen, L.S. Benner, T. Suzuki, K. Meguro and S. Tanaka, Editors. (Translated from Japanese with updates and revisions).

Proceedings of the International Precious Metals Institute Conferences, held once a year in June since 1979. Abstract books were published from 1976 (when IPMI was founded) to 1979.

Proceedings of the International Precious Metals institute Seminars, held several times a year since 1981. Some of the titles are: **Platinum Group Metals, Gold and Silver, Precious Metals - Sampling and Analysis, Precious Metals - Mine to Market, Precious Metals Recovery and Refining, Rhodium - Sampling and Analysis.**

Proceedings of the Process Mineralogy Sessions, held during the AIME Conferences since 1979 and published alternatively by The Metallurgical Society (TMS) and the Society of Mining Engineers (SME) of the American Institute of Mining Engineering (AIME) since 1981.

PRYOR, E.J., (1965), **Mineral Processing**, Elsevier Publishing Company Ltd., London.

RAGLAND, P.C., (1964), Autoradiographic investigation of naturally occurring materials: pp. 129-151 in **The Natural Radiation Environment,** The Chicago University Press, Adams and Lowder, Editors.

RAMDOHR, P., (1980), **The Ore Minerals and their Intergrowths,** Pergamon Press, Oxford.

RAU, E. (1985), Nonmetallic industrial minerals - Magnesium minerals: in **SME Mineral Processing Handbook**, published by the Society of Mining Engineers, N.L. Weiss, Editor.

RICHARDS, A.W., (Editor), (1988), Industries - Extraction and processing-zinc production: pp. 399-403 in **Encyclopaedia Britannica - Macropaedia**, vol. 21, Encyclopaedia Britannica, Inc., Chicago.

RICHARDS, R.H., and LOCKE, C.E., (1925), **Textbook of Ore Dressing**, McGraw-Hill Book Company, New York.

RILEY, J.F., (1974), Tetrahedrite-freibergite series with reference to the Mt. Isa Pb-Zn-Ag ore body: Mineralium Deposita, vol. 9, pp. 117-124.

RIZLEY, J.H., (Editor), (1988), Industries - Extraction and processing-magnesium: pp. 413-415 in **Encyclopaedia Britannica-Macropaedia,** vol. 21, Encyclopaedia Britannica, Inc., Chicago.

ROSE, T.K., (1986), **The Metallurgy of Gold**, Met-Chem Research Inc., Boulder, Colorado.

RUNCK, R.J., (1988), (Editor), Are there platinum reserves along the US-Canadian border?: Precious Metals News and Review, vol. 12, no. 9, p. 1.

SANTOSH, M., PHILIP, R., JACOB, M.K., and OMANA, P.K., (1992), Highly pure placer gold formation in the Nilambur Valley, Wynad Gold Fields, Southern India: Mineralium Deposita, vol. 27, pp. 336-339.

SARBUTT, K., (1983), Principles of cyanidation: notes distributed by McGill University during the Professional Development Seminar titled **Mineral Processing of Gold Ores.**

SAWYER, J.B.P., (1992), Assaying in resource evaluation: the need for a clear and open mind: Geological Society Special Publication no. 63, pp. 37-45.

SHADE, J.P., (1986), Nickel in service of society: paper presented at the 25th CIM annual conference of metallurgists held in Toronto.

SHAFFER, J.W., (1985), Alumina - Bauxite: in **SME Mineral Processing Handbook**, published by the Society of Mining Engineers, N.L. Weiss, Editor.

SCHEINER, B.J., LINDSTROM, E.R., and HENRIE, T.A., (1971), Processing refractory carbonaceous ores for gold recovery: Journal of Metallurgy, 23, 337-409.

SCHEINER, B.J., POOL, D.L., SJOBERG, J.J., and LINDSTROM, R.E., (1973), **Extraction of Silver from Refractory Ores,** US Department of the Interior, Bureau of Mines report of Investigations no. 7736.

SHELTON, J.E., (Editor), (1988), Industries - Extraction and processing-mercury: pp. 415-417 in **Encyclopaedia Britannica-Macropaedia**, vol. 21, Encyclopaedia Britannica, Inc., Chicago.

SHEPHARD, O.C., and DIETRICH, W.F., (1989), **Fire Assay**, Met-Chem Research Inc., Boulder, Colorado.

SCHLECHTEN, A.W., (1988), Industries - Extraction and processing-copper production: pp. 396-399 in **Encyclopaedia Britannica-Macropaedia**, vol. 21, Encyclopaedia Britannica, Inc., Chicago.

SCHOUTEN, C., (1962), **Determination Tables for Ore Microscopy**, Elsevier Publishing Company, New York.

SCHWELLNUS, J.S.I., HIEMSTRA, S.A., and GASPARRINI, Claudia, (1976), The Merensky Reef at the Atok platinum mine and its environs: Economic Geology, vol. 71, pp. 249-260.

SMIRNOV, S.S., (1951), The zone of oxidation of sulfidic mineral deposits: Izd. Akad. Nauk USSR, Moscow, p. 335. Translated and published in Boyle's **Gold - History and Genesis of Deposits (1987)**, Van Nostrand Reinhold Company, New York, pp. 569-575.

SMITH, E.A., (1987), **The Sampling and Assaying of the Precious Metals**, Met-Chem Research, Inc., Boulder, Colorado.

SMITH, S.L., (Editor), (1988), Industries - Extraction and processing-silver: pp. 420-422 in **Encyclopaedia Britannica**, vol. 21, Encyclopaedia Britannica, Inc., Chicago.

SMITH, S.W., (1908), in transactions of the Institute of Mineralogy and Metallurgy, vol. 17, p. 468.

SOBOLEV, S.N., (1981), Siberian diamonds - Vast kimberlite reserves: World Mining, October issue, pp. 55-57.

STEACY, H.R., and KAIMAN, S., (1978), Uranium minerals in Canada, their description, identification and field guides: pp. 107-140 in **MAC Short Course in Uranium Deposits: Their Mineralogy and Origin**, published by the Mineralogical Association of Canada, Toronto, M.M. Kimberly, Editor.

TAGGART, A.F. (1953), **Handbook of Mineral Dressing, Ores and Industrial Minerals**, John Wiley and Sons Publishing Company, New York.

THEYS, L.F., et al., (1985), Nickel and cobalt - Cobalt-bearing ores: in **SME Mineral Processing Handbook**, published by the Society of Mining Engineering, N.L. Weiss, Editor.

TREMBLAY, Remy, (1983), Laboratory testing for flowsheet development and optimization: notes distributed by McGill University during the Professional Development Seminar titled **Mineral Processing of Gold Ores.**

US BUREAU OF MINES MATERIAL SURVEY, (1951), **Antimony**, compiled for the Materials Office, National Security Resources Board.

VANCE, E.R., and ROUTCLIFFE, P., (1976), Heat treatment of some metamict allanites: Mineralogical Magazine, vol. 40, pp. 521-523.

VAN HORN, K.R., (Editor), (1988), Industries - Extraction and processing-aluminum production: pp. 388-396 in **Encyclopaedia Britannica-Macropaedia**, vol. 21, Encyclopaedia Britannica, Inc., Chicago..

VAN LOON, Jon, (1980), **Analytical Atomic Absorption: Selected Methods**, Academic Press, New York.

VASSILIOU, A.H., (1989), The form of occurrence of uranium in deposits associated with organic material: Economic Geology, vol. 75, pp. 609-617.

VASSILIOU, A.H., and KERR, P.F., (1973), Analytical data on nature of urano-organic deposits: American Association of Petroleum Geologists Bulletin, vol. 57, pp. 1291-1296.

VINCENT, J.D., (1985), Molybdenum - Introduction: in **SME Mineral Processing Handbook**, published by the Society of Mining Engineers, N.L. Weiss, Editor.

VINCENT, J.D., and WEISS, N.L., (1985), Copper - Introduction-production: in SME **Mineral Processing Handbook**, published by the Society of Mining Engineers, N.L. Weiss, Editor.

WADSWORTH, M. E., (1979), Hydrometallurgical processes: in **Rate Processes in Extractive Metallurgy**, Plenum publishing Company, H.Y. Sohn and M.E. Wadsworth, Editors.

WEIR, D.R., and BEREZOWSKY, R.M.G.S, (1984), Gold extractions from refractory concentrates: paper presented at the 14th CIM annual hydrometallurgy meeting, held in Timmins, Ontario.

WESTLAND, A.D., (1981), Inorganic chemistry of the platinum group elements: pp. 7-18 in **Platinum Group Elements: Mineralogy, Geology and Recovery**, CIM Special Volume no. 23.

WILLIAMS, J., (Editor), (1988), Industries - Extraction and processing-uranium: pp. 433-435 in **Encyclopaedia Britannica-Macropaedia**, vol. 21, Encyclopaedia Britannica, Inc., Chicago.

WILSON, W.P., (1985), Hydrometallurgy - Recovery from solutions-evaporation and crystallization: in **SME Mineral Processing Handbook**, published by the Society of Mining Engineers, N.L. Weiss, Editor.

WISE, E.M., (Editor), (1988), Industries - Extraction and processing-gold: pp. 410-413 in **Encyclopaedia Britannica-Macropaedia**, vol. 21, Encyclopaedia Britannica, Inc., Chicago.

WISE, E.M., (1988), Industries - Extraction and processing-nickel: pp. 417-419 in **Encyclopaedia Britannica-Macropaedia**, vol. 21, Encyclopaedia Britannica, Inc., Chicago.

WISE, E.M., (1988), Industries - Extraction and processing-platinum group metals: pp. 419-420 in **Encyclopaedia Britannica-Macropaedia**, vol. 21, Encyclopaedia Britannica, Inc., Chicago.

World Mine Production of Gold, 1986-1990, (1987), report published by the Gold Institute, Washington, D.C.

World Survey of Nonferrous Smelters, (1988), proceedings of the symposium on world survey of nonferrous smelters, sponsored by the Pyrometallurgical Committee of TMS, at the TMS-AIME annual meeting held in Phoenix.

YANNOPOULOS, J.C., (1991), **The Extractive Metallurgy of Gold:** Chapman and Hall, London.

YARAR, Baki, (1993), Alternatives to cyanide in the leaching and recovery of gold: an overview, in **Precious Metals Economics and Refining Technology**, Proceedings of a seminar of the International Precious Metals Institute held in Las Vegas.

YOUNG, R.S., (1988), Industries - Extraction and processing-cobalt: pp. 409-410 in **Encyclopaedia Britannica-Macropaedia**, vol. 21, Encyclopaedia Britannica, Inc., Chicago.

ZAMAN, S. (1985), Carbon, sulphur, mercury, miscellaneous sulfides - Graphite: in **SME Mineral Processing Handbook**, published by the Society of Mining Engineers, N.L. Weiss, Editor.

ZAMAN, S., (1985), Carbon, sulfur, mercury, miscellaneous sulfides - Antimony and arsenic: in **SME Mineral Processing Handbook**, published by the Society of Mining Engineers, N.L. Weiss, Editor.

ZAMAN, S., (1985), Oxides of heavy metals - Columbium and tantalum: in **SME Mineral Processing Handbook**, published by the Society of Mining Engineers, N. L. Weiss Editor.

ZAMAN, S., (1985), Carbon, sulfur, mercury, miscellaneous sulfides - Bismuth: in **SME Mineral Processing Handbook**, published by the Society of Mining Engineers, N.L. Weiss, Editor.

ZANDON, V.A., (1985), Potash - Introduction and potash beneficiation: in **SME Mineral Processing handbook**, published by the Society of Mining Engineers, N.L. Weiss, Editor.

ZHANG JIAN, LAN XINZHE, DING FENG, YUAN ZHICHANG, and YANG LONGQIAN, (1992), Leaching gold and silver by lime-sulfur synthetic solutions (LSSS), pp 289-398 in **Precious Metals 1992**, proceedings of the 16th International Precious Metals Institute Conference held in Scottsdale, Arizona, A.K. Mehta and R.M. Nadkarni, Editors.

Subject Index

MIX
Papier aus verantwortungsvollen Quellen
Paper from responsible sources
FSC® C105338

If you have any concerns about our products,
you can contact us on
ProductSafety@springernature.com

In case Publisher is established outside the EU,
the EU authorized representative is:
Springer Nature Customer Service Center GmbH
Europaplatz 3, 69115 Heidelberg, Germany

Printed by Libri Plureos GmbH
in Hamburg, Germany